Progress in Mathematics

Volume 271

Krzysztof Galicki • Santiago R. Simanca

Riemannian Topology and Geometric Structures on Manifolds

Birkhäuser
Boston • Basel • Berlin

Editors

Krzysztof Galicki
Department of Mathematics
University of New Mexico
Albuquerque, NM 87131
galicki@math.unm.edu

Santiago R. Simanca
Department of Mathematics
University of New Mexico
Albuquerque, NM 87131
santiago@math.unm.edu

ISBN 978-0-8176-4742-1   e-ISBN 978-0-8176-4743-8
DOI 10.1007/978-0-8176-4743-8

Library of Congress Control Number: 2008938278

Mathematics Subject Classification (2000): 58-06

Printed on acid-free paper

www.birkhauser.com

Progress in Mathematics

Riemannian Topology
and Geometric Structures
on Manifolds

Proceedings of the Conference on Riemannian Topology
and Geometric Structures on Manifolds*
In honor of Charles P. Boyer's 65th birthday
October 10-14, 2006
The University of New Mexico, Albuquerque, New Mexico

Krzysztof Galicki
Santiago R. Simanca
Editors

Birkhäuser, Boston

*The realization of this event and the preparation of these Proceedings were supported by the National Science Foundation grant DMS 0623676, the Efroymson Foundation, and an anonymous source.

This volume is dedicated to Charles P. Boyer
on the occasion of his 65th birthday

Preface

The articles in this volume originate either from lectures presented at the conference *Riemannian Topology and Geometric Structures on Manifolds* or from discussions directly surrounding the topics pertaining to this event. Held in Albuquerque from October 10 to October 14, 2006, the conference had three objectives: to provide a forum for the analysis of recent advances in the areas of positive sectional curvature, Kählerian and Sasakian geometry, and their interrelation to mathematical physics; to create a platform for the discussion of the growth that these fundamental ideas have experienced in recent past; and to examine the open problems of interest in these fields. The conference was used also as an occasion to celebrate Charles P. Boyer's 65th birthday in recognition of his academic achievements while on the faculty at the University of New Mexico.

The objectives of the conference are all reflected in this volume. The interrelation of the subjects is such that ordering the articles alphabetically by authors is a good as any way of presenting them all. With due deference to all outside contributors, this is exactly what I have chosen to do, simply leaving for last the one contribution to these proceedings written by the hosts. Be as it may, I hope that you all enjoy the ubiquitousness of the subjects and ideas in all of these articles.

I acknowledge the financial support of the National Science Foundation, the Efroymson Foundation, and that of an anonymous donor for the realization of the Riemannian Topology Conference and the preparation of these proceedings. At a time where support for mathematical sciences is dwindling, it would have been very hard to accomplish this much without their backing. In an area of the country where support for mathematics is next to nonexistent, it would have been impossible to do so without it.

I thank all the referees for their careful job in reviewing the contributions to this volume and, where appropriate, for their thoughtful suggestions for improvements of the contributions that they reviewed.

And I thank all the contributors to these proceedings. With submissions that they all could have sent for publication to excellent mathematics journals, they made possible the volume that I present here.

To the extent that I could properly say this volume was made in Albuquerque, I hope that its universal quality can be thought of as a testimony that mathematics can flourish anywhere, transcending any barrier that there may be. Mathematics is for and at the service of everyone.

January 2008   *Santiago R. Simanca*

———————————————————————

Last July 8 was a sad day for us. For on that day Krzysztof Galicki, the coeditor of these proceedings, suffered an accident while hiking in the Swiss Alps. He remained in a coma until his death in Albuquerque on September 24. Kris' fatal accident caused a deep feeling of loss among all of us. I present this volume in his memory.

Charles P. Boyer — An Autobiographical Sketch

Charles P. Boyer was born April 27, 1942, in Norristown, Pennsylvania. After working as an electronic technician for General Electric in the early 1960s, he graduated from Penn State University in 1966, and then enrolled in the graduate program in physics at Penn State where he completed his Ph.D in 1972. His dissertation used the representation theory of noncompact Lie groups in models of elementary particles. Charles' early work stressed the mathematics more than the physics by using the representation theory of Lie groups to study properties of the solutions to partial differential equations of mathematical physics much of which was done in collaboration with E. Kalnins and W. Miller Jr.

After postdoctoral positions at the then Centro de Investigaciones en Matemáticas Aplicadas y en Sistemas (CIMAS) of the Universidad Nacional Autónoma de México (UNAM) and Le Centre de Recherches Mathematiques of the University of Montreal, he emigrated permanently to Mexico where he became a researcher at UNAM. While in Mexico, Charles' interests changed to the area of differential geometry and topology, where he collaborated with Polish general relativist Jerzy Plebanski and Mexican topologist Samuel Gitler.

Charles spent a one-year sabbatical in the Mathematics Department of Harvard University. After the large devaluation of the Mexican peso in 1982, he returned to the United States taking a position at the then Clarkson College of Technology in upstate New York. He accepted a position at the University of New Mexico in 1988 where he is currently employed.

Charles is best known for the proof of the Atiyah–Jones conjecture with Jacques Hurtubise, Ben Mann, and Jim Milgram, as well as his work with Kris Galicki and others on Sasakian geometry. In particular, together with János Kollár, Boyer and Galicki used Sasakian geometry to show the existence of an abundance of Einstein metrics on odd-dimensional spheres including exotic spheres.

Charles is married to University of New Mexico Law Professor Margaret Montoya. They have two daughters, Diana and Alejandra, and Charles has a son, Chuck, who currently resides in Hawaii, where he is a professor of Spanish.

Charles is a lover of Mexican culture with many friends in Mexico. He retains strong ties to Mexico with frequent visits of both a professional and personal nature.

Charles P. Boyer's List of Publications

1. *The Matrix Elements of the Most Degenerate Representations of $SO_0(p,1)$.* J. Mathematical Phys. 12 (1971), 8, pp. 1599–1603.
2. *On the Decomposition $SO_0(p-1,1) \subset SO_0(p,1)$ for Most Degenerate Representations.* J. Mathematical Phys. 12 (1971), 10, pp. 2070–2075, with Farhad Ardalan.
3. *On the Supplementary Series of $SO_0(p,1)$.* J. Mathematical Phys. 14 (1973), 5, pp. 609–617.
4. *Quantum Field Theory on a Seven Dimensional Homogeneous Space of the Poincaré Group.* J. Mathematical Phys. 15 (1974), 7, pp. 1002–1024, with G.N. Fleming.

5. *Deformations of Inhomogeneous Classical Lie Algebras to the Algebras of the Linear Groups*. J. Mathematical Phys. 14 (1973), 12, pp. 1853–1859, with K.B. Wolf.

6. *sl(4,R) as a Spectrum-Generating Algebra for the Coulomb Systems*. Lettre a Nuovo Cimento, 8 (1973), 7, pp. 458–460, with K.B. Wolf.

7. *Deformations of Lie Algebras and Groups and Their Applications*. Rev. Mexicana Fís. 23 (1974), pp. 99–122.

8. *The Group Chain $SO_0(n, 1) \supset SO_0(1, 1) \times SO(n-1)$: A Complete Solution to the 'Missing Label' Problem*. J. Mathematical Phys. 15 (1974), 5, pp. 560–564, with K.B. Wolf.

9. *The Algebra and Group Deformations $I^m[SO(n) \times SO(m)] \supset SO(n,m)$, $I^m[U(n) \times U(m)] \supset U(n,m)$ and $I^m[Sp(n) \times Sp(m)] \supset Sp(n,m)$ for $1 \leq m < n$*. J. Mathematical Phys. 15 (1974), 12, pp. 2096–2101, with K.B. Wolf.

10. *The Maximal 'Kinematical' Invariance Group for an Arbitrary Potential*. Helv. Phys. Acta, 47 (1974), pp. 589–605.

11. *A Classification of Second Order Raising Operators for Hamiltonians in Two Variables*. J. Mathematical Phys. 15 (1974), 9, pp. 1484–1489, with W. Miller, Jr.

12. *Lie Theory and Separation of Variables. VI. The Equation $iU_t + \triangle_2 U = 0$*. J. Mathematical Phys. 16 (1975), 3, pp. 499–511, with E.G. Kalnins and W. Miller, Jr.

13. *Lie Theory and Separation of Variables. VII. The Harmonic Oscillator in Elliptic Coordinates and Ince Polynomials*. J. Mathematical Phys. 16 (1975), 3, pp. 512–517, with E.G. Kalnins and W. Miller, Jr.

14. *Canonical Transforms. III. Configuration and Phase Descriptions of Quantum Systems Possessing an $sl(2,R)$ Dynamical Algebra*. J. Mathematical Phys. 16 (1975), 7, pp. 1493–1502, with K.B. Wolf.

15. *The 2:1 Anisotropic Oscillator, Separation of Variables, and Symmetry Group in Bargmann Space*. J. Mathematical Phys. 16 (1975), 11, pp. 2215–2223, with K.B. Wolf.

16. *Conformal Symmetry of the Hamiltonian-Jacobi Equation and Quantization*. Nuovo Cimento B, 31 (1976), 2, pp. 195–210, with M. Penafiel.

17. *Lie Theory and Separation of Variables for the Equation $iU_t + \triangle_2 U - \left(\frac{\alpha}{x_1^2} + \frac{\beta}{x_2^2} \right) U = 0$*. SIAM J. Math. Anal. 7 (1976), 2, pp. 230–263.

18. *Symmetry and Separation of Variables for the Helmholtz and Laplace Equations*. Nagoya Math. J. 60 (1976), pp. 35–80, with E.G. Kalnins and W. Miller, Jr.

19. *Symmetries of Differential Equations in Mathematical Physics*. Group Theoretical Methods in Physics (Fourth Internat. Colloq., Nijmegen, 1975), pp. 425–434. Lecture Notes in Phys., Vol. 50, Springer, Berlin, 1976.

20. *Symmetries and Exterior Differential Forms*. Memorias del Simposio Internacional de Física-Matemática, Mexico City.

21. *Finite $SL(2,R)$ Representation Matrices of the D_k^+ Series for all Subgroup Reductions*. Rev. Mexicana Fís. 25 (1976), pp. 31–45, with K.B. Wolf.

22. *Symmetry Breaking Interactions for the Time Dependent Schrödinger Equation*. J. Mathematical Phys. 17 (1976), 8, pp. 1439–1461, with R.T. Sharp and P. Winternitz.

23. *Some Comments on Symplectic Structures in Complex Riemannian Geometry*. Group Theoretical Methods in Physics (Proc. Fifth Internat. Colloq., Univ. Montréal, Montreal, Que., 1976), pp. 261–265. Academic Press, New York, 1977.

24. *Heavens and their Integral Manifolds*. J. Mathematical Phys. 18 (1877), 5, pp. 1022–1031, with J.F. Plebanski.

25. *Symmetries of the Hamiltonian-Jacobi Equation*. J. Mathematical Phys. 18 (1977), 5, pp. 1032–1045, with E.G. Kalnins.

26. *Symmetry and Separation of Variables for the Hamilton-Jacobi Equation $W_t^2 - W_x^2 - W_y^2 = 0$*. J. Mathematical Phys. 19 (1978), 1, pp. 200–211, with E.G. Kalnins and W. Miller, Jr.

27. *A Relationship between Lie Theory and Continued Fraction Expansions for Special Functions*. Padé and rational approximation (Proc. Internat. Sympos., Univ. South Florida, Tampa, Fla., 1976), pp. 147–155. Academic Press, New York, 1977, with W. Miller, Jr.

28. *General Relativity and G-structures. I. General Theory and Algebraically Degenerate Spaces*. Rep. Math. Phys. 14 (1978), 1, 111–145, with J.F. Plebanski.

29. *R-Separable Coordinates for Three-Dimensional Complex Riemannian Spaces*. Trans. Am. Math. Soc. 242 (1978), pp. 355–376, with E.G. Kalnins and W. Miller, Jr.

30. *Separable Coordinates for Four-Dimensional Riemannian Spaces*. Comm. Math. Phys. 59 (1978), 3, pp. 285–302, with E.G. Kalnins and W. Miller, Jr.

31. *Formal Algebraic Models of Graded Differential Geometry*. J. Pure Appl. Algebra, 18 (1980), pp. 1–16.

32. *Complex General Relativity, H and HH spaces — A Survey of One Approach*. General relativity and gravitation, Vol. 2, pp. 241–281, Plenum, New York-London, 1980, with J.D. Finley, III and J.F. Plebanski.

33. *Graded G-Structures*. Group Theoretical Methods in Physics (W. Beiglbock, A. Bohn and E. Takasugi, Eds.) Springer, New York, 1979.

34. *Separation of Variables in Einstein Spaces*, Proceedings of the Second Marcel Grossmann Meeting on General Relativity, North-Holland, Amsterdam, 1982.

35. *Separation of Variables in Einstein Spaces. I. Two Ignorable and One Null Coordinates*. J. Phys. A 14 (1981), 7, pp. 1675–1684, with E.G. Kalnins and W. Miller, Jr.

36. *Separation of Variables in Einstein Spaces. II. No Ignorable Coordinates*. Gen. Relativity Gravitation 12 (1980), 9, pp. 733–741.

37. *Gravitational Instantons*. Proceedings of the UNAM Workshop on Gauge Theories (Univ. Nac. Autónoma México, Mexico City, 1979). Hadronic J. 4 (1980/81), 1, pp. 2–18.

38. *The Theory of G^∞-Supermanifolds*. Group Theoretical Methods in Physics (K.B. Wolf Ed.) Springer, New York (1980), with S. Gitler.

39. *Killing Vectors in Self-dual Euclidean Einstein Spaces*. J. Mathematical Phys. 23 (1982), 6, pp. 1126–1130, with J.D. Finley, III.

40. *On the Structure of Supermanifolds*. Contemporary Mathematics, Vol. 12, pp. 53–59 (1982).

41. *Completely Integrable Relativistic Hamiltonian Systems and Separation of Variables in Hermitian Hyperbolic Spaces*. J. Mathematical Phys. 24 (1983), 8, pp. 2022–2034, with E.G. Kalnins and P. Winternitz.

42. *The Theory of G^∞-Supermanifolds*. Trans. Am. Math. Soc. 285 (1984), 1, pp. 241–267, with S. Gitler.

43. *Separation of Variables for the Hamilton-Jacobi Equation on Complex Projective Spaces*. SIAM J. Math. Anal. 16 (1985), 1, pp. 93–109, with E.G. Kalnins and P. Winternitz.

44. *An Infinite Hierarchy of Conservation Laws and Nonlinear Superposition Principles for Self-dual Einstein Spaces*. J. Math. Phys. 26 (1985), 2, pp. 229–234, with J.F. Plebanski.

45. *The Geometry of Complex Self-dual Einstein Spaces*. Nonlinear phenomena (Oaxtepec, 1982), pp. 25–46, Lecture Notes in Phys., 189, Springer, Berlin, 1983.

46. *Stäckel-equivalent Integrable Hamiltonian Systems*. SIAM J. Math. Anal. 17 (1986), 4, pp. 778–797, with E.G. Kalnins and W. Miller.

47. *Conformally Self-dual Spaces and Maxwell's Equations*. Phys. Lett. A 106 (1984), 3, pp. 125–129, with J.F. Plebanski.

48. *Response to Query "On the Solvability of the Nonlinear P.D.E. $\theta_{xx}\theta_{yy} - \theta_{xy}^2 + \theta_{xp} + \theta_{yq} = 0$."* Notices Am. Math. Soc. 35, 380 (1985).

49. *Conformal Duality and Compact Complex Surfaces*. Math. Ann. 274 (1986), pp. 517–526.

50. *The Curved Twistor Construction for the Self-dual Einstein Equations — An Integrable System*. Systèmes dynamiques non linéaires: intégrabilité et comportement qualitatif, pp. 23–39, Sém. Math. Sup., 102, Presses Univ. Montréal, Montreal, QC, 1986.

51. *A Note on Hyper-Hermitian Four-Manifolds*. Proc. Am. Math. Soc. 102 (1988), pp. 157–164.

52. *Self-Dual and Anti-Self-Dual Hermitian Metrics on Compact Complex Surfaces*. Mathematics and general relativity (Santa Cruz, CA, 1986), pp. 105–114, Contemp. Math., 71, Am. Math. Soc., Providence, RI, 1988.

53. *Homology Operations on Instantons*. J. Differential Geom. 28 (1988), 3, pp. 423–465, with B. Mann.

54. *Monopoles, Non-Linear σ Models, and Two-Fold Loop Spaces.* Commun. Math. Phys. 115 (1988), 4, pp. 571–594, with B. Mann.

55. *Some Problems of Elementary Calculus in Superdomains.* Proceedings of the XXth National Congress of the Mexican Mathematical Society (Spanish) (Xalapa, 1987), pp. 111–143, Aportaciones Mat. Comun., 5, Soc. Mat. Mexicana, México, 1988, with O. A. Sánchez Valenzuela.

56. *Instantons and Homotopy.* Algebraic topology (Arcata, CA, 1986), pp. 87–102, Lecture Notes in Math., 1370, Springer, Berlin, 1989, with B. Mann.

57. *Symmetries of the Self Dual Einstein Equations. I. The Infinite Dimensional Symmetry Group and its Low Dimensional Subgroups.* J. Math. Phys. 30 (1989), 5, pp. 1081–1094. with P. Winternitz.

58. *Instantons, quaternionic Lie algebras and hyper-Khler geometry.* Lie theory, differential equations and representation theory (Montreal, PQ, 1989), pp. 119–136, Univ. Montréal, Montreal, QC, 1990, with B.M. Mann.

59. *Lie Supergroup Actions on Supermanifolds.* Trans. Am. Math. Soc. 323 (1991), 1, pp. 151–175, with O. A. Sánchez Valenzuela.

60. *On the Homology of SU(n) Instantons.* Trans. Am. Math. Soc. 323 (1991), 2, pp. 529–561, with B.M. Mann and D. Waggoner.

61. *The Atiyah Jones Conjecture.* Bull. Am. Math. Soc. (N.S.) 26 (1992), 2, pp. 317–321, with J.C. Hurtubise, B.M. Mann, and R.J. Milgram.

62. *The Hyperkähler Geometry of the ADHM Construction and Quaternionic Geometric Invariant Theory.* Proc. Sympos. Pure Math., 54, Part 2, pp. 45–83, Am. Math. Soc., Providence, RI, 1993, with B.M. Mann.

63. *The Topology of Instanton Moduli Spaces. I: The Atiyah-Jones Conjecture.* Ann. of Math. (2) 137 (1993), 3, pp. 561–609, with J.C. Hurtubise, B.M. Mann, and R.J. Milgram.

64. *Algebraic Cycles and Infinite Loop Spaces.* Invent. Math. 113 (1993), 2, pp. 373–388, with H.B. Lawson, P. Lima-Filho, B.M. Mann, and M.-L. Michelsohn.

65. *Quaternionic Reduction and Einstein Manifolds.* Comm. Anal. Geom. 1 (1993), 2, pp. 229–279, with K. Galicki, and B.M. Mann.

66. *3-Sasakian Manifolds.* Proc. Japan Acad. Ser. A Math. Sci. 69 (1993), 8, pp. 335–340, with K. Galicki and B.M. Mann.

67. *New Examples of Inhomogeneous Einstein Manifolds of Positive Scalar Curvature.* Math. Res. Lett. 1 (1994), 1, pp. 115–121, with K. Galicki and B.M. Mann.

68. *Topology of Holomorphic Maps into Generalized Flag Manifolds.* Acta Math. 173 (1994), 1, pp. 61–101, with J. Hurtubise, B.M. Mann, and R.J. Milgram.

69. *The Geometry and Topology of 3-Sasakian Manifolds.* J. Reine Angew. Math. 455 (1994), pp. 183–220, with K. Galicki and B.M. Mann.

70. *Some New Examples of Compact Inhomogeneous Hypercomplex Manifolds.* Math. Res. Lett. 1 (1994), 5, pp. 531–538, with K. Galicki and B.M. Mann.

71. *The Atiyah-Jones Conjecture.* Homotopy theory and its applications (Cocoyoc, 1993), pp. 51–56, Contemp. Math., 188, Am. Math. Soc., Providence, RI, 1995.

72. *Quaternionic Geometry and 3-Sasakian Manifolds.* Quaternionic structures in mathematics and physics (Trieste, 1994), pp. 7–24 (electronic), Int. Sch. Adv. Stud. (SISSA), Trieste, 1998, with K. Galicki and B.M. Mann.

73. *Hypercomplex Structures on Stiefel Manifolds.* Ann. Global Anal. Geom. 14 (1996), 1, pp. 81–105, with K. Galicki and B.M. Mann.

74. *On Strongly Inhomogeneous Einstein Manifolds.* Bull. London Math. Soc. 28 (1996), 4, pp. 401–408, with K. Galicki and B.M. Mann.

75. *The Twistor Space of a 3-Sasakian Manifold.* Internat. J. Math. 8 (1997), 1, pp. 31–60, with K. Galicki.

76. *Einstein manifolds of positive scalar curvature with arbitrary second Betti number.* Balkan J. Geom. Appl. 1 (1996), 2, pp. 1–7, with K. Galicki, B.M. Mann, and E.G. Rees.

77. *Compact 3-Sasakian 7-Manifolds with Arbitrary second Betti Number.* Invent. Math. 131 (1998), 2, pp. 321–344, with K. Galicki, B.M. Mann, and E.G. Rees.

78. *A Note on Smooth Toral Reductions of Spheres.* Manuscripta Math. 95 (1998), 2, pp. 149–158, with K. Galicki and B.M. Mann.

79. *Hypercomplex Structures from 3-Sasakian Structures.* J. Reine Angew. Math. 501 (1998), pp. 115–141, with K. Galicki and B.M. Mann.

80. *3-Sasakian Manifolds.* Surveys in differential geometry: essays on Einstein manifolds, pp. 123–184, Surv. Differ. Geom., VI, Int. Press, Boston, MA, 1999, with K. Galicki.

81. *A Note on Toric Contact Geometry.* J. Geom. Phys. 35 (2000), 4, pp. 288–298, with K. Galicki.

82. *On Sasakian-Einstein Geometry.* Internat. J. Math. 11 (2000), 7, pp. 873–909, with K. Galicki.

83. *Stability Theorems for Spaces of Rational Curves.* Internat. J. Math. 12 (2001), 2, pp. 223–262, with J. Hurtubise, and R.J. Milgram.

84. *Einstein manifolds and contact geometry.* Proc. Am. Math. Soc. 129 (2001), 8, pp. 2419–2430, with K. Galicki.

85. *New Einstein Metrics in Dimension Five.* J. Differential Geom. 57 (2001), 3, pp. 443–463, with K. Galicki.

86. *3-Sasakian Geometry, Nilpotent Orbits, and Exceptional Quotients.* Ann. Global Anal. Geom. 21 (2002), 1, pp. 85–110, with K. Galicki and P. Piccinni.

87. *Sasakian-Einstein Structures on $9\#(S^2 \times S^3)$.* Trans. Am. Math. Soc. 354 (2002), 8, pp. 2983–2996, with K. Galicki and M. Nakamaye.

88. *Rational Homology 5-Spheres with Positive Ricci Curvature.* Math. Res. Lett. 9 (2002), 4, pp. 521–528, with K. Galicki.

89. *Einstein Metrics on Rational Homology 7-Spheres.* Ann. Inst. Fourier (Grenoble) 52 (2002), 5, pp. 1569–1584, with K. Galicki and M. Nakamaye.

90. *On the Geometry of Sasakian-Einstein 5-Manifolds.* Math. Ann. 325 (2003), 3, pp. 485–524, with K. Galicki and M. Nakamaye.

91. *Sasakian Geometry, Homotopy Spheres, and Positive Ricci Curvature.* Topology 42 (2003), 5, pp. 981–1002, with K. Galicki and M. Nakamaye.

92. *New Einstein Metrics on $8\#(S^2 \times S^3)$.* Differential Geom. Appl. 19 (2003), 2, pp. 245–251, with K. Galicki.

93. *On Positive Sasakian Geometry.* Geom. Dedicata 101 (2003), pp. 93–102, with K. Galicki and M. Nakamaye.

94. *Toric Self-dual Einstein Metrics as Quotients.* Commun. Math. Phys. 253 (2005), 2, pp. 337–370, with D. Calderbank, K. Galicki, and P. Piccinni.

95. *Sasakian Geometry, Hypersurface Singularities, and Einstein Metrics.* Rend. Circ. Mat. Palermo (2) Suppl. 75 (2005), pp. 57–87, with K. Galicki.

96. *Einstein Metrics on Exotic Spheres in Dimensions 7, 11, and 15.* Experiment. Math. 14 (2005), 1, pp. 59–64, with K. Galicki, J. Kollár, and E. Thomas.

97. *Einstein Metrics on Spheres.* Ann. Math. (2) 162 (2005), 1, pp. 557–580, with K. Galicki and J. Kollár.

98. *On Eta-Einstein Sasakian Geometry.* Commun. Math. Phys. 262 (2006), 1, pp. 177–208, with K. Galicki and Paola Matzeu.

99. *Sasakian Geometry and Einstein Metrics on Spheres.* Perspectives in Riemannian geometry, pp. 47–61, CRM Proc. Lecture Notes, 40, Am. Math. Soc., Providence, RI, 2006, with K. Galicki.

100. *Einstein Metrics on Rational Homology Spheres.* J. Differential Geom. 74 (2006), 3, pp. 353–362, with K. Galicki.

101. *Highly Connected Manifolds with Positive Ricci Curvature.* Geom. Topol. 10 (2006), pp. 2219–2235, with K. Galicki.

102. *Constructions in Sasakian Geometry.* Math. Z. 257 (2007), 4, pp. 907–924, with K. Galicki and L. Ornea.

103. Sasakian Geometry. Oxford Mathematical Monographs. Oxford University Press MR2382957, with K. Galicki.

104. *Canonical Sasakian Metrics*. Commun. Math. Phys. (3) 279 (2008), pp. 705–733, with K. Galicki and S.R. Simanca.
105. *The Sasaki Cone and Extremal Sasakian Metrics*. These Proceedings, Birkhäuser, K. Galicki and S.R. Simanca Eds., with K. Galicki and S.R. Simanca.

Accepted for Publication

1. *Sasakian Geometry, Holonomy, and Supersymmetry*. Handbook of pseudo-Riemannian Geometry and Supersymmetry, European Mathematical Society (EMS) (to appear), with K. Galicki.

Contents

L^2-Cohomology of Spaces with Nonisolated Conical Singularities and Nonmultiplicativity of the Signature

Jeff Cheeger and Xianzhe Dai

Abstract We study from a mostly topological standpoint the L^2-signature of certain spaces with nonisolated conical singularities. The contribution from the singularities is identified with a topological invariant of the link fibration of the singularities. This invariant measures the failure of the signature to behave multiplicatively for fibrations for which the boundary of the fiber is nonempty. The result extends easily to cusp singularities and can be used to compute the L^2-cohomology of certain noncompact hyperkähler manifolds that admit geometrically fibered end structures.

1 Introduction

In this paper, we study the L^2-cohomology and L^2-signature for certain spaces with nonisolated conical singularities. We call these generalized Thom spaces. Appropriately formulated, our results extend easily to cusp singularities as well. Our main theorem identifies the contribution to the L^2-signature from a singular stratum with a topological invariant of the link fibration of the stratum. As an immediate application, we get a proof of the adiabatic limit formula of [13], in the case of odd dimensional fiber, without resorting to the quite nontrivial analytical results of [24]. This was actually one of original motivations. A second motivation was to study certain spaces with singularities that can be viewed as generalizations of Thom spaces.[1]

The L^2-cohomology of spaces with conical singularities has been studied extensively in [9, 10]; see also [9, 12] for the relation with intersection cohomology,

J. Cheeger
Courant Institute, New York, New York, USA

X. Dai
Department of Mathematics, University of California, Santa Barbara, California, USA

[1] This part of our work was done more than fifteen years ago but remained unpublished. We learned of the more recent connections with the Sen's conjecture from conversations with Tamas Hausel; see the end of this section.

K. Galicki and S.R. Simanca (eds.), *Riemannian Topology and Geometric Structures on Manifolds,* Progress in Mathematics 271, DOI 10.1007/978-0-8176-4743-8,
© Springer Science+Business Media LLC 2009

and [12, 26] for the Cheeger–Goresky–MacPherson conjecture. For case of the cusp singularities, see [31, 32], and for Zucker's conjecture, see [22, 28]. For hyperbolic manifolds, see [23, 25].

The singular spaces that we consider can be described as follows. Recall that a compact Riemannian manifold M with finite isolated conical singularities is modeled on the finite cone. That is, M is a compact topological space such that there are finite many points $p_1, \cdots, p_k$, so that $M \setminus \{p_1, \cdots, p_k\}$ is a smooth Riemannian manifold, and a neighborhood of each singular point p_i is isomorphic to a finite metric cone, $C_{[0,a]}(Z_i)$, on a closed Riemannian manifold Z_i. In addition to isolated conical singularities, we also allow finitely many closed singular strata of positive dimension, whose normal fibers are of metric conical type. A prominent feature of these spaces is that the link fibration of a singular stratum need not be trivial. The discussion in this generality is necessary because we wish to identify the contribution to the L^2-signature from the singular strata in terms of global topological invariants of the link fibration.

A neighborhood of a singular stratum of positive dimension can be described as follows. Let

$$Z^n \to M^m \xrightarrow{\pi} B^l \tag{1.1}$$

denote a fibration of closed oriented smooth manifolds. Denote by $C_\pi M$ the mapping cylinder of π. This is obtained by attaching a cone to each of the fibers. Indeed, we have

$$C_{[0,a]}(Z) \to C_\pi M \to B. \tag{1.2}$$

The space $C_\pi M$ also comes with a natural quasi-isometry class of metrics. A metric can be obtained by choosing a submersion metric on M:

$$g_M = \pi^* g_B + g_Z.$$

Then, on the nonsingular part of $C_\pi M$, we take the metric,

$$g_1 = dr^2 + \pi^* g_B + r^2 g_Z, \tag{1.3}$$

and complete it.

The general class of spaces with nonisolated conical singularities as above can be described as follows. A space X in the class will be of the form

$$X = X_0 \cup X_1 \cup \cdots \cup X_k, \tag{1.4}$$

where X_0 is a compact smooth manifold with boundary, and each X_i (for $i = 1, \ldots, k$) is the associated mapping cylinder, $C_{\pi_i} M_i$, for some fibration, (M_i, π_i), as above. We require that the restriction of the metric to X_i is quasi-isometric to one of the form (1.3). Spaces with more complicated singularities can be obtained by iterating this construction, namely, by allowing the base and fiber of the fibration, (1.1), to be closed manifolds with nonisolated conical singularities.

Consider again the space, $C_\pi M$, in (1.2). By coning off the boundary, ∂M, we obtain what we call a generalized Thom space T. Thus, $T = C_\pi M \cup_M C(M)$ is a stratified space with two singular strata, namely, B and a single point.

The metric on T is constructed as follows. Equip $C(M)$ with the conical metric

$$g_2 = dr^2 + r^2 g_M.$$

Perturb g_1, g_2 near $r = 1$ so that they can be glued together so as to obtain a smooth metric, g, on T. We will call (T, g) a *generalized Thom space*; see the example below. Clearly, a different choice of g_M will give rise to a metric quasi-isometric to g.

Example 1. Let $\xi \xrightarrow{\pi} B$ be a vector bundle of rank k. Then we have the associated sphere bundle:

$$S^{k-1} \to S(\xi) \xrightarrow{\pi} B.$$

The generalized Thom space constructed out of this fibration coincides with the usual Thom space equipped with a natural metric.

We now introduce the topological invariant that gives the contribution to the L^2-signature for each singular strata. In [13], in studying adiabatic limits of eta invariants, the second author introduced a global topological invariant associated with a fibration. (For adiabatic limits of eta invariants, see also [5, 6, 11, 30].) Let (E_r, d_r) be the E_r-term with differential, d_r, of the Leray spectral sequence of (1.1). Define a pairing

$$
\begin{aligned}
E_r \otimes E_r &\to \mathbf{R} \\
\phi \otimes \psi &\mapsto \langle \phi \cdot d_r \psi, \xi_r \rangle
\end{aligned}
,
$$

where ξ_r is a basis for E_r^m naturally constructed from the orientation. In case $m = 4k - 1$, when restricted to $E_r^{\frac{m-1}{2}}$, this pairing becomes symmetric. We define τ_r to be the signature of this symmetric pairing and put

$$\tau = \sum_{r \geq 2} \tau_r.$$

It is shown in [13] that, unlike the case of fibrations whose fibers are closed manifolds, when the fibers have nonempty boundary, the signature does not always behave multiplicatively, even in a generalized sense; compare [1, 27]. The failure of such multiplicative behavior is intrinsically measured by the τ invariant of the associated boundary fibration; see [13].

Bismut and Cheeger studied related questions by introducing spaces with conical singularities as a technical tool; see [3, 4]. They showed that if one closes up the fibration of manifolds with boundary by attaching cones to the boundary of each fiber, then for the corresponding fibration of manifolds with singularities, then the L^2-signature does in fact behave multiplicatively. One reason for studying generalized Thom spaces is to understand the difference between the approaches of [13] and [3, 4].

In this chapter, we restrict attention to the case in which the fiber Z of (1.1) is either odd dimensional or its middle dimensional L^2-cohomology vanishes. Furthermore, we make the same assumptions for the links of the isolated conical singular points of the base and the fiber. (The general case requires the introduction of an "ideal boundary condition" as in [10, 11].) The result of [9] shows that $H_{(2)}^*(T)$, the L^2-cohomology of T, is finite dimensional and the Strong Hodge theorem holds. In fact, $H_{(2)}^*(T)$ agrees with the middle intersection cohomology of Goresky and MacPherson [14, 15]. Consequently, the L^2-signature of T is a topological invariant. Here, in defining the signature, we take the natural orientation on $C_\pi M$ and glue to $C(M)$ with the reverse of its natural orientation, in order to obtain the orientation on T.

Theorem 1.1. *The L^2-signature of the generalized Thom space T is equal to $-\tau$:*

$$\text{sign}_{(2)}(T) = -\tau.$$

Let $\tau(X_i)$ denote the τ invariant for the fibration associated with X_i. Theorem 1.1 combined with Novikov additivity of the signature yields the following. (Note the reversing of the orientation.)

Corollary 1.2. *For the space X of the form (1.4), the L^2-signature is given by*

$$\text{sign}_{(2)}(X) = \text{sign}(X_0) + \sum_{i=1}^{k} \tau(X_i).$$

Example 2. Consider again the sphere bundle of a vector bundle,

$$S^{k-1} \to S(\xi) \xrightarrow{\pi} B.$$

Let Φ denote the Thom class and χ the Euler class. Then the Thom isomorphism gives

$$H^*(D(\xi), S(\xi)) \otimes H^*(D(\xi), S(\xi)) \to \qquad \mathbf{R}$$

$$\uparrow \pi^*(\cdot) \cup \Phi \qquad \qquad \uparrow \pi^*(\cdot) \cup \Phi$$

$$H^*(B) \qquad \otimes \qquad H^*(B) \qquad \to \qquad \mathbf{R}$$

$$\phi \qquad \qquad \psi \qquad \to [\phi \cup \psi \cup \chi][B].$$

Thus, $\text{sign}(D(\xi)) = -\text{sign}_{(2)}(T)$ is the signature of this bilinear form on $H^*(B)$. Since in this case, the spectral sequence degenerates at E_2, and $d_2 \psi = \psi \cup \chi$, it follows that the invariant, $\text{sign}(D(\xi))$, agrees with τ. According to Theorem 1.1, the same result is still true even if the sphere bundle does not arise from a vector bundle.

In spirit, our proof of Theorem 1.1 follows Example 2. Thus, we first establish an analog of Thom's isomorphism theorem in the context of generalized Thom spaces. In part, this consists of identifying the L^2-cohomology of (T, g) in terms of the spectral sequence of the original fibration. The Mayer–Vietoris argument as in [9] shows that

$$H^i_{(2)}(T) = \begin{cases} H^i_{(2)}(C_\pi M, M), & i > \frac{m+1}{2} \\ \mathrm{Im}\,(H^i_{(2)}(C_\pi M, M) \to H^i_{(2)}(C_\pi M)), & i = \frac{m+1}{2} \\ H^i_{(2)}(C_\pi M), & i < \frac{m+1}{2} \end{cases}.$$

(Recall that $m = \dim M$ is the dimension of the total space of the fibration and $n = \dim Z$ is the dimension of the fiber.)

Let $E_r(M) = \oplus E_r^{p,q}(M)$, $d_r^{p,q} : E_r^{p,q}(M) \to E_r^{p+r,q-r+1}(M)$, denote the Leray spectral sequence of the fibration (1.1). (For some of the notation in the following theorem, we refer to Section 4.)

Theorem 1.3. *The following are isomorphisms.*

$$H^k(C_\pi(M), M) \cong \oplus_{p+q=k, q \geq (n+3)/2} \left[\mathrm{Im}\,(d^{p,q-1}_{q-(n-1)/2})^* \oplus \right.$$
$$\left. \oplus \mathrm{Im}\,(d^{p,q-1}_{q-(n-1)/2+1})^*) \oplus \cdots \oplus E^{p,q-1}_\infty(M)\right],$$

and

$$H^k(C_\pi(M)) \cong \oplus_{p+q=k, q \leq (n-1)/2} \left[\mathrm{Im}\,d^{k-(n+3)/2,(n+1)/2}_{(n+3)/2-q} \oplus \right.$$
$$\left. \oplus \mathrm{Im}\,d^{k-(n+5)/2,(n+3)/2}_{(n+5)/2-q} \oplus \cdots \oplus E^{p,q}_\infty(M)\right].$$

Moreover, in terms of these identifications, the map,

$$H^k(C_\pi(M), M) \to H^k(C_\pi(M)),$$

is given by $\oplus d_r$.

Remark. It can be shown that Theorem 1.1, Corollary 1.2, and Theorem 1.3 have extensions to the case of iterated conical singularities.

From the standpoint of index theory, the L^2-signature is of particular interest. For the case of fibrations with smooth fibers, it was considered in [4].

Let Z and B of (1.1) be closed smooth manifolds. Let A_M denote the signature operator on M with respect to the metric g_M. In addition, let $A_{M,\varepsilon}$ denote the signature operator on M with respect to the metric $g_{M,\varepsilon}$, where

$$g_{M,\varepsilon} = \varepsilon^{-1}\pi^* g_B + g_Z.$$

Define the $\tilde{\eta}$-form as in [3,5]. Let the modified, L-form, $\mathcal{L}$, be defined as in [4]. Let R^B denote the curvature of B. According to [4] the following holds.

Theorem 1.4 (Bismut–Cheeger). *If the fiber of (1.1) is odd dimensional, then*

$$\mathrm{sign}_{(2)}(T) = -\lim_{\varepsilon \to 0} \eta(A_{M,\varepsilon}) + \int_B \mathcal{L}\left(\frac{R^B}{2\pi}\right) \wedge \tilde{\eta}.$$

Remark. Since the smooth part of T is diffeomorphic to $(0,1) \times M$, its contribution to the index formula vanishes. In the above formula, the first term arises from the isolated conical point, and the second term arises from the singular stratum B.

Combining Theorem 1.4 with our result on the L^2-signature, Theorem 1.1, we recover the following adiabatic limit formula of [13]; see also [5, 6, 10, 30].

Corollary 1.5. *With the same assumptions as in Theorem 1.4,*

$$\lim_{\varepsilon \to 0} \eta(A_{M,\varepsilon}) = \int_B \mathcal{L}\left(\frac{R^B}{2\pi}\right) \wedge \tilde{\eta} + \tau.$$

Note that if (1.1) actually bounds a fibration of manifolds with boundary, then the above adiabatic limit formula is a consequence of the signature theorem of Atiyah–Patodi–Singer [2] and the Families Index Theorem for manifolds with boundary of Bismut–Cheeger [3].[2] On the other hand, it seems difficult to decide whether every fibration (with odd dimensional fibers) actually bounds and it is generally believed that this is not the case. The method of attaching cones enables one to avoid this issue.

In the case in which (1.1) consists of closed smooth manifolds, in place of a cone, one can attach a cusp to each fiber. The metric at infinity of a locally symmetric space of rank one is of this type. Essentially because the Poincaré lemma for metric cusps gives the same calculation as for metric cones, similar results hold in this case.

The study of the L^2-cohomology of the type of spaces with conical singularities discussed here turns out to be related to work on the L^2-cohomology of noncompact hyperkähler manifolds that is motivated by Sen's conjecture; see, e.g., [16, 17]. Hyperkähler manifolds often arise as moduli spaces of (gravitational) instantons and monopoles, and so-called S-duality predicts the dimension of the L^2-cohomology of these moduli spaces (Sen's conjecture). Many of these spaces can be compactified to given a space with nonisolated conical singularities. In such cases, our results can be applied. We would also like to refer the reader to the work of Hausel–Hunsicker–Mazzeo [16], which studies the L^2-cohomology and L^2-harmonic forms of noncompact spaces with fibered geometric ends and their relation to the intersection cohomology of the compactification. Various applications related to Sen's conjecture are also considered there.

In the general case, i.e., with no dimension restriction on the fiber, the L^2-signature for generalized Thom spaces is discussed in [19]. In particular, Theorem 1.1 is proved for the general case in [19]. However, one of ingredients there is the adiabatic limit formula of [13], rather than the direct topological approach taken here. As mentioned earlier, one of our original motivations was to give a simple topological proof of the adiabatic limit formula. In [18], the methods and techniques introduced in our old unpublished work are used in the more general situation to derive a very interesting topological interpretation for the invariant τ_r. This circumstance provided additional motivation for us to write up this work for publication.

---

[2] In this case, the invariant, τ, enters because it measures the nonmultiplicativity of the signature as in [13].

2 Review of L^2-Cohomology

We begin by reviewing the basic properties of L^2-cohomology; for details, see [9]. Let (Y, g) denote an open (possibly incomplete) Riemannian manifold. We denote by $[g]$ the quasi-isometry class of g; i.e., the collection of Riemannian metrics g' on M such that for some positive constant c,

$$\frac{1}{c} g \leq g' \leq cg.$$

Let $\Omega^i = \Omega^i(Y)$ denote the space of C^∞ i-forms on Y and $L^2 = L^2(Y)$ the L^2 completion of Ω^i with respect to the L^2-metric induced by g. Define d to be the exterior differential with the domain

$$\text{dom} \, d = \{\alpha \in \Omega^i(Y) \cap L^2(Y); \, d\alpha \in L^2(Y)\}.$$

Put $\Omega^i_{(2)}(Y) = \Omega^i(Y) \cap L^2(Y)$. As usual, let δ denote the formal adjoint of d. In terms of a choice of local orientation for Y, we have $\delta = \pm * d *$, where $*$ is the Hodge star operator. We define the domain of δ by

$$\text{dom} \, \delta = \{\alpha \in \Omega^i(Y) \cap L^2(Y); \, \delta\alpha \in L^2(Y)\}.$$

Note that d, δ have well-defined strong closures $\bar{d}$, $\bar{\delta}$. That is, $\alpha \in \text{dom} \, \bar{d}$ and $\bar{d}\alpha = \eta$ if there is a sequence $\alpha_j \in \text{dom} \, d$ such that $\alpha_j \to \alpha$ and $d\alpha_j \to \eta$ in L^2.

Usually, the L^2-cohomology of Y is defined by

$$H^i_{(2)}(Y) = \ker d_i / \text{Im} \, d_{i-1}.$$

One can also define the L^2-cohomology using the closure $\bar{d}$. Put

$$H^i_{(2), \#}(Y) = \ker \bar{d}_i / \text{Im} \, \bar{d}_{i-1}.$$

In fact, the natural map,

$$\iota_{(2)} : H^i_{(2)}(Y) \longrightarrow H^i_{(2), \#}(Y),$$

is always an isomorphism.

In general, the image of $\bar{d}$ need not be closed. The reduced L^2-cohomology is defined by

$$\bar{H}^i_{(2)}(Y) = \ker \bar{d}_i / \overline{\text{Im} \, \bar{d}_{i-1}}.$$

The space of L^2-harmonic i-forms $\mathcal{H}^i_{(2)}(Y)$ is the space,

$$\mathcal{H}^i_{(2)}(Y) = \{\theta \in \Omega^i \cap L^2; d\theta = \delta\theta = 0\}.$$

When Y is oriented, the Hodge star operator induces the Poincaré duality isomorphism

$$* : \mathcal{H}^i_{(2)}(Y) \rightarrow \mathcal{H}^{n-i}_{(2)}(Y), . \tag{2.1}$$

Remark. Some authors define the space of harmonic forms differently, using the Hilbert spaces adjoint of $\bar{d}$; see for example, [29]. In this case, the Hodge star operator does not necessarily leave invariant the space of harmonic forms.

Clearly, there is a natural map

$$\mathcal{H}^i_{(2)}(Y) \rightarrow H^i_{(2)}(Y) . \tag{2.2}$$

The question of when this map is an isomorphism is of crucial interest. The most basic result here is the Kodaira decomposition,

$$L^2 = \mathcal{H}^i_{(2)} \oplus \overline{d\Lambda^{i-1}_0} \oplus \overline{\delta\Lambda^{i+1}_0},$$

which leaves invariant the subspaces of smooth forms. It follows then that

$$\ker \bar{d}_i = \mathcal{H}^i_{(2)} \oplus \overline{d\Lambda^{i-1}_0}.$$

Adopting the terminology of [9], we will say that the Strong Hodge Theorem holds if the natural map (2.2) is an isomorphism. By the above discussion, if Im $\bar{d}$ is closed, then the map in (2.2) is surjective. In particular, this holds if the L^2-cohomology is finite dimensional.

On the other hand, if we assume that Stokes' theorem holds for Y in the L^2 sense, i.e.,

$$\langle \bar{d}\alpha, \beta \rangle = \langle \alpha, \bar{\delta}\beta \rangle \tag{2.3}$$

for all $\alpha \in \operatorname{dom} \bar{d}$, $\beta \in \operatorname{dom} \bar{\delta}$, or equivalently, for all $\alpha \in \operatorname{dom} d$, $\beta \in \operatorname{dom} \delta$, then one has

$$\mathcal{H}^i_{(2)}(Y) \perp \operatorname{Im} \bar{d}_{i-1},$$

and hence,

$$\mathcal{H}^i_{(2)}(Y) \perp \overline{\operatorname{Im} \bar{d}_{i-1}} .$$

Thus, (2.2) is injective in this case. Moreover,

$$\mathcal{H}^i_{(2)}(Y) \cong \bar{H}^i_{(2)}(Y), \tag{2.4}$$

and

$$H^i_{(2)}(Y) = \bar{H}^i_{(2)}(Y) \oplus \overline{\operatorname{Im} \bar{d}_{i-1}} / \operatorname{Im} \bar{d}_{i-1}. \tag{2.5}$$

Here, by the closed graph theorem, the last summand is either 0 or infinite dimensional.

To summarize, if the L^2-cohomology of Y is finite dimensional and Stokes' theorem holds on Y in the L^2-sense, then the L^2-cohomology of Y is isomorphic to the space of L^2-harmonic forms and therefore, when Y is orientable, Poincaré duality holds as well. Consequently, the L^2 signature of Y is well-defined in this case.

Next, we recall the relative de Rham theory [7] and relative L^2-cohomology. Let $f : S \to Y$ denote a map between manifolds. Define a complex, $(\Omega^*(f), d)$, by

$$\Omega^p(f) = \Omega^p(Y) \oplus \Omega^{p-1}(S), \quad d(\omega, \theta) = (d\omega, f^*(\omega) - d\theta). \qquad (2.6)$$

Clearly, $d^2 = 0$, and hence, the corresponding cohomology $H^*(f)$ is well defined.

Put $\alpha(\theta) = (0, \theta)$, $\beta(\omega, \theta) = \omega$, and consider the short exact sequence,

$$0 \to \Omega^{p-1}(S) \overset{\alpha}{\to} \Omega^p(f) \overset{\beta}{\to} \Omega^p(Y) \to 0, \qquad (2.7)$$

where the differential of the complex $\Omega^{*-1}(S)$ is $-d$. There is an induced a long exact sequence on the cohomology,

$$\cdots \to H^p(f) \overset{\beta^*}{\to} H^p(Y) \overset{f^*}{\to} H^p(S) \overset{\alpha^*}{\to} H^{p+1}(f) \to \cdots . \qquad (2.8)$$

If S is a submanifold of M and $i : S \to Y$ is the inclusion map, we define the relative cohomology, $H^*(Y, S)$, to be $H^*(i)$. To define the relative L^2-cohomology $H^*_{(2)}(Y, S)$, we assume further that S has trivial normal bundle in Y and that the metric in a neighborhood of S is quasi-isometric to the product metric. Then $H^*_{(2)}(Y, S)$ can be defined as the cohomology of the complex,

$$\Omega^*_{(2)}(Y, S) = \Omega^*_{(2)}(Y) \oplus \Omega^{*-1}_{(2)}(S),$$

with $\mathrm{dom}(d) = \{(\omega, \theta) \mid d\omega \in L^2, \ d\theta \in L^2.\}$. It follows that the long exact sequence for the pair (Y, S) is also valid in L^2-cohomology.

We note that L^2-cohomology is quasi-isometry invariant and conformally invariant in the middle dimension. Also, Künneth formula holds for the L^2-cohomology. Furthermore, given an open cover $\{U_\alpha\}_\alpha$, the Mayer–Vietoris principle holds for the L^2-cohomology, provided there is a constant, C, such that there is a partition of unity $\{f_\alpha\}$ subordinate to $\{U_\alpha\}$, such that $|df_\alpha| \leq C$, for all α . Hence, the Leray spectral sequence in L^2-cohomology is valid for a fibration, if such a partition of unity, subordinate to a trivializing open cover, can be found on the base.[3] Clearly, this holds for the fibrations considered here.

3 L^2-Cohomology of Generalized Thom Spaces

In this section, we begin to specialize to the case of generalized Thom spaces. Thus, we retain the notation of (1.1)–(1.3). The results in this section are special cases, and in some instances, refinements, of those that hold for more general stratified pseudomanifolds; see [9] and also [29]. For completeness and later purposes, we provide a somewhat detailed account.

---

[3] This can be shown by the usual double complex construction as in [7].

Let N be a Riemannian manifold, possibly incomplete. For simplicity, we assume that $m = \dim N$ is odd. Further, we assume that for N, the L^2-cohomology is finite dimensional and that Stokes' theorem holds in the L^2 sense. Then for the finite cone, $C_{[0,1]}(N)$, over N, we have the following facts from [9].

(1) The L^2-cohomology is finite dimensional and Stokes' theorem holds in the L^2 sense. Consequently, the Strong Hodge Theorem holds for the cone.

(2) For $i \leq m/2$, the restriction map induces an isomorphism,

$$H_{(2)}^i(C_{[0,1]}(N)) \cong H_{(2)}^i(N),$$

and for $i \geq (m+1)/2$,
$$H_{(2)}^i(C_{[0,1]}(N)) = 0.$$

The above results are consequences of the following lemmas; for proofs, see [9].

Lemma 3.1. *Let θ be an i-form on N that is in L^2. Let $\tilde{\theta}$ denote the extension of θ to $C_{[0,1]}(N)$ so that $\tilde{\theta}$ is radially constant. Then $\tilde{\theta} \in L^2(C_{[0,1]}(N))$ if and only if $i < (m+1)/2$.*

When there is no danger of confusion, we will just write θ for $\tilde{\theta}$.

For some $a \in (0,1)$, define the homotopy operator K^0 as follows. If $\alpha = \phi + dr \wedge \omega$ is an i-form and $i < (m+1)/2$, then

$$K^0 \alpha = \int_a^r \omega.$$

If $i \geq (m+1)/2$, then

$$K^0 \alpha = \int_0^r \omega.$$

Lemma 3.2 (Poincaré lemma). *For $\alpha \in \mathrm{dom}\, \bar{d}$,*

$$(\bar{d}K^0 + K^0\bar{d})\alpha = \alpha - \alpha(a), \quad i < (m+1)/2$$

and

$$(\bar{d}K^0 + K^0\bar{d})\alpha = \alpha, \quad i \geq (m+1)/2.$$

We now turn to the case of fibrations whose fibers are cones.

Theorem 3.3. *The L^2-cohomology of $C_\pi M$ is finite dimensional, and Stokes' theorem holds in the L^2 sense. Hence, the Strong Hodge Theorem holds.*

Proof. By the Mayer–Vietoris principle, verifying finite dimensionality reduces to verifying finite dimensionality for a product fibration. Since the L^2-cohmology is a quasi-isometry invariant, we can use the product metric. Then the Künneth formula yields the desired result.

It was proved in [9] that if the L^2-Stokes theorem holds locally, then it holds globally. Further, the validity of the L^2-Stokes theorem is a quasi-isometry invariant.

Thus, once again we can reduce to a product situation, and the validity of the L^2-Stokes theorem follows from its validity for $C_{[0,1]}(Z)$ and B.   □

Finally, we give a vanishing statement for L^2-cohomology that is a direct generalization of the one that holds in the case in which the base consists of a single point; [9]. Put $\dim B = \ell$, $\dim Z = n$.

Lemma 3.4.

$$H^i_{(2)}(C_\pi M) = 0 \qquad \text{for} \quad i \geq l + \frac{n+1}{2}.$$

Consequently,

$$H^{m+1}_{(2)}(C_\pi M, M) \simeq H^m(M) \cong \mathbf{R}.$$

Proof. Assuming the first statement, the second follows from the long exact sequence

$$\to H^i_{(2)}(C_\pi M, M) \to H^i_{(2)}(C_\pi M) \to H^i_{(2)}(M) \to H^{i+1}_{(2)}(C_\pi M, M) \to .$$

To prove the first statement, we make use of K_0, the chain homotopy operator for cones. By defining K^0 fiberwise, we can naturally extend K^0 to a cone bundle. If $\alpha = \phi + dr \wedge \omega$ and $\deg \alpha \geq l + \frac{n+1}{2}$, put

$$K^0 \alpha = \int_0^r \omega.$$

Clearly, from what was shown for the case of a single cone, if $\alpha \in L^2(C_\pi M)$, then $K^0 \alpha \in L^2(C_\pi M)$. Now we verify

$$(\bar{d} K^0 + K^0 \bar{d}) \alpha = \alpha.$$

Again, everything is local on B and quasi-isometric invariant, therefore, we can check it for a product fibration with product metric. In this case

$$\bar{d} = \bar{d}_{C_{[0,1]}(Z)} + d_B,$$

and $(\bar{d}_{C(Z)} K^0 + K^0 \bar{d}_{C(Z)}) \alpha = \alpha$. Hence it suffices to check

$$(d_B K^0 + K^0 d_B) \alpha = 0.$$

Note that α can be written as linear combinations of forms of type, $(\phi + dr \wedge \omega) \otimes \tau_B$, where $\phi + dr \wedge \omega$ lives on $C_{[0,1]}(Z)$. Then

$$d_B(K^0 \alpha) = d_B \left(\int_0^r \omega \otimes \tau_B \right) = (-1)^{\deg \omega} \left(\int_0^r \omega \right) \otimes d\tau_B,$$

$$K^0 d_B \alpha = (-1)^{\deg \omega + 1} \left(\int_0^r \omega \right) \otimes d\tau_B.$$

These terms cancel one another.   □

4 Spectral Sequences

In this section, we recall some basics concerning spectral sequences; a general reference is [7].

A filtered differential complex (K,d) is a differential complex that comes with a filtration by subcomplexes

$$K = K_0 \supset K_1 \supset K_2 \supset \cdots. \tag{4.1}$$

A graded filtered complex, (K,d), has in addition, a grading $K = \oplus_{n \in \mathbb{Z}} K^n$. In this case, the filtration (4.1) induces a filtration on each K^n: $K_p^n = K^n \cap K_p$.

One has the following general result; see [7].

Theorem 4.1. *Let (K,d) denote a graded filtered differential complex such that the induced filtration on each K^n has finite length. Then the short exact sequence,*

$$0 \to \oplus K_{p+1} \to \oplus K_p \to \oplus K_p/K_{p+1} \to 0,$$

induces a spectral sequence, $(E_r^{p,q},\ d_r)$, which converges to the cohomology group $H^*(K,d)$.

In fact, if $d: K^n \to K^{n+1}$ is of degree one, then

$$E_r^{p,q} = \frac{d^{-1}(K_{p+r}^{p+q+1}) \cap K_p^{p+q}}{d^{-1}(K_{p+r}^{p+q+1}) \cap K_{p+1}^{p+q} + K_p^{p+q} \cap d(K_{p-r+1}^{p+q-1})}, \tag{4.2}$$

where the differential, d_r, is naturally induced by d.

We now recall Serre's filtration and the Leray spectral sequence of the fibration (1.1). A p-form ω is in F^i if

$$\omega(U_1, U_2, \ldots, U_p) = 0,$$

whenever $p - i + 1$ of the tangent vectors, U_j, are vertical. If the fibration is equipped with a connection, there is a splitting

$$TM = T^H M \oplus T^V M \cong \pi^* TB \oplus T^V M.$$

Thus, $\Lambda^* M = \pi^* \Lambda^* B \otimes \Lambda^* T^V M$. Formally, we can use $a(y,z)\, dy^\alpha \wedge dz^\beta$ with y the local coordinates of B, z local coordinates of Y to indicate such a splitting. With this convention, Serre's filtration can be simply described as

$$F^i = \{a(y,z)\, dy^\alpha \wedge dz^\beta : |\alpha| \geq i\}.$$

The global definition shows that this filtration is independent of the particular choice of local coordinates.

Definition. The horizontal degree of ω is at least i, h-deg $\omega \geq i$, or equivalently, the vertical degree of ω is at most $p - i$ ($p = \deg \omega$), v-deg $\omega \leq p - i$, if $\omega \in F^i$.

Serre's filtration, together with the grading given by the degree of differential form, gives rise to the Leray spectral sequence of the fibration, (E_r, d_r), which converges to the cohomology of the total space M. Moreover, the E_2 term is given by

$$E_2 = \oplus E_2^{p,q}, \quad E_2^{p,q} = H^p(B, \mathcal{H}^q(Z)), \tag{4.3}$$

where $\mathcal{H}^q(Z)$ denotes the flat bundle over B, for which the fiber over $b \in B$ is the q-th L^2-cohomology of the fiber $\pi^{-1}(b)$.

Now assume that both the base B and the vertical tangent bundle $T^V M$ are oriented. When the fibration is equipped with a submersion metric, we can identify $\mathcal{H}^q(Z)$ with the bundle of fiberwise L^2-harmonic q-forms. This induces a fiberwise metric on $\mathcal{H}^q(Z)$. Note that $\mathcal{H}^q(Z)$ is a flat vector bundle over B. Via Hodge theory for cohomology with coefficients in a flat bundle, we obtain an inner product on $E_2^{p,q} = H^p(B, \mathcal{H}^q(Z))$.

The E_3 term of the spectral sequence is given by the cohomology groups of the E_2 term with respect to the differential d_2. Therefore, finite dimensional Hodge theory gives us

$$E_2 = E_3 \oplus \mathrm{Im}\, d_2 \oplus \mathrm{Im}(d_2)^*, \tag{4.4}$$

when the adjoint is defined with respect to the inner product introduced above. This in turn, induces an inner product on the E_3 terms. By iterating this construction, we obtain inner products on each of the E_r, with the associated Hodge decomposition.

The above discussion extends to fibrations whose base and fibers are closed manifolds with conical singularities, if we replace the usual cohomology with the L^2-cohomology.

5 The Generalized Thom Isomorphism

This section is devoted to the proof of Theorem 1.3, which is a generalization of the Thom isomorphism theorem. As explained in the introduction, the statement consists of identifying the L^2-cohomology of $C_\pi M$ and $(C_\pi M, M)$ in terms of the Leray spectral sequence of M. Additionally, the map,

$$H^*_{(2)}(C_\pi M, M) \to H^*_{(2)}(C_\pi M),$$

is identified in terms of the differential of the spectral sequence. We begin with $H^*_{(2)}(C_\pi M)$.

Note that since $C_\pi M$ is fibered over B,

$$C_{[0,1]}(Z) \to C_\pi M \to B,$$

we also have a spectral sequence, $(E_r^{p,q}(C_\pi M) \bar{d}_r^{p,q})$, converging to the L^2-cohomology of $C_\pi M$. On the other hand, the pullback of the inclusion $i : M \to C_\pi M$,

$$i^* : \Omega^*(C_\pi M) \to \Omega^*(M),$$

is filtration preserving. Hence, i^* induces homomorphism

$$i_r : E_r^{p,q}(C_\pi M) \rightarrow E_r^{p,q}(M)$$

which commutes with the differentials.

Now

$$E_2^{p,q}(C_\pi M) = H^p(B, \mathcal{H}^q(C(Z))) = \begin{cases} 0, & q > n/2 \\ H^p(B, \mathcal{H}^q(Z)) = E_2^{p,q}(M), & q < n/2 \end{cases}.$$

It follows that $E_r^{p,q}(C_\pi M) = 0$, for all $q > n/2$ and $r \geq 2$. Moreover, by the Poincaré Lemma 3.2, the identification here is given by i_2, i.e.,

$$i_2^{p,q} = \begin{cases} 0, & q > n/2 \\ \text{Ident}, & q < n/2 \end{cases}.$$

It follows that

$$\bar{d}_2^{p,q} = \begin{cases} 0, & q > n/2 \\ d_2^{p,q}, & q < n/2 \end{cases},$$

and thus

$$E_3^{p,q}(C_\pi M) = E_3^{p,q}(M), \quad \text{if } q < (n-1)/2,$$

$$E_3^{p,(n-1)/2}(C_\pi M) = \ker d_2^{p,(n-1)/2} = E_3^{p,(n-1)/2}(M) \oplus \operatorname{Im} d_2^{p-2,(n+1)/2}.$$

This implies that for $q < (n-1)/2$, $i_3^{p,q} = \text{Ident}$.

Also, because $i_3^{p,(n-1)/2}$ is induced by

$$\text{Ident}: E_2^{p,(n-1)/2}(C_\pi M) \rightarrow E_2^{p,(n-1)/2}(M),$$

one sees that $i_3^{p,(n-1)/2}$ is the natural projection,

$$E_3^{p,(n-1)/2}(C_\pi M) = \ker d_2^{p,(n-1)/2} \rightarrow \frac{\ker d_2^{p,(n-1)/2}}{\operatorname{Im} d_2^{p-2,(n+1)/2}} = E_3^{p,(n-1)/2}(M).$$

Hence, because i_r commutes with $\bar{d}_r$, d_r, we get

$$\bar{d}_3^{p,q} = d_3^{p,q}, \qquad \text{if } q < (n-1)/2,$$

and

$$\bar{d}_3^{p,(n-1)/2} \,|\, \operatorname{Im} d_2^{p-2,(n+1)/2} = 0,$$

$$\bar{d}_3^{p,(n-1)/2} \,|\, E_3^{p,(n-1)/2}(M) = d_3^{p,(n-1)/2}.$$

It follows that

$$E_4^{p,q}(C_\pi M) = E_4^{p,q}(M) \qquad \text{if } q < (n-1)/2 - 1,$$

$$E_4^{p,(n-3)/2}(C_\pi M) = \ker d_3^{p,(n-3)/2}$$
$$= E_4^{p,(n-3)/2}(M) \oplus \operatorname{Im} d_3^{p-3,(n+1)/2},$$

$$E_4^{p,(n-1)/2}(C_\pi M) = \ker d_3^{p,(n-1)/2} \oplus \operatorname{Im} d_2^{p-2,(n+1)/2}$$
$$= E_4^{p,(n-1)/2}(M) \oplus \operatorname{Im} d_3^{p-3,(n+3)/2} \oplus \operatorname{Im} d_2^{p-2,(n+1)/2}.$$

By an inductive argument, we obtain the following proposition, in which some of the $\operatorname{Im} d$ summands are obviously zero.

Proposition 5.1. *There are equalities,*

$$E_\infty^{p,(n-1)/2}(C_\pi M) = E_\infty^{p,(n-1)/2}(M) \oplus \operatorname{Im} d_2^{p-2,(n+1)/2} \oplus \operatorname{Im} d_3^{p-3,(n+3)/2} \oplus \cdots$$
$$\cdots \oplus \operatorname{Im} d_l^{p-l,(n+1)/2+l-2},$$

$$E_\infty^{p,(n-3)/2}(C_\pi M) = E_\infty^{p,(n-3)/2}(M) \oplus \operatorname{Im} d_3^{p-3,(n+1)/2} \oplus \operatorname{Im} d_4^{p-4,(n+3)/2} \oplus \cdots$$
$$\cdots \oplus \operatorname{Im} d_l^{p-l,(n+1)/2+l-3},$$

$$\vdots$$

$$E_\infty^{p,0}(C_\pi M) = E_\infty^{p,0}(M) \oplus \operatorname{Im} d_{(n+3)/2}^{p-(n+3)/2,(n+1)/2} \oplus \operatorname{Im} d_{(n+5)/2}^{p-(n+5)/2,(n+3)/2} \oplus \cdots$$
$$\cdots \oplus \operatorname{Im} d_l^{p-l,l-1}.$$

Similarly, we have:

Proposition 5.2. *For the relative cohomology,*

$$E_\infty^{p,(n+3)/2}(C_\pi M, M) = E_\infty^{p,(n+1)/2}(M) \oplus \operatorname{Im}(d_2^{p,(n+1)/2})^* \oplus \operatorname{Im}(d_3^{p,(n+1)/2})^* \oplus \cdots$$
$$\cdots \oplus \operatorname{Im}(d_l^{p,(n+1)/2})^*,$$

$$E_\infty^{p,(n+5)/2}(C_\pi M, M) = E_\infty^{p,(n+3)/2}(M) \oplus \operatorname{Im}(d_3^{p,(n+3)/2})^* \oplus \operatorname{Im}(d_4^{p,(n+3)/2})^* \oplus \cdots$$
$$\cdots \oplus \operatorname{Im}(d_l^{p,(n+3)/2})^*,$$

$$\vdots$$

$$E_\infty^{p,n+1}(C_\pi M, M) = E_\infty^{p,n}(M) \oplus \operatorname{Im}(d_{(n+3)/2}^{p,n})^* \oplus \operatorname{Im}(d_{(n+5)/2}^{p,n})^* \oplus \cdots$$
$$\cdots \oplus \operatorname{Im}(d_l^{p,n})^*.$$

Proof. Recall that

$$\Omega_{(2)}^*(C_\pi M, M) = \Omega_{(2)}^*(C_\pi M) \oplus \Omega_{(2)}^{*-1}(M).$$

Clearly, the map,

$$\Omega_{(2)}^{*-1}(M) \rightarrow \Omega_{(2)}^*(C_\pi M, M)$$
$$\theta \quad \rightarrow \quad (0, \theta),$$

is filtration preserving and commuting with differentials. Hence, this map induces homomorphisms

$$j_r: (E_r^{p,q-1}(M), d_r) \to (E_r^{p,q}(C_\pi M, M), \tilde{d}_r).$$

Moreover,

$$E_2^{p,q}(C_\pi M, M) = \begin{cases} H^p(B, \mathcal{H}^{q-1}(Z)) = E_2^{p,q-1}(M), & q > n/2+1 \\ 0, & q < n/2+1 \end{cases}.$$

In terms of this identification,

$$j_2^{p,q} = \begin{cases} \text{id} & q > n/2+1 \\ 0 & q < n/2+1 \end{cases}.$$

It follows that

$$\tilde{d}_2^{p,q} = \begin{cases} d_2^{p,q-1} & q > (n+3)/2 \\ 0 & q \le (n+3)/2 \end{cases},$$

and thus

$$E_3^{p,q}(C_\pi M, M) = E_3^{p,q-1}(M), \text{ if } q > (n+3)/2,$$

$$E_3^{p,(n+3)/2}(C_\pi M, M) = \frac{E_2^{p,(n+1)/2}(M)}{\text{Im} \, d_2} = E_3^{p,(n+1)/2}(M) \oplus \text{Im}\,(d_2^{p,(n+1)/2})^*.$$

This implies that $j_3^{p,q} = \text{Ident}$, for $q > (n+3)/2$.

Again, because $j_3^{p,(n+3)/2}$ is induced by

$$\text{Ident}: E_2^{p,(n+1)/2}(M) \to E_2^{p,(n+3)/2}(C_\pi M, M),$$

one sees that $j_3^{p,(n+3)/2}$ is the natural inclusion. It follows that

$$\tilde{d}_2^{p,q} = \begin{cases} d_2^{p,q-1} & q > (n+3)/2 \\ 0 & q \le (n+3)/2 \end{cases},$$

and therefore,

$$E_4^{p,q}(C_\pi M, M) = E_4^{p,q}(M), \qquad \text{if } q > (n+5)/2,$$

$$E_4^{p,(n+5)/2}(C_\pi M, M) = E_4^{p,(n+3)/2}(M) \oplus \text{Im}\,(d_3^{p,(n+3)/2})^*,$$

$$E_4^{p,(n+3)/2}(C_\pi M, M) = E_4^{p,(n+1)/2}(M) \oplus \text{Im}\,(d_3^{p,(n+1)/2})^* \oplus \text{Im}\,d_2^{p,(n+1)/2})^*.$$

By proceeding inductively, the desired result follows.   □

We are now ready to prove Theorem 1.3

Proof. From the general theory of the spectral sequence, the filtration on $\Omega^*_{(2)}(C_\pi M)$ induces a filtration on $H^*_{(2)}(C_\pi M)$,

$$H^*_{(2)}(C_\pi M) \supset F^0 H^*_{(2)}(C_\pi M) \supset F^1 H^*_{(2)}(C_\pi M) \supset \cdots \supset 0,$$

such that

$$E^{p,q}_\infty(C_\pi M) \cong F^p H^{p+q}_{(2)}(C_\pi M)/F^{p+1} H^{p+q}_{(2)}(C_\pi M).$$

Therefore,

$$H^k(C_\pi M) \cong \oplus_{p+q=k} E^{p,q}_\infty(C_\pi M)$$

$$\cong \oplus_{p+q=k, q \le (n-1)/2} [E^{p,q}_\infty(M) \oplus \mathrm{Im}(d^{k-(n+3)/2,(n+1)/2}_{(n+3)/2-q})$$

$$\oplus \mathrm{Im}(d^{k-(n+5)/2,(n+3)/2}_{(n+5)/2-q}) \oplus \cdots].$$

Similarly

$$H^k(C_\pi(M), M)$$

$$\cong \oplus_{p+q=k, q \ge (n+3)/2} [E^{p,q-1}_\infty(M) \oplus \mathrm{Im}(d^{p,q-1}_{q-(n-1)/2})^* \oplus \mathrm{Im}(d^{p,q-1}_{q-(n-1)/2+1})^* \oplus \cdots].$$

We claim that in terms of this identification, the map,

$$H^k(C_\pi(M), M) \to H^k(C_\pi(M)),$$

is given by applying appropriate d_r's to the appropriate factors. To verify that this is the case, we need to trace back the isomorphisms. It suffices to look at, for example, $\mathrm{Im}(d^{p,(n+1)/2}_2)^*$.

Let $[\theta] \in \mathrm{Im}(d^{p,(n+1)/2}_2)^* \subset E^{p,(n+1)/2}_2(M)$. In this case, θ can be represented by a $(p+(n+1)/2)$-form, such that

$$\theta \in F^p, \ d\theta \in F^{p+2}.$$

Therefore, v-deg $d\theta \le p + (n+1)/2 - (p+2) \le (n-1)/2$. By Lemma 3.1, $d\theta$ can be extended to an L^2-form on $C_\pi(M)$. To make $[0, \theta]$ an element of $E^{p,(n+1)/2}_\infty(C_\pi(M), M)$, we modify its representative slightly:

$$(d\theta, \theta) \in [0, \theta] \text{ and } [d\theta, \theta] \in E^{p,(n+1)/2}_\infty(C_\pi(M), M).$$

This implies that

$$(d\theta, \theta) \in F^p H^{p+(n+1)/2}_{(2)}(C_\pi M, M),$$

which is mapped to

$$d\theta \in F^p H^{p+(n+1)/2}_{(2)}(C_\pi M),$$

which, via the identification with the spectral sequence terms, is

$$d\theta \,|\, M = d_M \theta.$$

This is exactly $d_2[\theta]$. The rest of the terms can be treated in exactly the same fashion.

$\square$

6 L^2-Signatures of Generalized Thom Spaces

Assume that $\dim T = m+1$ is divisible by 4.

By definition, the L^2-signature, $\mathrm{sign}_{(2)}(T)$, of the generalized Thom space, (T,g), is the signature of the pairing,

$$H_{(2)}^{(m+1)/2}(T) \otimes H_{(2)}^{(m+1)/2}(T) \to \mathbf{R},$$

induced by wedge product and integration.

The L^2-signature of $C_\pi M$ as a (singular) manifold with boundary, $\partial(C_\pi M) = M$, is, by definition, the signature of the (possibly degenerate) pairing

$$H_{(2)}^{(m+1)/2}(C_\pi M, M) \otimes H_{(2)}^{(m+1)/2}(C_\pi M, M) \to \mathbf{R}, \tag{6.1}$$

defined via the map

$$H^k(C_\pi(M), M) \to H^k(C_\pi(M)), \tag{6.2}$$

and the (nondegenerate) pairing

$$H_{(2)}^{m+1-k}(C_\pi M, M) \otimes H_{(2)}^{k}(C_\pi M) \to \mathbf{R}, \tag{6.3}$$

given by

$$([\omega, \theta], \, [\alpha]) \to \int_{C_\pi M} \omega \wedge \alpha - \int_M \theta \wedge \alpha. \tag{6.4}$$

As in [9], one has

$$\mathrm{sign}_{(2)}(T) = \mathrm{sign}_{(2)}(C_\pi M).$$

Thus, to prove Theorem 1.1, we just have to compute the L^2-signature of $C_\pi M$. The computation is in the spirit of Example 2 of the introduction.

Proof of Theorem 1.1. The outline of the proof is as follows. First, we consider the intersection matrix with respect to the block decompositions in terms of $E_\infty^{r,s}(C_\pi M, M)$ and $E_\infty^{p,q}(C_\pi M)$ and show that the matrix is lower antidiagonal. Then we consider each of these antidiagonal blocks and show that they are also antidiagonal with respect to the block decompositions given by the Thom isomorphism. Finally, we identify the pairings along the antidiagonals and show that they give rise to the τ invariant.

Consider the pairing between $H_{(2)}^k(C_\pi M)$ and $H_{(2)}^{m+1-k}(C_\pi M, M)$. Let $\ell = \dim B$, $n = \dim Z$. (In what follows below, we make a change of index.)

First of all, we show that in terms of the identifications,

$$H^k(C_\pi(M))$$
$$\cong \oplus_{p+q=k, q\leq (n-1)/2}[E_\infty^{p,q}(M) \oplus \operatorname{Im} d_{(n+3)/2-q}^{k-(n+3)/2,(n+1)/2} \oplus \operatorname{Im} d_{(n+5)/2-q}^{k-(n+5)/2,(n+3)/2} \oplus \cdots,]$$

and

$$H^{m+1-k}(C_\pi(M), M)$$
$$\cong \oplus_{p'+q'=k, q'\leq (n-1)/2}[E_\infty^{l-p',n-q'}(M) \oplus \operatorname{Im} (d_{(n+3)/2-q'}^{l-p',n-q'})^* \oplus \operatorname{Im} (d_{(n+5)/2-q'}^{l-p',n-q'})^* \oplus \cdots]$$

the pairing has the form of the block matrix,

$$\begin{pmatrix} 0 & \cdots & 0 & * \\ 0 & \cdots & * & * \\ \vdots & & \vdots & \vdots \\ * & \cdots & * & * \end{pmatrix},$$

with the blocks on the antidiagonal coming from the pairing between $E_\infty^{p,q}(M)$ and $E_\infty^{l-p,n-q}(M)$, $E_{(n+3)/2-q}^{p,q}(M) \supset \operatorname{Im} d_{(n+3)/2-q}^{k-(n+3)/2,(n+1)/2}$ and $E_{(n+3)/2-q}^{l-p,n-q}(M) \supset \operatorname{Im} d_{q-(n-1)/2}^{l-p,n-q})^*$, etc.

To see this, we observe that in terms of the identification,

$$H^k(C_\pi M) \cong \oplus_{p+q=k} E_\infty^{p,q}(C_\pi M),$$

and

$$H^{m+1-k}(C_\pi M, M) \cong \oplus_{p'+q'=k} E_\infty^{l-p',n+1-q'}(C_\pi M, M),$$

the pairing is of the form

$$\begin{pmatrix} 0 & \cdots & 0 & * \\ 0 & \cdots & * & * \\ \vdots & & \vdots & \vdots \\ * & \cdots & * & * \end{pmatrix},$$

with the entries on the antidiagonal coming from the pairing between $E_\infty^{l-p,n+1-q}(C_\pi M, M)$ and $E_\infty^{p,q}(C_\pi M)$. This is a consequence of the following formal properties of the spectral sequence:

$$E_\infty^{p,k-p} = F^p H^k / F^{p+1} H^k,$$

$$F^p(C_\pi M, M) \cdot F^{p'}(C_\pi M) \subset F^{p+p'}((C_\pi M, M)),$$

and

$$F^{p+p'} = 0, \text{ if } p + p' > l.$$

We now consider the pairings between $E_\infty^{l-p,n+1-q}(C_\pi M, M)$ and $E_\infty^{p,q}(C_\pi M)$. For this we need to once again look at the isomorphisms in Propositions 5.1 and 5.2. As we have shown before, an element $[\theta] \in E_r^{l-p,n-q}(M)$ lifts to $[d\theta, \theta] \in E_\infty^{l-p,n+1-q}(C_\pi(M), M)$, where $r = (n+3)/2 - q + k$ with k a non-negative integer. On the other hand, for $q \le (n-1)/2$, an element, $[\omega] \in E_\infty^{p,q}(M)$, $\operatorname{Im} d_{(n+3)/2-q}^{k-(n+3)/2,(n+1)/2}$, $\cdots$, lifts to $[\omega] \in E_\infty^{p,q}(C_\pi M)$. Hence, by tracing through the isomorphisms, we find that the pairing is given by

$$\langle [\theta], [\omega] \rangle = \int_{C_\pi M} d\theta \wedge \omega - \int_M \theta \wedge \omega.$$

Now

$$\text{h-}\deg d\theta \wedge \omega \ge \text{h-}\deg d\theta + \text{h-}\deg \omega \ge l+2,$$

which implies

$$\omega \wedge d\theta = 0.$$

Thus,

$$\langle [\theta], [\omega] \rangle = -\int_M \theta \wedge \omega.$$

If both $[\theta]$ and $[\omega]$ are elements of E_r, i.e., if both are at the same level, this yields

$$\langle [\theta], [\omega] \rangle = -\langle [\theta] \cdot [\omega], \xi_r \rangle.$$

We now show that if $[\theta]$ and $[\omega]$ are in different levels of the spectral sequence, then the pairing,

$$\int_M \theta \wedge \omega,$$

is of the desired triangular form. We do it for

$$[\theta] \in \operatorname{Im} d_{(n+3)/2-q}^{k-(n+3)/2,(n+1)/2}, \quad [\omega] \in \operatorname{Im} (d_{(n+5)/2-q}^{l-p,n-q})^*.$$

The remaining cases are similar.

By assumption $\theta = d\alpha$, for some $\alpha \in F^{k-(n+3)/2}$. Therefore, by the L^2-Stokes' theorem,

$$\int_M \theta \wedge \omega = (-1)^k \int_M \alpha \wedge d\omega.$$

Since

$$\text{h-}\deg \alpha \wedge d\omega \ge k - (n+3)/2 + l - p + (n+5)/2 - q = l+1,$$

it follows that $\alpha \wedge d\omega = 0$ as claimed.

To compute the signature of $C_\pi M$, we note that the pairing (6.1) factors through the pairing (6.3), via the map in (6.2). It is known that the radical for this pairing is

$$\operatorname{Im}(H^{k-1}(M) \to H^k(C_\pi M, M)),$$

which in terms of the Leray spectral sequence is given by

$$\oplus_{p'+q'=k, q' \leq (n-1)/2} E_\infty^{l-p', n-q'}(M).$$

It follows that we can choose our decomposition so that the pairing will look like

$$\begin{pmatrix} 0 & & \cdots & & 0 \\ & \begin{pmatrix} 0 & \cdots & 0 & * \\ 0 & \cdots & * & * \\ \vdots & & \vdots & \vdots \\ * & \cdots & * & * \end{pmatrix} & \\ 0 & & & & \\ & & & & \\ * & & \cdots & & * \end{pmatrix},$$

with the entries on the antidiagonal given by the nondegenerate pairing

$$
\begin{array}{ccccc}
\mathrm{Im}\,(d_r^{p,q})^* & \otimes & \mathrm{Im}\,(d_r^{p,q})^* & \to & \mathbf{R} \\
\cap & & \cap & & \\
E_r^{p,q} & \otimes & E_r^{p,q} & \to & \mathbf{R} \\
\varphi & & \psi & \to & -\langle \varphi \cdot d_r \psi, \xi_r \rangle
\end{array}
\tag{6.5}
$$

for $[\frac{n+r}{2}] \leq q \leq [\frac{n+r}{2}]+r-2$ and $p+q = \frac{m-1}{2}$. Now, one can deform the matrix, without changing its signature, to one whose entries are all zero except on the antidiagonal. Consequently, $\mathrm{sign}_{(2)}(T)$ is given by the signature of the pairing (6.5) restricted to the direct sum of the $E_r^{p,q}$, with $[\frac{n+r}{2}] \leq q \leq [\frac{n+r}{2}]+r-2$ and $p+q = \frac{m-1}{2}$.

To get the final result, we observe the following symmetry of the pairing on $E_r^{\frac{m-1}{2}} = \oplus_{p+q=\frac{m-1}{2}} E_r^{p,q}$. Recall the inner product, $(\,,\,)_r$, on E_r, defined at the end of Section 4. Also, there is a natural basis, $\xi_r \in E_r^{l,n}$ (i.e., a volume element) constructed from the orientation. Define the finite dimensional star operator, $\star_r$, by

$$\langle \varphi \cdot \psi, \xi_r \rangle = (\varphi, \star_r \psi)_r.$$

In fact, $\star_2$ coincides with the usual $*$ operator when E_2 is identified with the harmonic differential forms on the base with values in the harmonic differential forms along the fibers. Then τ_r, the signature of the pairing on $E_r^{\frac{m-1}{2}}$, is exactly the signature of the self-adjoint operator, $\star_r d_r$, on $E_r^{\frac{m-1}{2}}$. Now,

$$\star_r d_r : E_r^{p,q} \to E_r^{l-p-r, n-q+r-1}.$$

Moreover $q = n-q+r-1$ iff $q = \frac{n+r-1}{2}$. It follows that we have the decomposition,

$$E_r^{\frac{m-1}{2}} = A_0 \oplus A_1 \oplus A_2,$$

where if r is even,

$$A_0 = E_r^{(l-r)/2, (n+r-1)/2},$$

and otherwise, $A_0 = 0$.

Also,

$$A_1 = \sum_{p+q=\frac{m-1}{2}, q<\frac{n+r-1}{2}} E_r^{p,q},$$

and

$$A_2 = \sum_{p+q=\frac{m-1}{2}, q>\frac{n+r-1}{2}} E_r^{p,q}.$$

With respect to this decomposition, the operator $\star_r d_r$ restricts to $\star_r d_r$ on A_0 and has the form

$$\begin{pmatrix} 0 & T \\ T^* & 0 \end{pmatrix}.$$

Hence, the spectrum of this operator is symmetric about 0. This shows that only the term, $E_r^{(l-r)/2,(n+r-1)/2}$, contributes to the signature, τ_r, and in fact, only when r is even. This completes the proof of Theorem 1.1.   □

7 L^2-Cohomology of Hyperkähler Manifolds

In recent years, there has been considerable interest in L^2-harmonic forms on noncompact moduli spaces arising in gauge theory. Typically, these spaces come equipped with hyperkähler structures. In [17], using Gromov's Kähler hyperbolicity trick as adapted by Jost-Zuo [20], (see also [8]), Hitchen showed that for complete hyperkähler manifolds, with one of the Kähler forms having linear growth, the L^2-harmonic forms are all concentrated in the middle dimension. Moreover, these L^2-harmonic forms are either all self-dual or all anti–self-dual. Thus, on these hyperkähler manifolds, the L^2-cohomology is completely determined by the L^2-signature.

Theorem 7.1 (Hitchin). *Let M be a complete hyperkähler manifold of real dimension $4k$ such that one of the Kähler forms $\omega_i = d\beta$ where β has linear growth. Then any L^2-harmonic forms are primitive and of the type (k,k) with respect to all complex structures. Therefore, they are anti–self-dual when k is odd and self-dual when k is even.*

The complete hyperkähler manifolds that are ALE have been classified by Kronheimer [21]. The general classification remains open. Thus far, all known examples come with a fibered (or more generally, stratified) geometric structure at infinity. In such cases, our results can be applied. More precisely, consider a complete hyperkähler manifold, M, of real dimension $4k$, such that

$$M = M_0 \cup M_1, \cup \cdots \cup M_r, \tag{7.1}$$

where M_0 is a compact manifold with boundary and each of M_i, $1 \leq i \leq r$, is a geometrically fibered end ([16]). By this we mean that there is a fibration

$$Z_i \to Y_i \xrightarrow{\pi} B_i$$

such that $M_i = [1, \infty) \times Y_i$, and the metric is quasi-isometric to the fibered cusp metric,

$$g_{M_i} = dr^2 + \pi^* g_{B_i} + e^{-2r} g_{Z_i}, \tag{7.2}$$

respectively, the fibered boundary metric

$$g_{M_i} = dr^2 + r^2 \pi^* g_{B_i} + g_{Z_i}. \tag{7.3}$$

Such metrics appear in the ALF and ALG gravitational instantons, for example, Taub-NUT space.

Theorem 7.2. *With the same hypothesis as above, we have*

$$\mathrm{sign}_{(2)}(M) = \mathrm{sign}(M_0) + \sum_{i=1}^{r} \tau(M_i),$$

Proof. We note that our result on the conical singularity extends easily to cusp singularities. This is because it depends only on two ingredients: Lemma 3.1 on the radially constant forms and the Poincaré lemma, Lemma 3.2. Both of these are true for cusp singularities; see [31, 32]. Also, the fibered conical end is conformally equivalent to a fibered cusp end. Because the middle dimensional L^2-cohomology is conformally invariant, the theorem follows. $\square$

Acknowledgment The second author would like to thank Tamas Hausel, Eugenie Hunsicker, and Rafe Mazzeo for very stimulating conversations. We also thank the referee for useful suggestions. During the preparation of this work, the first author was supported by NSF grant DMS 0105128, while the second author was supported by NSF grant DMS 0707000.

References

1. M.F. Atiyah. The signature of fibre-bundles. In *Global Analysis*. Princeton University Press, 1969, pp. 73–84.
2. M.F. Atiyah, V.K. Patodi, and I.M. Singer. Spectral asymmetry and riemannian geometry. I, II, III. *Math. Proc. Cambridge Philos. Soc.*, 77(1975):43–69, 78(1975):405–432, 79(1976):71–99.
3. J.-M. Bismut and J. Cheeger. Families index for manifolds with boundary, superconnections, and cones. I, II. *J. Funct. Anal.*, 89:313–363, 90:306–354, 1990.
4. J.-M. Bismut and J. Cheeger. Remarks on famlies index theorem for manifolds with boundary. eds. Blaine Lawson and Kitti Tenanbaum. *Differential Geometry*. Pitman Monogr. Surveys Pure Appl. Math., 52, Longman Sci. Tech., Harlow, 1991, pp. 59–83.
5. J.-M. Bismut and J. Cheeger. η-invariants and their adiabatic limits. *J. Am. Math. Soc.*, 2:33–70, 1989.
6. J.-M. Bismut and D.S. Freed. The analysis of elliptic families I, II. *Commun. Math. Phys.*, 106:159–167, 107:103–163, 1986.
7. R. Bott and L. Tu. *Differential Forms in Algebraic Topology*. Graduate Text in Mathematics, 82. Springer-Verlag, New York Berlin, 1982.

8. J. Cao and X. Frederico. Kähler parabolicity and the Euler number of compact manifolds of non-positive sectional curvature. *Math. Ann.*, 319(3):483–491, 2001.

9. J. Cheeger. On the Hodge theory of Riemannian pseudomanifolds. *Geometry of the Laplace Operator*, Proc. Sympos. Pure Math., XXXVI, Am. Math. Soc., Providence, RI, 1980, pp. 91–146.

10. J. Cheeger. Spectral geometry of singular Riemannian spaces. *J. Diff. Geom.*, 18:575–657, 1983.

11. J. Cheeger. Eta invariants, the adiabatic approximation and conical singularities. *J. Diff. Geom.*, 26:175–211, 1987.

12. J. Cheeger, M. Goresky, and R. MacPherson. L^2-cohomology and intersection homology of singular algebraic varieties. *Seminar on Differential Geometry*, Ann. of Math. Stud., 102, Princeton Univ. Press, Princeton, N.J, 1982, pp. 303–340.

13. X. Dai. Adiabatic limits, nonmultiplicativity of signature, and Leray spectral sequence. *J. Am. Math. Soc.*, 4:265–321, 1991.

14. M. Goresky and R. MacPherson. Intersection homology theory. *Topology*, 19:135–162, 1980.

15. M. Goresky and R. MacPherson. Intersection homology II. *Invent. Math.*, 71:77–129, 1983.

16. T. Hausal, E. Hunsicker, and R. Mazzeo. The hodge cohomology of gravitational instantons. *Duke Math. J.* 122, no. 3:485–548, 2004.

17. N. Hitchin. L^2-cohomology of hyperkähler quotients. *Comm. Math. Phys.*, 211(1):153–165, 2000.

18. E. Hunsicker. Hodge and signature theorems for a family of manifolds with fibration boundary. *Geometry and Topology*, 11:1581–1622, 2007.

19. E. Hunsicker, R. Mazzeo. Harmonic forms on manifolds with edges. *Int. Math. Res. Not.* no. 52, 3229–3272, 2005.

20. J. Jost and K. Zuo. Vanishing theorems for L^2-cohomology on infinite coverings of compact Kähler manifolds and applications in algebraic geometry. *Comm. Anal. Geom.*, 8(1):1–30, 2000.

21. P. Kronheimer. A Torelli-type theorem for gravitational instantons. *J. Diff. Geom.*, 29(3):685–697, 1989.

22. E. Looijenga. L^2-cohomology of locally symmetric varieties. *Compositio Math.*, 67:3–20, 1988.

23. R. Mazzeo. The Hodge cohomology of a conformally compact metric. *J. Diff. Geom.*, 28:309–339, 1988.

24. R. Mazzeo and R. Melrose. The adiabatic limit, Hodge cohomology and Leray's spectral sequence for a fibration. *J. Diff. Geom., 31,* no. 1, 185–213, 1990.

25. R. Mazzeo and R. Phillips. Hodge theory on hyperbolic manifolds. *Duke Math. J.*, 60:509–559, 1990.

26. W. Pardon and M. Stern. L^2-$\bar{\partial}$-cohomology of complex projective varieties. *J. Am. Math. Soc.*, 4(3):603–621, 1991.

27. J. Serre, S. Chern, F. Hirzebruch. On the index of a fibered manifold. *Proc. AMS*, 8:587–596, 1957.

28. L. Saper and M. Stern. L^2-cohomology of arithmetic varieties. *Ann. Math.*, 132(2):1–69, 1990.

29. L. Saper and S. Zucker. An introduction to L^2-cohomology. *Several Complex Variables and Complex Geometry*, Proc. Sympos. Pure Math., 52, Part 2, Am. Math. Soc., Providence, RI, 519–534, 1991.

30. E. Witten. Global gravitational anomalies. *Commun. Math. Phys.*, 100:197–229, 1985.

31. S. Zucker. Hodge theory with degenerating coefficients. L_2 cohomology in the Poincará metric. *Ann. Math. (2)*, 109:415–476, 1979.

32. S. Zucker. L_2 cohomology of warped products and arithmetic groups. *Invent. Math.*, 70:169–218, 1982.

Hirzebruch Surfaces and Weighted Projective Planes

Paul Gauduchon

Abstract For any positive integer, we show that the standard self-dual orbifold Kähler structure of the weighted projective surface $\mathbb{P}_{1,1,k}$ can be realized as a limit of the Hirzebruch surface $\mathbb{F}_k$, equipped with a sequence of Calabi extremal Kähler metrics whose Kähler classes tend to the boundary of the Kähler cone, and that this collapsing process is compatible with the natural toric structures of $\mathbb{P}_{1,1,k}$ and $\mathbb{F}_k$.

In reference to [25], nontrivial (geometrically) ruled surfaces of genus zero are usually called *Hirzebruch surfaces*. The first Hirzebruch surface $\mathbb{F}_1$ is well-known to be the blow-up of the complex projective plane at one point; more generally, the k-th Hirzebruch surface $\mathbb{F}_k$ is the blow-up of the *weighted projective plane* $\mathbb{P}_k^2$ of weight $\boldsymbol{k} = (1,1,k)$ at its (unique) singular point, cf., e.g., [19]. The aim of this article is to show that, for any fixed positive integer k, the weighted projective plane $\mathbb{P}_k^2$, equipped with its *standard self-dual orbifold Kähler metric* — cf. Section 1 — can be viewed as a limit of the Hirzebruch surface $\mathbb{F}_k$, when the latter is equipped with a sequence of Calabi extremal Kähler metrics whose Kähler classes tend to the boundary of the Kähler cone. Moreover, we show that this limiting — or collapsing — process fits nicely with the natural toric structures of $\mathbb{F}_k$ and $\mathbb{P}_k^2$.

Notice that our construction can be regarded as an illustration of the general *weak compactness theorem* recently established by X. Chen and B. Weber in [16], cf. also [15].

In order to make this paper reasonably self-contained, we included a somewhat detailed exposition of the Bochner-flat Kähler metrics of weighted projective spaces in general (Section 1), of Calabi extremal Kähler metrics on Hirzebruch surfaces (Section 2), and of their toric structures (Section 3). The limiting process itself is firstly described in the toric setting in Section 3, then, in a more precise formulation — cf. Theorem 2 — in Section 4.

P. Gauduchon
École Polytechnique-UMR 7640 du CNRS, Palaiseau, France

K. Galicki and S.R. Simanca (eds.), *Riemannian Topology and Geometric Structures on Manifolds,* Progress in Mathematics 271, DOI 10.1007/978-0-8176-4743-8,
© Springer Science+Business Media LLC 2009

1 The Bochner-Flat Metric of Weighted Projective Spaces

Let $a = (a_1, \ldots, a_{m+1})$ be any $(m+1)$-uple of positive integers, with $\gcd(a_1, \ldots, a_{m+1}) = 1$. The weighted (complex) projective space of weight a, denoted by $\mathbb{P}_a^m$, is defined as the quotient of $\mathbb{C}^{m+1} \setminus \{0\}$ by the weighted $\mathbb{C}^*$-action defined by

$$\zeta \cdot (z_1, \ldots, z_{m+1}) = (\zeta^{a_1} z_1, \ldots, \zeta^{a_{m+1}} z_{m+1}). \tag{1}$$

Alternatively, $\mathbb{P}_a^m$ can be realized as the quotient of the unit sphere $S^{2m+1} \subset \mathbb{C}^{m+1}$ by the S^1-action obtained by restricting the above $\mathbb{C}^*$-action to S^1. Together with the standard (flat) CR-structure of S^{2m+1}, the generator of the weighted S^1-action makes S^{2m+1} into a Sasakian manifold; according to [33], the induced (orbifold) Kähler metric on $\mathbb{P}_a^m$ is *Bochner-flat*, cf. [18] for details[1]; we also refer the reader to the beautiful forthcoming book by C. Boyer and K. Galicki [8] for complete information concerning orbifolds in general, Sasakian structures, etc.

In this paper, Bochner-flat metrics on weighted projective spaces $\mathbb{P}_a^m$ constructed as above and suitably rescaled will be referred to as *standard Bochner-flat Kähler metrics* — *standard self-dual Kähler metrics* if $m = 2$ — or, simply, *standard metrics*. The overall scaling is chosen in such a way that, in the case when all a_j's are equal to 1 — $\mathbb{P}_a^m$ is then the usual complex projective space $\mathbb{P}^m$ — the standard metric be the Fubini–Study metric of (constant) holomorphic sectional curvature equal to 2, see Section 2.

For any $\mathbf{z} = (z_1, \ldots, z_{m+1})$ in $\mathbb{C}^{m+1} \setminus \{0\}$, the corresponding element of $\mathbb{P}_a^m$ will be denoted $[\mathbf{z}] = (z_1 : \cdots : z_{m+1})$ (beware that the "homogeneous coordinates" $(z_1 : \cdots : z_{m+1})$ are relative to the chosen weight a, so that $(z_1 : \cdots : z_{m+1}) = (\zeta^{a_1} z_1 : \ldots : \zeta^{a_{m+1}} z_{m+1})$ for any ζ in $\mathbb{C}^*$).

For each index $j = 1, \ldots, m+1$, denote by $U^{(j)}$ the open affine subset of $\mathbb{P}_a^m$ defined by the condition $z_j \neq 0$. For definiteness, assume that $j = m+1$; then, any element of $U^{(m+1)}$ has a unique representative in $\mathbb{C}^{m+1}$ of the form $(w_1, \ldots, w_m, 1)$, and the m-uple $(w_1, \ldots, w_m)$ is then uniquely defined up to a a_{m+1}-th root of 1. Moreover, for any m-uple $w_1, \ldots, w_m$, there exists a unique positive number, $s = s(w_1, \ldots, w_m)$, such that $s \cdot (w_1, \ldots, w_m, 1) = (s^{a_1} w_1, \ldots, s^{a_m} w_m, s^{a_{m+1}})$ belongs to the unit sphere S^{2m+1} in $\mathbb{C}^{m+1}$, namely the unique positive root of the equation

$$\sum_{i=1}^{m} s^{2a_i} |w_i|^2 + s^{2a_{m+1}} = 1. \tag{2}$$

Notice that s can be continuously extended to $\mathbb{P}_a^m$ by setting $s = 0$ on $\mathbb{P}_a^m \setminus U^{(m+1)}$.

From now on, we denote by $u = (u_1, \ldots, u_{m+1})$ any representative of an element of $\mathbb{P}_a^m$ in the unit sphere S^{2m+1}; we then have $\sum_{j=1}^{m+1} |u_j|^2 = 1$ and

$$|u_{m+1}| = s^{a_{m+1}} \tag{3}$$

---

[1] A proof of the existence of Bochner-flat Kähler — or *Bochner–Kähler* — metrics on weighted projective spaces, as well as an explicit description of these metrics, first appeared in Bryant's seminal paper [9].

on the affine subspace $U^{(m+1)}$. The Kähler form of the standard metric of $\mathbb{P}^m_a$ has then the following expression on $U^{(m+1)}$:

$$\omega_a = -\frac{1}{a_{m+1}} dd^c \log |u_{m+1}| \tag{4}$$

whereas the Ricci form and the scalar curvature are given by

$$\rho_a = \left(\sum_{j=1}^{m+1} a_j\right) \omega_{FS} + \frac{(m+2)}{2} dd^c \log \sum_{j=1}^{m+1} a_j |u_j|^2, \tag{5}$$

and

$$s_a = 4(m+1) \sum_{j=1}^{m+1} a_j - 2(m+1)(m+2) \frac{\sum_{j=1}^{m+1} a_j^2 |u_j|^2}{\sum_{j=1}^{m+1} a_j |u_j|^2}, \tag{6}$$

respectively, for any representative $(u_1,\ldots,u_{m+1})$ in S^{2m+1}, cf., e.g., Propositions 4 and 5 in [18].

Each weighted projective space $\mathbb{P}^m_a$, equipped with its standard metric, can be made into a toric Kähler orbifold in the following manner. Let $\mathbb{T}^{m+1} = S^1 \times \ldots \times S^1$ — $(m+1)$ factors — be the standard torus acting in the standard way on $S^{2m+1} \subset \mathbb{C}^{m+1}$. The weighted action of S^1 on S^{2m+1} can be viewed as the restriction of this action to S^1, via the embedding $j_a : S^1 \to \mathbb{T}^{m+1}$ defined by $j_a(\zeta) = (\zeta^{a_1},\ldots,\zeta^{a_{m+1}})$. The quotient torus $\mathbb{T}^m_a := \mathbb{T}^{m+1}/j_a(S^1)$ then acts effectively on $\mathbb{P}^m_a$; by its very construction, the standard metric of $\mathbb{P}^m_a$ is preserved by this action; moreover, this action is Hamiltonian. In order to get an explicit expression of the momentum map $\mu_a : \mathbb{P}^m_a \to (\mathfrak{t}^m_a)^*$, we first identify the Lie algebra, $\mathfrak{t}^{m+1}$, of $\mathbb{T}^{m+1}$ with $\mathbb{R}^{m+1}$ via the isomorphism $\mathbb{T}^{m+1} = \mathbb{R}^{m+1}/\mathbb{Z}^{m+1}$ induced by the map $(\theta_1,\ldots,\theta_{m+1}) \to (e^{2\pi i\theta_1},\ldots,e^{2\pi i\theta_{m+1}})$ from $\mathbb{R}^{m+1}$ to $\mathbb{T}^{m+1}$; we then denote by $e_1,\ldots,e_{m+1}$ the natural basis of $\mathfrak{t}^{m+1} = \mathbb{R}^{m+1}$, by $e_1^*,\ldots,e_{m+1}^*$ the dual basis of $(\mathfrak{t}^{m+1})^*$, by $\xi_1,\ldots,\xi_{m+1}$ the associated coordinates on $(\mathfrak{t}^{m+1})^*$; finally, the Lie algebra, $\mathfrak{t}^m_a$, of $\mathbb{T}^m_a$ is identified with the quotient of $\mathfrak{t}^{m+1} = \mathbb{R}^{m+1}$ by the real line generated by $e_a = \sum_{r=1}^{m+1} a_r e_r$ — in particular, the defining lattice, Λ_a, of $\mathbb{T}^m_a$ in $\mathfrak{t}^m_a$ is the quotient of $\mathbb{Z}^{m+1}$ by the sublattice $\mathbb{Z} e_a$ — whereas the dual $(\mathfrak{t}^m_a)^*$ is identified with the hyperplane $\sum_{j=1}^{m+1} a_j \xi_j = 0$ in $(\mathfrak{t}^{m+1})^*$. With this notation, the momentum map μ_a has the following expression:

$$\mu_a([u]) = \sum_{j=1}^{m+1} \left(\frac{|u_j|^2}{\sum_{r=1}^{m+1} a_r |u_r|^2} - \frac{1}{(m+1)a_j} \right) e_j^* \tag{7}$$

for any $[u]$ in $\mathbb{P}^m_a$, represented by $(u_1,\ldots,u_{m+1})$ in S^{2m+1}, cf., e.g., [18, Section 2.4]. The image of μ_a in $(\mathfrak{t}^m_a)^*$ is then the simplex, Δ_a, of vertices $p_1,\ldots,p_{m+1}$, with $p_j = (-\frac{1}{(m+1)a_1},\ldots,\frac{m}{(m+1)a_j},\ldots,-\frac{1}{(m+1)a_{m+1}})$; equivalently, Δ_a is determined in $(\mathfrak{t}^m_a)^*$ by the set of inequalities:

$$\ell_j(\xi) := \xi_j + \lambda_j \geq 0, \tag{8}$$

$j = 1, \ldots, m+1$, with $\lambda_j = \frac{1}{(m+1)a_j}$. In particular, for each facet, $F_j = \{\xi \in \Delta_a \mid \ell_j(\xi) = 0\}$, the corresponding (inward) normal of F_j in $\mathfrak{t}_a$ is $[e_j]$, the class of $e_j \mod \mathbb{R}\, e_a$.

In general, if (V, Γ) is a real n-dimensional vector space equipped with a lattice $\Gamma \cong \mathbb{Z}^n$, a (compact) convex polytope, Δ, in the dual vector space V^* equipped with the dual lattice Γ^*, is called a *Delzant polytope* with respect to Γ if it satisfies the following conditions [17]:

$\mathbb{D}_1$. Δ is *simple*: exactly n facets meet at each vertex;

$\mathbb{D}_2$. Δ is *rational* with respect to Γ: for each facet, the (inward) normal, in $\mathfrak{t}$, can be chosen in the lattice Γ;

$\mathbb{D}_3$. Δ satisfies the *Delzant condition*: for each vertex, the (inward) normals to the n adjacent facets can be chosen so as to form a $\mathbb{Z}$-basis of Γ.

In the current situation, the simplex Δ_a clearly meets conditions $\mathbb{D}_1$ and $\mathbb{D}_2$, as all $[e_j]$'s belong to Λ_a, but does not satisfy the Delzant condition $\mathbb{D}_3$, except in the case when all a_j's are equal to 1; more precisely, Δ_a satisfies the Delzant condition at the vertex p_j if and only if $a_j = 1$. Moreover, each facet F_j is labeled, in the sense of [28], by $d_j = \gcd(a_1, \ldots, \hat{a}_j, \ldots, a_{m+1})$, meaning that $[e_j]/d_j$ still belongs to Λ_a and is primitive there.

The interior, Δ_a^0, of Δ_a is the image by μ_a of the open subset, U, of elements of $\mathbb{P}_a^m$ represented by $(u_1, \ldots, u_{m+1})$ in S^{2m+1}, with $u_j \neq 0$ for all j's; equivalently, $U = \cap_{j=1}^{m+1} U^{(j)}$. By (4), the restriction of ω_a to U is equal to $-\frac{1}{a_j} dd^c \log|u_j|$ for *all* j's; it can then be rewritten as $\omega_a = dd^c F$, with

$$F = -\sum_{j=1}^{m+1} \frac{1}{(m+1)a_j} \log|u_j|. \tag{9}$$

Now, via the momentum map μ_a, the ℓ_j's defined by (8) can be viewed as functions defined on $\mathbb{P}_a^m$, and we then have

$$\ell_j([u]) = \frac{|u_j|^2}{\sum_{r=1}^{m+1} a_r |u_r|^2}, \tag{10}$$

hence

$$\log|u_j| = \frac{1}{2}\left(\log \ell_j([u]) - \log \ell_0([u])\right), \tag{11}$$

by setting $\ell_0 := \sum_{r=1}^{m+1} \ell_r$.

Similarly, the Kähler potential F on U can be viewed as a function defined on Δ_a^0; in view of (9) and (11), it has there the following expression:

$$F = \frac{1}{2}\left(\lambda_0 \log \ell_0 - \sum_{j=1}^{m+1} \lambda_j \log \ell_j\right), \tag{12}$$

with $\lambda_0 := \sum_{r=1}^{m+1} \lambda_r$, cf. [1, Theorem 3].

According to the general theory of toric Kähler manifolds developed by V. Guillemin [23], cf. also [1], the induced metric, g_{Δ_a}, on Δ_a^0 is of Hessian type, i.e., is of the form

$$g_{\Delta_a} = D^0 d G, \tag{13}$$

where G — the *symplectic potential* of the toric Kähler structure — is a real function defined on $\Delta_a^0 \subset (\mathfrak{t}_a^m)^*$ and where D^0 denotes the standard flat connection on $(\mathfrak{t}_a^m)^*$. The symplectic potential G is deduced from the Kähler potential F by the *Legendre transformation*:

$$G = \langle \tau, \xi \rangle - F, \tag{14}$$

where τ is a $\mathfrak{t}_a^m$-valued function on Δ_a^0 — equivalently, a $\mathbb{T}_a^m$-invariant $\mathfrak{t}_a^m$-valued function defined on U — called the *dual momentum map*.[2] This is defined in terms of the complexified action of $\mathbb{T}_a^m$ on $\mathbb{P}_a^m$ induced by the standard holomorphic action of $\mathbb{T}_\mathbb{C}^{m+1} = \mathbb{C}^* \times \ldots \times \mathbb{C}^*$ $((m+1)$ factors) on $\mathbb{C}^{m+1}$ and depends on the choice of a base point of U. If the latter is chosen to be $[\mathbf{1}] = (1 : 1 : \ldots : 1)$, $\tau = \sum_{j=1}^{m+1} \tau_j [e_j]$ is determined by

$$[\mathbf{z}] = (e^{\tau_1([\mathbf{z}])} : \ldots : e^{\tau_{m+a}([\mathbf{z}])}) \mod \mathbb{T}_a^m; \tag{16}$$

equivalently,

$$\tau([\mathbf{z}]) = \sum_{j=1}^{m+1} \log |z_j| [e_j], \tag{17}$$

for any $[\mathbf{z}] = (z_1 : \ldots : z_{m+1})$ in U; in view of (11), we infer

$$\tau = \frac{1}{2} \sum_{j=1}^{m+1} \left(\log \ell_j - \log \ell_0 \right) [e_j] \tag{18}$$

as a $\mathfrak{t}_a^m$-valued function on Δ_a^0; we thus have

$$\langle \tau, \xi \rangle = \frac{1}{2} \sum_{j=1}^{m+1} (\log \ell_j - \log \ell_0)(\ell_j - \lambda_j). \tag{19}$$

By (14) and (13), the symplectic potential and the induced metric on Δ_a^0 have then the following expressions:

$$G = \frac{1}{2} \left(\sum_{j=1}^{m+1} \ell_j \log \ell_j - \ell_0 \log \ell_0 \right), \tag{20}$$

---

[2] In general, the Kähler potential F, the symplectic potential G, the induced metric g_{Δ_a}, and the dual momentum map τ, all regarded as defined on Δ_a^0 via the momentum map μ_a, are related to each other by (14) and by:

$$dF = \langle d\tau, \xi \rangle, \qquad dG = \langle \tau, d\xi \rangle, \qquad g_{\Delta_a} = \langle d\tau \otimes d\xi \rangle, \tag{15}$$

cf. [23].

and

$$g_{\Delta_a} = \frac{1}{2} \left(\sum_{j=1}^{m+1} \frac{d\ell_j \otimes d\ell_j}{\ell_j} - \frac{d\ell_0 \otimes d\ell_0}{\ell_0} \right), \tag{21}$$

cf. [9, Section 4.4.6].

2 Calabi Extremal Kähler Metrics on Hirzebruch Surfaces

In this section, we briefly recall the construction by E. Calabi [10] of extremal Kähler metrics on *Hirzebruch surfaces*. Our treatment closely follows [22, Chapter 9] and [4]. For detail, in particular for statements and formulae here given without proof, we refer the reader to these references and to Calabi's original paper [10].

In general, a (complex) ruled surface is the total space of the projective bundle $M = \mathbb{P}(E)$ associated to a holomorphic rank 2 vector bundle E over a compact Riemann surface Σ, endowed with its natural complex structure. When Σ is the complex projective line $\mathbb{P}^1 = \mathbb{P}(\mathbb{C}^2)$, E splits as the sum of two holomorphic line bundles. Without loss of generality, we can then assume that $M = \mathbb{P}(1 \oplus L)$, where 1 stands for the trivial line bundle $\mathbb{P}^1 \times \mathbb{C}$, and L denotes a holomorphic line bundle over $\mathbb{P}^1$ of nonpositive degree $-k$, for some nonnegative integer k. If $k = 0$, L is the trivial line bundle and M is then the product $M = \mathbb{P}^1 \times \mathbb{P}^1$. If $k > 0$, $L = \Lambda^k$, where $\Lambda = \mathcal{O}(-1)$ denotes the tautological line bundle over $\mathbb{P}^1$: for any y in $\mathbb{P}^1$, the fiber Λ_y is y itself, viewed as a complex line[3] in $\mathbb{C}^2$. The resulting ruled surface $\mathbb{P}(1 \oplus L)$ is then the k-th *Hirzebruch surface*, here denoted by $\mathbb{F}_k$ or simply M. An element of $\mathbb{F}_k$ is then determined by an element y of $\mathbb{P}^1$, viewed as a complex line in $\mathbb{C}^2$, and by a complex line in the complex 2-plane $\mathbb{C} \oplus y^{\otimes k}$.

Alternatively, $M = \mathbb{F}_k$ is the compactification of the total space of $L = \Lambda^k$ obtained by adding a *point at infinity* to each fiber. The union of these point at infinity, denoted by Σ_∞, is called the *infinity section*, whereas the (image of the) zero section of L is denoted by Σ_0. We then have

$$M = \Sigma_0 \cup M_0 \cup \Sigma_\infty, \tag{22}$$

where $M_0 = L \setminus \Sigma_0$ is the total space of the $\mathbb{C}^*$-principal bundle of non-zero elements of L. When $k = 1$, $M_0 = \Lambda \setminus \Sigma_0$ is naturally identified with $\mathbb{C}^2 \setminus \{0\}$; in general, M_0 is covered by $\mathbb{C}^2 \setminus \{0\}$ via the k-fold covering map

$$u \mapsto u^{\otimes k} \tag{23}$$

from $\mathbb{C}^2 \setminus \{0\} = \Lambda \setminus \Sigma_0$ to $M_0 = \Lambda^k \setminus \Sigma_0$.

For any (compact) complex manifold M, denote by $H(M)$ the identity component of the group of diffeomorphisms of M that preserve the complex structure:

---

[3] Here, and in the whole paper, *complex line* is a short cut for *one-dimensional complex vector space*; similar convention for *complex 2-plane*, etc.

$H(M)$ is then a connected complex Lie group. In particular, $H(\mathbb{P}^1) = PGl(2,\mathbb{C}) = Gl(2,\mathbb{C})/\mathbb{C}^*$, and the action of $PGl(2,\mathbb{C})$ on $\mathbb{P}^1$ is covered by an (effective) action of $Gl(2,\mathbb{C})/\mu_k$ on Λ^k for any positive integer k, where μ_k denotes the group of k-th roots of 1; this action extends to a holomorphic action of $Gl(2,\mathbb{C})/\mu_k$ on $\mathbb{F}_k$, which is transitive on M_0, on Σ_0, and on Σ_∞.

The full group $H(\mathbb{F}_k)$ is then the semidirect product $Gl(2,\mathbb{C})/\mu_k \ltimes H^0(\mathbb{P}^1,(\Lambda^k)^*)$, where $H^0(\mathbb{P}^1,(\Lambda^k)^*) = \odot^k(\mathbb{C}^2)^*$ denotes the space of holomorphic sections of the dual line bundle $(\Lambda^k)^*$, whereas the action of $H^0(\mathbb{P}^1,(\Lambda^k)^*)$ on $\mathbb{F}_k$ is defined as follows: For any α in $H^0(\mathbb{P}^1,(\Lambda^k)^*)$ and for any element of $\mathbb{F}_k$ represented by y in $\mathbb{P}^1$ and by the complex line generated by the pair $(z,u^{\otimes k})$ in the complex 2-plane $\mathbb{C} \oplus y^{\otimes k}$, the image of this element by α is the complex line in $\mathbb{C} \oplus y^{\otimes k}$ generated by the pair $(z + \langle \alpha, u^{\otimes k} \rangle, u^{\otimes k})$. Within this identification, $U(2)/\mu_k \subset Gl(2,\mathbb{C})/\mu_k$ is a maximal compact subgroup of $H(\mathbb{F}_k)$.

The tautological line bundle Λ comes naturally equipped with a fiberwise Hermitian inner product h determined by the usual Hermitian inner product of $\mathbb{C}^2$. The inner product h induces a fiberwise Hermitian inner product on $L = \Lambda^k$ for any positive integer k. Denote by ∇ the corresponding Chern connection and by R^∇ the curvature of ∇; then, $R^\nabla = -ik\,\omega_0$, where ω_0 is the Kähler form of a Kähler metric, g_0, of constant sectional curvature 2 on $\mathbb{P}^1$. We denote by $|\cdot|_h$ the Euclidean distance to the origin with respect to h in each fiber of L and by t the function on M_0 defined by

$$t = \log|\cdot|_h, \tag{24}$$

so that

$$dd^c t = k\,\pi^* \omega_0, \tag{25}$$

where π denotes the natural projection of M on $\mathbb{P}^1$ (here restricted to M_0). The natural S^1-action on L, determined by the multiplication in each fiber by an element of S^1, extends to a holomorphic S^1-action on M, which is free on M_0 and reduces to the identity on Σ_0 and Σ_∞ (this action actually coincides with the restriction of the action of the group $U(2)/\mu_k$ to its center). Denote by T the generator of this action; on M_0 we have

$$dt(T) = 1, \qquad dt(JT) = 0, \qquad dt_{|H^\nabla} = 0, \tag{26}$$

where J denotes the natural complex structure of M, and H^∇ denotes the horizontal distribution on L induced by ∇.

On M, we consider Kähler metrics whose restriction to M_0 admit a Kähler potential, F, which factors through t. These metrics will be referred to as *admissible*. They are clearly invariant under the action of $U(2)/\mu_k$; conversely, any $U(2)/\mu_k$ Kähler metric on $\mathbb{F}_k$ is admissible, cf., e.g., [22, Proposition 9.3.1]. For simplicity, the notation will not distinguish F, function on M_0, and the corresponding function of t, defined on $\mathbb{R}$; in particular, $F',F'' \ldots$ will denote the derivatives of F with respect to t, still regarded as functions on M_0; the same convention will be used for any function on M_0 that factors through t.

For any admissible Kähler metric on M_0, of Kähler potential $F = F(t)$, we set: $\psi = \psi(t) = F'(t)$; the corresponding Kähler form, ω_ψ, has then the following expression:

$$\omega_\psi = dd^c F = \psi \, dd^c t + \psi' \, dt \wedge d^c t. \tag{27}$$

Conversely, the 2-form defined by (27) is the Kähler form of a Kähler metric on M_0 if and only if ψ and its derivative ψ' are positive. Moreover, by (25), $dd^c t$ is well-defined, whereas $e^{2t} dt \wedge d^c t$, resp. $e^{-2t} dt \wedge d^c t$, smoothly extends to Σ_0, resp. to Σ_∞, when t tends to $-\infty$, resp. $+\infty$; it follows that ω_ψ and the corresponding metric, g_ψ, extend to M if and only if ψ satisfied the following properties:

$\mathbb{A}_1$. Near $t = -\infty$, $\psi(t) = \Phi_-(e^{2t})$, where Φ_- is smoothly defined at 0, with $\Phi_-(0) = a > 0$ and $\Phi'_-(0) > 0$.

$\mathbb{A}_2$. Near $t = +\infty$, $\psi(t) = \Phi_+(e^{-2t})$, where $\Phi+$ is smoothly defined at 0, with $\Phi + (0) = b > a > 0$ and $\Phi'_+(0) < 0$.

The Kähler class $[\omega_\psi]$ in $H^2(M, \mathbb{R})$ is then

$$\Omega_{a,b} = 2\pi(-a\Sigma_0 + b\Sigma_\infty), \tag{28}$$

where Σ_0, Σ_∞ here stand for their homology classes in $H_2(M, \mathbb{Z})$, implicitly identified with elements of $H^2(M, \mathbb{Z})$ by Poincaré duality.

In particular, ψ extends to the whole of $\mathbb{F}_k$ and is there a momentum for the natural S^1-action with respect to ω_ψ, meaning that

$$\iota_T \omega_\psi = -d\psi. \tag{29}$$

Since T preserves the whole Kähler structure, it is a Hamiltonian Killing vector field and ψ is then a *Killing potential* with respect to g_ψ.

The momentum ψ maps M to the closed interval $[a, b]$. It may be convenient to substitute the *momentum profile*,[4] Θ, defined on the open interval (a, b) in the following manner: for any x in $[a, b]$, $\Theta(x) = \psi'(\psi^{-1}(x))$ (since ψ is an increasing function from $\mathbb{R}$ to (a, b), its inverse, ψ^{-1}, is a well-defined (increasing) function from (a, b) to $\mathbb{R}$). The momentum profile is positive on (a, b) and the boundary conditions $\mathbb{A}_1$-$\mathbb{A}_2$ for ψ readily imply the following boundary conditions for Θ:

$\mathbb{B}$. Θ smoothly extends to the closed interval $[a, b]$, with

$$\begin{aligned} \Theta(a) &= 0, & \Theta(b) &= 0, \\ \Theta'(a) &= 2, & \Theta'(b) &= -2. \end{aligned} \tag{30}$$

Conversely, the momentum profile determines the Kähler metric in the following sense:

Proposition 1. *Any function Θ smoothly defined on $[a, b]$, which is positive on (a, b) and satisfies the boundary condition $\mathbb{B}$, is the momentum profile of an admissible*

---

[4] The momentum profile has been first introduced and advocated by A. Hwang and M. Singer in [27].

Kähler metric defined on M, for a momentum function ψ uniquely defined up to transformations of the form $\psi(t) \mapsto \tilde{\psi}(t) = \psi(t + t_0)$, for any real number t_0.

Proof. (Sketch). We first recover the function t, up to an additive constant, via the differential equation $\frac{dt}{dx} = \frac{1}{\Theta(x)}$ on the open interval (a,b); since Θ is positive on (a,b), we get x as an increasing function of t, which is the function ψ we are looking for; finally, the boundary condition $\mathbb{B}$ for Θ ensures that ψ satisfies the boundary conditions $\mathbb{A}_1$-$\mathbb{A}_2$ (more detail can be found in [10] or [22]).   □

In general, the Ricci form of a Kähler manifold (M, J, g, ω) has the following local expression: $\rho = -\frac{1}{2}dd^c \log \frac{v_g}{v_0}$, where v_g denotes the volume form of g and v_0 the volume form of the flat Kähler metric determined by any local holomorphic chart. For any admissible Kähler metric g_ψ on M_0, the volume form is $v_{g_\psi} = \frac{\omega_\psi \wedge \omega_\psi}{2} = \psi\psi' dd^c t \wedge dt \wedge d^c t$, whereas a holomorphic chart is provided by the covering map (23) from $\mathbb{C}^2 \setminus \{0\} = \Lambda \setminus \Sigma_0$ to M_0; the Kähler form of the associated flat Kähler metric is then $\omega_0 = \frac{1}{4}dd^c e^{\frac{2t}{k}} = e^{\frac{2t}{k}}(\frac{1}{2k}dd^c t + \frac{1}{k^2}dt \wedge d^c t)$, so that $v_0 = \frac{\omega_0 \wedge \omega_0}{2} = \frac{e^{\frac{4t}{k}}}{k^3}dd^c t \wedge dt \wedge d^c t$. On M_0, the Ricci form of (g_ψ, ω_ψ) is then written as

$$\rho = dd^c w, \tag{31}$$

where the *Ricci potential* $w = w(t)$ is given by

$$w = w(t) = \kappa t - \frac{1}{2}\log \psi(t) - \frac{1}{2}\log \psi'(t), \tag{32}$$

by setting (cf. Remark 2 below)

$$\kappa = \frac{2}{k}, \tag{33}$$

whereas the scalar curvature $s = \frac{2\rho \wedge \omega_\psi}{v_{g_\psi}}$ is given by

$$s = \frac{2(w'\psi)'}{\psi\psi'}. \tag{34}$$

Via the momentum map ψ, the scalar curvature can be read as a function defined on the open interval (a,b) and has then the following expression in terms of the momentum profile Θ:

$$s = \frac{2\kappa - (x\Theta(x))''}{x}. \tag{35}$$

Notice that the derivatives in the rhs of (34) are relative to t, whereas the derivatives in the rhs of (35) are relative to x.

A Kähler metric is extremal if its scalar curvature is a Killing potential, i.e., the momentum of a Hamiltonian Killing vector field [10]. For any admissible Kähler metric g_ψ on M_0, the scalar curvature s factors through t, hence is a Killing potential if and only if it is an affine function of the Killing potential ψ:

$$s = \alpha\psi + \beta, \tag{36}$$

for two real numbers α, β. By using (35), (36) can be expressed as the differential equation

$$(x\Theta(x))'' = -\alpha x^2 - \beta x + 2\kappa, \tag{37}$$

for the momentum profile $\Theta = \Theta(x)$. This readily integrates to:

$$\Theta(x) = \frac{P(x)}{x}, \tag{38}$$

with

$$P(x) = -\frac{\alpha}{12}x^4 - \frac{\beta}{6}x^3 + \kappa x^2 - \frac{\gamma}{6}x - \frac{\delta}{12}, \tag{39}$$

where γ, δ are constants. Moreover, the boundary condition $\mathbb{B}$ for Θ is equivalent to the following conditions for P:

$$\begin{aligned} P(a) = 0, \quad & P(b) = 0, \\ P'(a) = 2a, \quad & P'(b) = -2b. \end{aligned} \tag{40}$$

Equivalently, the (normalized) coefficients $\alpha, \beta, \gamma, \delta$ of P are given by

$$\begin{aligned}
\alpha &= \frac{12(2(a+b) - \kappa(b-a))}{(b-a)(a^2 + 4ab + b^2)}, \\
\beta &= \frac{12(-(a^2 + b^2) + \kappa(b^2 - a^2))}{(b-a)(a^2 + 4ab + b^2)}, \\
\gamma &= ab\beta, \\
\delta &= a^2 b^2 \alpha.
\end{aligned} \tag{41}$$

Conversely, the conditions (40), together with $P''(0) = 2\kappa$, unambiguously determine a polynomial of degree equal to or less than 4 for any pair a, b with $0 < a < b$, i.e., for any Kähler class $\Omega_{a,b}$ on $\mathbb{F}_k$. This polynomial, denoted by $P_{\Omega_{a,b}}$, will be called the *Calabi polynomial* or, according to [4], the *extremal polynomial* of $\Omega_{a,b}$. It can also be written in the following form:

$$P_{\Omega_{a,b}}(x) = (x-a)(b-x)Q_{a,b}(x), \tag{42}$$

with $Q_{a,b} = Ax^2 + Bx + C$ and

$$\begin{aligned}
A &= \frac{2(a+b) - \kappa(b-a)}{(b-a)(a^2 + 4ab + b^2)}, \\
B &= \frac{4ab + \kappa(b^2 - a^2)}{(b-a)(a^2 + 4ab + b^2)}, \\
C &= abA.
\end{aligned} \tag{43}$$

The Calabi polynomial $P_{\Omega_{a,b}}$ is well defined for any Kähler class, independently of the existence of any extremal Kähler metric in $\Omega_{a,b}$. On the other hand, in view of

(38) and Proposition 1, $\Omega_{a,b}$ *does* admit an (admissible) extremal Kähler metric if and only if $P_{\Omega_{a,b}}$ is *positive* on the open interval (a,b).[5] By using (42)–(43), it is easily checked that this condition is satisfied for any $0 < a < b$ and for any positive integer k (we then have $0 < \kappa \le 2$).

We conclude that on any Hirzebruch surface $\mathbb{F}_k$, any Kähler class $\Omega_{a,b}$ contains an admissible extremal Kähler metric, which is determined by

$$\Theta(x) = \frac{P_{\Omega_{a,b}}(x)}{x}, \tag{44}$$

or, equivalently, by

$$\psi \psi' = P_{\Omega_{a,b}}(\psi), \tag{45}$$

hence is uniquely defined up to the transformations $t \mapsto t + t_0$, for any real number t_0, i.e., the holomorphic transformations of $M = \mathbb{P}(1 \oplus L)$ induced by the multiplication in each fiber of L by e^{t_0}.

According to a celebrated theorem of E. Calabi [11], any extremal Kähler metric on $\mathbb{F}_k$ can be made $U(2)/\mu_k$-invariant — hence admissible — by a holomorphic transformation in $H(\mathbb{F}_k)$. We then eventually get:

Theorem 1 (E. Calabi [10,11]). *For any positive integer k, the Hirzebruch surface $\mathbb{F}_k$ admits an extremal Kähler metric in each Kähler class, unique up to the action of $H(\mathbb{F}_k)$.*

Remark 1. In general, the uniqueness issue for extremal Kähler metrics in a Kähler class of a (compact, connected) complex manifold proved to be a highly nontrivial question, even in the particular case of Kähler–Einstein metrics, solved by S. Bando and T. Mabuchi [6]. Uniqueness of extremal metric in a *rational* Kähler class on a projective algebraic manifold M was proved by S. Donaldson [20], cf. also [7], with the assumption that $H(M) = \{1\}$ (extremal metrics are then of constant scalar curvature); this condition has then been removed by T. Mabuchi [29]. Finally, uniqueness of extremal metric up to the action of $H(M)$ was proved in full generality by X. X. Chen-G. Tian [12–14].

Remark 2. The above construction can be extended to any *Hirzebruch-like* ruled surfaces $\mathbb{P}(1 \oplus L)$ over a compact Riemann surface Σ of any genus g, for any holomorphic line bundle L of negative degree $-k$. All arguments and formulae in this section still hold in this extended context, by only substituting

$$\kappa = \frac{2(1-g)}{k} \tag{46}$$

to (33). In particular, admissible Kähler metrics are defined in the same way, Kähler classes are still parameterized by pairs of real numbers a,b with $0 < a < b$, each Kähler class $\Omega_{a,b}$ determines an extremal polynomial $P_{\Omega_{a,b}}$ — by (40) and $P''_{\Omega_{a,b}}(0) = 2\kappa$ — and we conclude as before that $\Omega_{a,b}$ contains an admissible extremal metric

---

[5] A similar theorem holds in a much more general setting, cf. [4], see also Remark 2 below.

if and only if $P_{\Omega_{a,b}}$ is positive on the open interval (a,b). The main point here —
first observed by C. W. Tønnesen-Friedman [32] — is that if $g > 1$, the condition
that $P_{\Omega_{a,b}}$ be positive on (a,b) is only satisfied for "small" Kähler classes; more
precisely, for any $g > 1$ and any $k > 0$, define c_κ as the unique positive root greater
than 1 of the equation

$$(\kappa x^2 + 4x - \kappa)^2 - 4x((2-\kappa)x + (2+\kappa))^2 = 0 \tag{47}$$

where κ is defined by (46); then [32] a Kähler class $\Omega_{a,b}$ on $\mathbb{F}_k$ admits an admissible extremal Kähler metric if and only if $b/a < c_\kappa$. It turns out that Kähler classes
$\Omega_{a,b}$ such that $b/a \geq c_\kappa$ admit no extremal Kähler metric. This has firstly been
proved by Apostolov–Calderbank–Gauduchon–Tønnesen-Friedman [4], by using
Chen–Tian's work [12–14], then, independently and by a different approach rely-
ing on Donaldson's work, in particular on [21], by G. Székelyhidi [30, 31].

Further generalizations of the Calabi construction can be found in several places,
in particular in [26] and in [3, 4].

3 Hirzebruch Surfaces as Toric Kähler Manifolds

On any Hirzebruch surface $\mathbb{F}_k$, the natural S^1-action can be combined with the natural toric structure of $\mathbb{P}^1$ to produce a Hamiltonian $\mathbb{T}^2$-action on $\mathbb{F}_k$ and make it into
a toric Kähler manifold for any admissible Kähler metric.

The natural toric structure of $\mathbb{P}^1$ is determined by the (effective) S^1-action defined, in terms of homogeneous coordinates, by

$$\zeta \cdot (y_1 : y_2) = (\zeta y_1 : y_2), \tag{48}$$

for any ζ in S^1 and any $y = (y_1 : y_2)$ in $\mathbb{P}^1$. We denote by Z the generator of this
action on $\mathbb{P}^1$. This action has a natural lift on the tautological line bundle Λ, whose
expression on $\Lambda \setminus \Sigma_0 = \mathbb{C}^2 \setminus \{0\}$ is

$$\zeta \cdot (u_1, u_2) = (\zeta u_1, u_2), \tag{49}$$

for any $u = (u_1, u_2)$ in $\mathbb{C}^2 \setminus \{0\}$. The generator of this action, denoted by $\hat{Z}$, has then
the following expression:

$$\hat{Z}(u) = (iu_1, 0). \tag{50}$$

On the other hand, we have that

$$\hat{Z} = \tilde{Z} + \mu_Z T, \tag{51}$$

where $\tilde{Z}$ denotes the horizontal lift of Z on Λ, with respect to the Chern connection
∇, and where μ_Z is a momentum of Z relative to ω_0. By definition, μ_Z is *the* momentum determined by the natural lift of the S^1-action on Λ (here and henceforth

we identify μ_Z with $\pi^*\mu_Z$). For any u in $\mathbb{C}^2 \setminus \{0\}$, $\tilde{Z}(u)$ is the part of $\hat{Z}(u)$ that is Hermitian-orthogonal to the complex line $\mathbb{C}u$. From (50) and (51), we thus get:

$$\mu_Z = \frac{|u_1|^2}{|u_1|^2 + |u_2|^2}, \tag{52}$$

and

$$\tilde{Z} = \frac{1}{|u_1|^2 + |u_2|^2}\left(|u_2|^2 iu_1, -|u_1|^2 iu_2\right). \tag{53}$$

From (52)–(53) we infer that the square norm of Z with respect to g_0 is given by

$$|Z|^2_{g_0} = 2\mu_Z(1 - \mu_Z). \tag{54}$$

The S^1-action (48) has a natural lift on $L = \Lambda^k$ for each k, whose generator is $\hat{Z} = \tilde{Z} + k\mu_Z T$. This action extends to the whole of M. The generator of the extended action will be denoted by K_2.

Lemma 1. *For any admissible Kähler form ω_ψ on M, K_2 is a Hamiltonian vector field of momentum σ_2 given by*

$$\sigma_2 = k\psi\mu_Z. \tag{55}$$

Moreover, K_2 preserves J and the Kähler metric g_ψ determined by ω_ψ.

Proof. On M_0, $K_2 = \hat{Z} = \tilde{Z} + k\mu_Z T$, whereas ω_ψ is given by (27); by using (25) and (26), we get

$$\iota_{K_2}\omega_\psi = k\psi\pi^*(\iota_Z\omega_0) + k\mu_Z\psi' \iota_T(dt \wedge d^c t)$$
$$= -k\psi d\mu_Z - k\psi'\mu_Z dt$$
$$= -k d(\psi\mu_Z).$$

Since ψ and μ_Z are defined on the whole of M, we thus get $\iota_{K_2}\omega_\psi = -d\sigma_2$ on M, with σ_2 given by (55). Then $\mathcal{L}_{K_2}\omega_\psi = d(\iota_{K_2}\omega_\psi) = 0$ and σ_2 is a momentum of K_2 with respect to ω_ψ; this proves the first assertion. The S^1-action on Λ described by (49) is clearly holomorphic, and so is the induced action on Λ^k for any k: the generator K_2 of this action thus preserves J, hence also $g_\psi = \omega_\psi(\cdot, J\cdot)$. $\square$

The momentum σ_2 is clearly T-invariant, meaning that T and K_2 commute and that the corresponding momenta, ψ and σ_2 Poisson-commute for any admissible Kähler form ω_ψ. For convenience, T and its momentum ψ relative to ω_ψ will be renamed K_1 and σ_1, respectively: $K_1 = T$ and K_2 are then the generators of a Hamiltonian action of the torus $\mathbb{T}^2$ on M, which makes M into a toric Kähler surface for any admissible Kähler structure. We denote by μ the corresponding momentum, as a map from M to $\mathfrak{t}^*$, the dual of the Lie algebra $\mathfrak{t}$ of $\mathbb{T}^2$. Denote by X_1, X_2 the generators of $\mathfrak{t}$ determined by $K_1 = T, K_2$, respectively, and by x_1, x_2 the associated linear coordinates in $\mathfrak{t}^*$. For any x in M, $\langle\mu(x), X_1\rangle = \sigma_1(x) = \psi(t)$ and

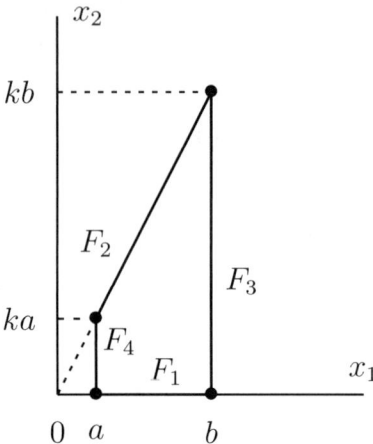

Fig. 1 Momentum polytope $\Delta_{a,b}$ for the Hirzebruch surface $\mathbb{F}_k$.

$\langle \mu(x), X_2 \rangle = \sigma_2(x) = k\,\psi(t)\,\mu_Z(x)$, and the image of the momentum map in $\mathfrak{t}^*$ has the shape shown in Fig. 1.

It is a convex polytope, denoted by $\Delta_{a,b}$, whose boundary is the right-angled trapezoid with vertices at the points $(a,0),(b,0),(a,ka),(b,kb)$. Equivalently, $\Delta_{a,b}$ is the convex polytope in the (x_1, x_2)-plane defined by

$$
\begin{aligned}
x_2 &\geq 0, \\
k x_1 - x_2 &\geq 0, \\
-x_1 + b &\geq 0, \\
x_1 - a &\geq 0
\end{aligned}
\tag{56}
$$

(the above picture is the case when $k = 2$). We denote by F_1, F_2, F_3, F_4 the facets determined by $x_2 = 0$, $-x_1 + b = 0$, $k x_1 - x_2 = 0$, $x_1 - a = 0$, respectively; their (inward) normals in the dual space $\mathfrak{t}$ are then

$$
v_1 = (0,1), \qquad v_2 = (k,-1), \qquad v_3 = (-1,0), \qquad v_4 = (1,0) \tag{57}
$$

respectively.

For any admissible metric g_ψ in $\Omega_{a,b}$, let H be the 2×2-matrix such that $H_{ij} = g_\psi(K_i, K_j)$, $i, j = 1, 2$. Via the momentum map $\mu = (\sigma_1, \sigma_2)$, H can be regarded as defined on $\Delta_{a,b}$; we then have:

Lemma 2. *For any admissible metric g_ψ,*

$$
H = \begin{pmatrix} \Theta(x_1) & \dfrac{x_2\,\Theta(x_1)}{x_1} \\[2ex] \dfrac{x_2\,\Theta(x_1)}{x_1} & 2x_2\left(1 - \dfrac{x_2}{k x_1}\right) + \dfrac{x_2^2\,\Theta(x_1)}{x_1^2} \end{pmatrix} \tag{58}
$$

where, we recall, Θ denotes the momentum profile of g_ψ. In particular

$$\det H = 2\,\Theta(x_1)\,x_2\left(1 - \frac{x_2}{k\,x_1}\right). \tag{59}$$

Proof. From (27), we readily infer $g_\psi(K_1,K_1) = g_\psi(T,T) = \psi'(t) = \Theta(\sigma_1)$, whereas $g_\psi(K_1,K_2) = g_\psi(T,\tilde{Z}) + k\mu_Z g_\psi(T,T) = \frac{\sigma_2}{\sigma_1}\Theta(\sigma_1)$, as $g_\psi(\tilde{Z},T) = 0$. Finally, $g_\psi(K_2,K_2) = \psi\,|Z|^2_{k\,\omega_0} + k^2\,\mu_{\tilde{Z}}^2\,\psi'$. By (54), this is equal to $2k\psi\mu_Z(1-\mu_Z) + k^2\mu_{\tilde{Z}}^2\psi' = 2\sigma_2(1 - \frac{\sigma_2}{k\psi}) + \frac{\sigma_2^2\,\Theta(\sigma_1)}{\sigma_1^2}$. $\qquad\square$

Remark 3. We easily check that H satisfies the boundary conditions in [3, Proposition 1], namely

$$H_{|F_i}(v_i,\cdot) = 0 \qquad dH_{|F_i}(v_i,v_i) = 2v_i, \qquad i = 1,\ldots,4, \tag{60}$$

if and only the boundary condition $\mathbb{B}$ for Θ are fulfilled.

The inverse of the matrix H has the following expression

$$H^{-1} = \begin{pmatrix} \frac{1}{\Theta(x_1)} + \frac{kx_2}{2x_1(kx_1-x_2)} & \frac{-k}{2(kx_1-x_2)} \\ \frac{-k}{2(kx_1-x_2)} & \frac{kx_1}{2x_2(kx_1-x_2)} \end{pmatrix}. \tag{61}$$

This is the matrix, with respect to the (x_1,x_2)-coordinates, of the induced metric $g_{\Delta_{a,b}}$ on the interior, $\Delta_{a,b}^0$, of $\Delta_{a,b}$; by setting $x = x_1, y = \frac{x_2}{kx_1}$ — recall that $\frac{\sigma_2}{k\sigma_1} = \mu_Z$ — we then get:

$$g_{\Delta_{a,b}} = \frac{dx\otimes dx}{\Theta(x)} + kx\frac{dy\otimes dy}{2y(1-y)}. \tag{62}$$

Notice that in the above expression, $\frac{dx\otimes dx}{\Theta(x)}$ is the induced metric of the admissible Kähler metric of $\mathbb{F}_k$ on the open interval (a,b), via the momentum map ψ, whereas $\frac{dy\otimes dy}{2y(1-y)}$ is the induced metric of the standard Kähler metric of $\mathbb{P}^1$ of curvature 2 on the open interval $(0,1)$, via the momentum map μ_Z.

As we already know, cf. Section 1, H^{-1} is the Hessian of the symplectic potential of $\mathbb{F}_k$ as a toric Kähler manifold; from (61) we easily deduce that the symplectic potential, call it $G_{a,b}$, has the following expression:

$$G_{a,b}(x_1,x_2) = \int^{x_1}\left(\int^{s_2}\frac{ds_1}{\Theta(s_1)}\right)ds_2$$
$$+ \frac{1}{2}(kx_1 - x_2)\log(kx_1 - x_2) - \frac{k}{2}x_1\log x_1 + \frac{1}{2}x_2\log x_2. \tag{63}$$

By (15), the components, τ_1, τ_2 of the dual momentum map τ are then given by:

$$\tau_1 = \frac{\partial G_{a,b}}{\partial x_1} = \int^{x_1} \frac{ds}{\Theta(s)} + \frac{k}{2} \log(k x_1 - x_2) - \frac{k}{2} \log x_1$$

$$\tau_2 = \frac{\partial G_{a,b}}{\partial x_2} = -\frac{1}{2} \log(k x_1 - x_2) + \frac{1}{2} \log x_2. \tag{64}$$

From (14), we infer the following expression of the Kähler potential $F_{a,b}$:

$$F_{a,b} = -G_{a,b} + x_1 \frac{\partial G_{a,b}}{\partial x_1} + x_2 \frac{\partial G_{a,b}}{\partial x_2}$$

$$= -\int^{x_1} \left(\int^{s_2} \frac{ds_1}{\Theta(s_1)} \right) ds_2 + x_1 \int^{x_1} \frac{ds}{\Theta(s)}. \tag{65}$$

As we already knew, the Kähler potential only depends on $\sigma_1 = \psi$. As a matter of fact, it readily follows from the definition of the momentum profile Θ that $\int^{\sigma_1} \frac{ds}{\Theta(s)}$ coincides, up to a constant, with the function t defined by (24), viewed as a function of $\psi = \sigma_1$. By differentiating (65) along t, we get $\frac{dF_{ab}}{dt} = -t \frac{d\psi}{dt} + t \frac{d\psi}{dt} + \psi = \psi$, which was the original definition of ψ.

We now restrict our attention to the case when the admissible Kähler metric (g_ψ, ω_ψ) is extremal in $\Omega_{a,b}$. As recalled in Section 2, this is the case when the momentum profile is given by $\Theta(x) = \frac{P_{\Omega_{a,b}}(x)}{x}$, where $P_{\Omega_{a,b}}(x) = (x - a)(b - x) Q_{a,b}(x)$ is the Calabi polynomial of $\Omega_{a,b}$, cf. (44). When $\frac{b}{a}$ is large enough, $Q_{a,b}$ has two distinct real roots $r_\pm$, both negative, given by

$$r_\pm = \frac{-(2abk + b^2 - a^2) \pm ((b-a)^4 + 12ab(b^2 - a^2)k - 4ab(a^2 + ab + b^2)k^2)^{\frac{1}{2}}}{2(a+b)k - 2(b-a)} \tag{66}$$

The function t, viewed as a function defined on the momentum (x_1, x_2)-plane, actually a function of x_1 alone, has then the following expression:

$$t(x_1) = \int^{x_1} \frac{dx}{\Theta(x)} = \int^{x_1} \frac{x\, dx}{P_{\Omega_{a,b}}(x)}$$

$$= \frac{1}{2} \log(x_1 - a) - \frac{1}{2} \log(b - x_1)$$

$$+ \frac{(b-a)}{2(r_+ - r_-)} \log(x_1 - r_+) - \frac{(b-a)}{2(r_+ - r_-)} \log(x_1 - r_-) \tag{67}$$

$$- \frac{(k-1)}{2} \log(k-1) + \frac{1}{2} \log k$$

(the constant $-\frac{(k-1)}{2} \log(k-1) + \frac{1}{2} \log k$ has been added for further convenience; we agree that it is equal to 0 for $k = 1$). Notice that (67) determines x_1 as a well-defined (smooth, increasing) function of t, hence determines unambiguously an admissible extremal Kähler metric in the Kähler class $\Omega_{a,b}$.

The Kähler potential $F_{a,b}$, whose general expression as a function defined on the (x_1, x_2)-plane is given by (65), then specializes to:

$$
\begin{aligned}
F_{a,b}(x_1) = {} & \frac{a}{2} \log (x_1 - a) - \frac{b}{2} \log (b - x_1) \\
& + \frac{(b-a)\, r_+}{2(r_+ - r_-)} \log (x_1 - r_+) - \frac{(b-a)\, r_-}{2(r_+ - r_-)} \log (x_1 - r_-).
\end{aligned}
\tag{68}
$$

We now fix $b > 0$ and make a tend to 0. The Calabi polynomial $P_{a,b}$ tends to the polynomial $P_{0,b}$ defined by

$$
P_{0,b}(x) = \frac{2}{b^2 k} x^2 (b - x)((k-1)x + b).
\tag{69}
$$

In particular, the root r_+ tends to 0, whereas r_- tends to $-\frac{b}{(k-1)}$ if $k > 1$ and tends to $-\infty$ like $\frac{-b^2}{2a}$ if $k = 1$.

If $k > 1$, the Kähler potential $F_{a,b}$ given by (68) tends to

$$
F_{0,b}(x_1) = \frac{b}{2} \Big(\log ((k-1)x_1 + b) - \log (b - x_1) \Big)
\tag{70}
$$

up to an additive constant. If $k = 1$, we substitute $F_{a,b} - \frac{(b-a)\, r_-}{2(r_+ - r_-)} \log a$, which tends to

$$
F_{0,b}(x_1) = -\frac{b}{2} \log (b - x_1)
\tag{71}
$$

when a tends to 0 up to an additive constant. The expression (70) then holds in both cases.

Similarly, the rhs of (67) tends to the function t_b defined by

$$
\begin{aligned}
t_b(x_1) = {} & \frac{k}{2} \log x_1 - \frac{1}{2} \log (b - x_1) \\
& - \frac{(k-1)}{2} \log ((k-1)x_1 + b) + \frac{1}{2} \log k.
\end{aligned}
\tag{72}
$$

In this limiting process, the momentum polytope $\Delta_{a,b}$ tends, in the (x_1, x_2)-plane, to the triangle $\Delta_{0,b}$ whose vertices are the points $(0,0)$, $(b,0)$ and (b, kb) (Fig. 2): this triangle is defined by the set of inequalities:

$$
\begin{aligned}
\ell_1 &:= x_2 + \lambda_1 \geq 0, \\
\ell_2 &:= k x_1 - x_2 + \lambda_2 \geq 0, \\
\ell_3 &:= -x_1 + \lambda_3 \geq 0,
\end{aligned}
\tag{73}
$$

with $\lambda_1 = \lambda_2 = 0$, $\lambda_3 = b$.

Except in the case when $k = 1$, $\Delta_{0,b}$ is no longer a Delzant polytope with respect to the lattice $\mathbb{Z}^2$. Indeed, the former face F_4 collapsed to the new vertex $(0,0)$, whereas the former facets F_1 and F_2 now meet at $(0,0)$, with normals $v_1 = (0,1)$ and

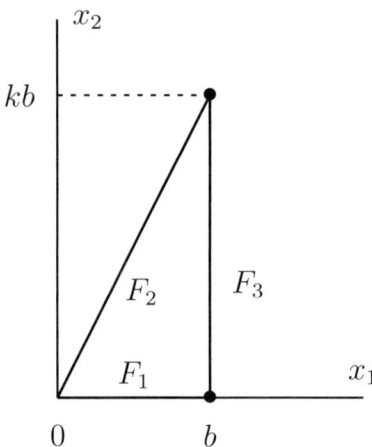

Fig. 2 Momentum polytope $\Delta_{0,b}$ for the weighted projective surface $\mathbb{P}_{1,1,k}$.

$v_2 = (k, -1)$, which do not generate the lattice $\mathbb{Z}^2$ any more if $k > 1$. It is nevertheless a simple rational polytope, and we easily recognize the momentum polytope Δ_k of the complex weighted projective space $\mathbb{P}_k^2$ of weight $\boldsymbol{k} = (1, 1, k)$ equipped with a self-dual Kähler metric with some scaling depending of b (the collapsing of the facet F_4 then corresponds with the collapsing of Σ_0 in $\mathbb{F}_k$, cf. Section 4 below).

It is easy to check that we recover the standard metric of $\mathbb{P}_k^2$ if and only if $b = \frac{1}{k}$. For this choice of b and with the notation of Section 1, the identification of $\Delta_{0, \frac{1}{k}}$, in the (x_1, x_2)-plane, with Δ_k, in the hyperplane $\xi_1 + \xi_2 + k\xi_3 = 0$ of the (ξ_1, ξ_2, ξ_3)-space, is given by the affine map

$$\xi_1 = x_2 - \frac{1}{3}, \qquad \xi_2 = kx_1 - x_2 - \frac{1}{3}, \qquad \xi_3 = -x_1 + \frac{2}{3k}; \tag{74}$$

in particular, the vertices $(0,0)$, $(\frac{1}{k}, 0)$, $(\frac{1}{k}, 1)$ of $\Delta_{0, \frac{1}{k}}$ are mapped to the vertices $p_3 = (-\frac{1}{3}, -\frac{1}{3}, \frac{2}{3k})$, $p_2 = (-\frac{1}{3}, \frac{2}{3}, -\frac{1}{3k})$, $p_1 = (\frac{2}{3}, -\frac{1}{3}, -\frac{1}{3k})$ of Δ_k, respectively.

By using (12) and the values of ℓ_j and λ_j given in (73), we get the following expression for the Kähler potential, F_k, of the standard self-dual metric of $\mathbb{P}_k^2$ on the interior of Δ_k:

$$F_k = \frac{b}{2}\left(\log\left((k-1)x_1 + b\right) - \log\left(b - x_1\right) \right). \tag{75}$$

By comparing with (70), we conclude that the pairs $(F_{a, \frac{1}{k}}, \Delta_{a, \frac{1}{k}})$ converge to the pair (F_k, Δ_k) when a tends to 0, meaning that the (admissible) Calabi extremal metric on $\mathbb{F}_k$ in the Kähler class $\Omega_{a, \frac{1}{k}}$ tends, in a sense that will be made more precise in Section 4, to the standard self-dual Kähler metric of the weighted projective plane $\mathbb{P}_k^2$.

4 The Weighted 2-Plane $\mathbb{P}^2_k$ as a Blow-Down of $\mathbb{F}_k$

By definition, the weighted projective plane $\mathbb{P}^2_k$ is the quotient of $\mathbb{C}^3 \setminus \{0\}$ by the $\mathbb{C}^*$-action of weight $\boldsymbol{k} = (1,1,k)$, defined by $\zeta \cdot (z_1, z_2, z_3) = (\zeta z_1, \zeta z_2, \zeta^k z_3)$, see Section 1.

For any $\mathbf{z} = (z_1, z_2, z_3)$ in $\mathbb{C}^3 \setminus \{0\}$, the isotropy group of the $\mathbb{C}^*$-action at $\mathbf{z}$ is the identity except if $\mathbf{z}$ belongs to the $\mathbb{C}^*$-orbit defined by $z_1 = z_2 = 0$, when the isotropy group is μ_k, the group of the k-th roots of 1. The weighted projective plane $\mathbb{P}^2_k$ has then a unique orbifold singularity at the point $(0:0:1)$, modeled on $\mathbb{C}^3/\mu_k$.

The link between the weighted projective plane $\mathbb{P}^2_k$ and the Hirzebruch surface $\mathbb{F}_k$ is provided by the following proposition:

Proposition 2. *For any positive integer k, $\mathbb{F}_k$ can be realized as the subvariety of the product $\mathbb{P}^2_k \times \mathbb{P}^1$ defined by the equation*

$$z_1 y_2 - z_2 y_1 = 0, \tag{76}$$

for $[\mathbf{z}] = (z_1 : z_2 : z_3)$ in $\mathbb{P}^2_k$ and $y = (y_1 : y_2)$ in $\mathbb{P}^1$.

Proof. Denote by $\mathbb{V}_k$ the subvariety of $\mathbb{P}^2_k$ defined by (76). The identification of $\mathbb{V}_k$ with $\mathbb{F}_k$ goes as follows. Recall that $\mathbb{F}_k = \mathbb{P}(1 \oplus \Lambda^k)$, where Λ denotes the tautological line bundle over $\mathbb{P}^1$: an element of $\mathbb{F}_k$ over y in $\mathbb{P}^1 = \mathbb{P}(\mathbb{C}^2)$ is then a complex line in the complex 2-plane $\mathbb{C} \oplus y^{\otimes k}$. To any pair $([\mathbf{z}] = (z_1 : z_2 : z_3), y = (y_1 : y_2))$ in $\mathbb{V}_k \subset \mathbb{P}^2_k \times \mathbb{P}^1$ we then associate the complex line in $\mathbb{C} \oplus y^{\otimes k}$ generated by $\left(z_3, \lambda\left((y_1, y_2)^{\otimes k}\right) \right)$, for any choice of a generator (y_1, y_2) of the complex line $y = (y_1 : y_2)$, where λ is determined by:

$$z_1^k = \lambda \, y_1^k, \qquad z_2^k = \lambda \, y_2^k. \tag{77}$$

$\square$

Notice that by the map from $\mathbb{P}^2_k \times \mathbb{P}^1$ to $\mathbb{P}^2 \times \mathbb{P}^1$ determined by the canonical map

$$(z_1 : z_2 : z_3) \mapsto (x_1 = z_1^k : x_2 = z_2^k : x_3 = z_3) \tag{78}$$

from $\mathbb{P}^2_k$ to $\mathbb{P}^2$ and the identity map on $\mathbb{P}^1$, $\mathbb{V}_k$ is mapped biholomorphically to the submanifold of $\mathbb{P}^2 \times \mathbb{P}^1$ defined by

$$x_1 y_2^k - x_2 y_1^k = 0, \tag{79}$$

which is the original definition of $\mathbb{F}_k$ in [25].

The above identification of $\mathbb{V}_k$ with $\mathbb{F}_k$ commutes with the projection from $\mathbb{V}_k$ to $\mathbb{P}^1$ induced by the second projection and the natural projection from $\mathbb{F}_k$ to $\mathbb{P}^1$. Moreover, in this identification, the projective line $\{p_0\} \times \mathbb{P}^1$ in $\mathbb{V}_k$ is identified with the zero section Σ_0, whereas the projective line of pairs of the form $([\mathbf{z}] = (z_1 : z_2 : 0), y = (z_1 : z_2))$ in $\mathbb{P}^2_k \times \mathbb{P}^1$ is identified with the infinity section Σ_∞.

Denote by q the map from $\mathbb{F}_k = \mathbb{V}_k \subset \mathbb{P}_k^2 \times \mathbb{P}^1$ to $\mathbb{P}_k^2$ induced by the projection to the first factor. Then, $\{p_0\} \times \mathbb{P}^1$ is mapped to the singular point p_0, whereas q maps $\mathbb{V}_k \setminus (\{p_0\} \times \mathbb{P}^1) = \mathbb{F}_k \setminus \Sigma_0$ biholomorphically to $\mathbb{P}_k^2 \setminus \{p_0\}$. The k-th Hirzebruch surface $\mathbb{F}_k$ is then a blow-up of $\mathbb{P}_k^2$ at the singular point p_0 — hence a desingularization of $\mathbb{P}_k^2$ — whereas $\mathbb{P}_k^2$, as an algebraic variety, is realized as the blow-down of $\mathbb{F}_k$ along the zero section Σ_0.

According to the above discussion, the blow-down map $q : \mathbb{F}_k \to \mathbb{P}_k^2$ has the following expression: for any element $x = (z : \lambda\,((y_1, y_2)^{\otimes k}))$ of $\mathbb{F}_k$ in the fiber $\pi^{-1}(y) = \mathbb{P}(\mathbb{C} \oplus y^{\otimes k})$ — see above — we have

$$q(x) = (z_1 = \lambda^{\frac{1}{k}} y_1 : z_2 = \lambda^{\frac{1}{k}} y_2 : z_3 = z) \tag{80}$$

in $\mathbb{P}_k^2$ for any choice of a generator (y_1, y_2) of y in $\mathbb{C}^2$; the inverse q^{-1} of q, from $\mathbb{P}_k^2 \setminus \{p_0\}$ to $\mathbb{F}_k \setminus \Sigma_0$ is then given by

$$q^{-1}([\mathbf{z}]) = (z_3 : ((z_1, z_2))^{\otimes k}), \tag{81}$$

for $[\mathbf{z}] = (z_1 : z_2 : z_3)$ in $\mathbb{P}_k^2 \setminus \{p_0\}$.

In particular, the function t defined on $M_0 = \mathbb{F}_k \setminus (\Sigma_0 \cap \Sigma_\infty)$ by (24) can be viewed as a function defined on $q(M_0) \subset \mathbb{P}_k^2$ and has there the following expression

$$t([\mathbf{z}]) = \frac{k}{2} \log (|z_1|^2 + |z_2|^2) - \frac{1}{2} \log |z_3|^2, \tag{82}$$

for any $[\mathbf{z}] = (z_1 : z_2 : z_3)$ in $q(M_0)$ (notice that $q(M_0) = U^{(3)} \setminus \{p_0\}$, i.e., is defined in $\mathbb{P}_k^2$ by the two conditions $(z_1, z_2) \neq (0, 0)$ and $z_3 \neq 0$; in terms of the orbifold coordinates w_1, w_2 on the affine part $U^{(3)}$ of $\mathbb{P}_k^2$, we then have

$$t(w_1, w_2) = \frac{k}{2} \log (|w_1|^2 + |w_2|^2); \tag{83}$$

alternatively,

$$t([u]) = \frac{k}{2} \log (1 - |u_3|^2) - \frac{1}{2} \log |u_3|^2, \tag{84}$$

for any $[u]$ in $q(M_0)$, represented by an element $u = (u_1, u_2, u_3)$ of S^5.

The blow-down map q is equivariant under the $\mathbb{T}^2$-action on $\mathbb{F}_k$ described in Section 2 and the $\mathbb{T}_k^2$-action on $\mathbb{P}_k^2$ described in Section 1, over an appropriate isomorphism from $\mathbb{T}^2$ to $\mathbb{T}_k^2$. More precisely, from the explicit formulation (80), we infer that the $\mathbb{T}^2$-action on $\mathbb{F}_k \setminus \Sigma_0$, when read on $\mathbb{P}_k^2 \setminus \{p_0\}$, has the following form:

$$(\zeta_1, \zeta_2) \cdot (z_1 : z_2 : z_3) = (\zeta_2 z_1 : z_2 : \zeta_1^{-1} z_3) \tag{85}$$

for any (ζ_1, ζ_2) in $\mathbb{T}^2 = S^1 \times S^1$ and any $(z_1 : z_2 : z_2)$ in $\mathbb{P}_k^2 \setminus \{p_0\}$. This shows that q is equivariant over the isomorphism

$$(\zeta_1, \zeta_2) \mapsto (\zeta_2, 1, \zeta_1^{-1}) \quad \mathrm{mod}\ j_k(S^1) \tag{86}$$

from $\mathbb{T}^2$ to $\mathbb{T}_k^2 = \mathbb{T}^3 / j_k(S^1)$. We can then arrange that the corresponding momenta be related as in (74) for any admissible Kähler metric on $\mathbb{F}_k$ and its image on $\mathbb{P}_k^2 \setminus \{p_0\}$ by the blow-down map q.

The discussion of Sections 3 and 4 can be summarized by the following statement:

Theorem 2. *For any positive integer k, let $\mathbb{F}_k = \mathbb{P}(1 \oplus \Lambda^k)$ be the k-th Hirzebruch surface and $\mathbb{P}_k^2$ the weighted projective plane of weight $k = (1, 1, k)$. Denote by q the blow-down map from $\mathbb{F}_k$ to $\mathbb{P}_k^2$, which maps the zero section Σ_0 of $\mathbb{F}_k$ to the singular point, p_0, of $\mathbb{P}_k^2$. For any Kähler class $\Omega_{a,b}$ on $\mathbb{F}_k$, with $0 < a < b$ and $\frac{b}{a}$ sufficiently large, denote by $g_{\psi_{a,b}}$ the Calabi admissible extremal Kähler metric in $\Omega_{a,b}$ determined by (67). Denote by $\tilde{g}_{\psi_{a,b}}$ the induced extremal Kähler metric on $\mathbb{P}_k^2 \setminus \{p_0\}$, via the blow-down map q restricted to $\mathbb{F}_k \setminus \Sigma_0$. Fix $b = \frac{1}{k}$ and let a tend to 0. Then, $\tilde{g}_{\psi_{a,\frac{1}{k}}}$ tends to the standard self-dual Kähler metric of $\mathbb{P}_k^2 \setminus \{p_0\}$, uniformly on each compact set, with all its derivatives.*

Proof. By putting together (72) and (84) — recall that t is independent of the chosen Kähler class $\Omega_{a,b}$, hence of the parameter a — we infer that the momentum map $\sigma_1 = \psi_{a,\frac{1}{k}}$, viewed as a function on $\mathbb{P}_k^2 \setminus \{p_0\}$, tends to

$$x_1 = \frac{1}{k} - \frac{|u_3|^2}{(1 + (k-1)|u_3|^2)} \tag{87}$$

on the affine part $U^{(3)} \cap (\mathbb{P}_k^2 \setminus \{p_0\})$, hence on $\mathbb{P}_k^2 \setminus \{p_0\}$. From (70), we conclude that the Kähler potential of $\tilde{g}_{a,\frac{1}{k}}$ tends to the function

$$\tilde{F}_{\lim} = -\frac{1}{k} \log |u_3|. \tag{88}$$

By (4), this is a Kähler potential of the standard metric of $\mathbb{P}_k^2$ on the affine part $U^{(3)}$. On each compact set of $\mathbb{P}_k^2 \setminus \{p_0\}$, t is bounded above and below: the convergence of all functions considered above, and of all their derivatives, are then uniform. $\square$

Remark 4. The group $Gl(2, \mathbb{C}) / \mu_k$ acts effectively and holomorphically on $\mathbb{P}_k^2$ by
$[A] \cdot (z_1 : z_2 : z_3) = (az_1 + bz_2 : cz_1 + dz_2 : z_3)$, for any $[A] = \begin{pmatrix} a & b \\ c & d \end{pmatrix}$ mod μ_k; this action fixes the singular point p_0 and is transitive on the projective line $q(\Sigma_\infty) = \mathbb{P}_k^2 \setminus U^{(3)}$ and on the open set $q(M_0) = U^{(3)} \setminus \{p_0\}$. Moreover, the induced action of $U(2) / \mu_k$ preserves the standard metric of $\mathbb{P}_k^2$. The blow-down map q is clearly equivariant with respect to this action and to the action of $Gl(2, \mathbb{C}) / \mu_k$ on $\mathbb{F}_k$, cf. Section 2. Then, $\mathbb{T}_k^2$, as a subgroup of $U(2) / \mu_k$, is the image of $\mathbb{T}^2$ in the map from $\mathbb{T}^2$ to $U(2) / \mu_k$ defined by

$$(\zeta_1, \zeta_2) \mapsto \begin{pmatrix} \zeta_1^{\frac{1}{k}} \zeta_2 & 0 \\ 0 & \zeta_1^{\frac{1}{k}} \end{pmatrix} \tag{89}$$

which is a directly deduced from (85).

Remark 5. The Ricci potential $w = w(t)$ of any admissible Kähler metric on M_0, whose expression is given in (32), can be rewritten, as a function of $x = x_1$ on $\Delta^0_{a,b}$, in the following manner:

$$w = w(x_1) = \frac{2}{k} t(x_1) - \frac{1}{2} \log x_1 - \frac{1}{2} \log \Theta(x_1), \qquad (90)$$

where $t = t(x_1)$ is given by (67). For an admissible *extremal* Kähler metric, this reduces to

$$w(x_1) = \frac{2}{k} t(x_1) - \frac{1}{2} \log P_{\Omega_{a,b}}(x_1), \qquad (91)$$

where $P_{\Omega_{a,b}}$ is the Calabi polynomial of the chosen Kähler class $\Omega_{a,b}$. When b is fixed and a tends to 0, the rhs of (91) tends to $\frac{2}{k} t_b(x_1) - \frac{1}{2} \log P_{0,b}(x_1)$, where t_b and $P_{0,b}$ are given by (72) and (69), respectively; the Ricci potential then tends to a limit, $w_{\lim}$, given by:

$$w_{\lim}(x_1) = -\frac{(k+2)}{2k} \log (b - x_1) - \frac{(3k-2)}{2k} \log ((k-1)x_1 + b). \qquad (92)$$

When $b = \frac{1}{k}$, x_1 is given by (87), and the rhs of (92) then reduces to

$$w_{\lim}(|u_3|) = -\frac{(k+2)}{2k} \log |u_3|^2 + 2 \log (1 + (k-1)|u_3|^2); \qquad (93)$$

by (5), this is the Ricci potential of the standard metric of $\mathbb{P}^2_k$ on the affine open set $U^{(3)}$. Similarly, by (41), the scalar curvature $s = \alpha x_1 + \beta$ tends to $s_{\lim} = \frac{12k(1-(k-1)|u_3|^2)}{1+(k-1)|u_3|^2}$, which, by (6), is the expression of the scalar curvature of the standard metric of $\mathbb{P}^2_k$ on $U^{(3)}$. If W^- denotes the anti–self-dual Weyl tensor — equivalently, the Bochner tensor — of $g_{\psi_{a,b}}$, we have that $|W^-|^2_{g_{\psi_{a,b}}} = \frac{s^2_{\bar{g}_{\psi_{a,b}}}}{24 \psi^4_{a,b}}$, where $s_{\bar{g}_{\psi_{a,b}}} = \frac{\delta}{\psi_{a,b}} + \gamma$ denotes the scalar curvature of the associated *dual Kähler metric*, cf. [2,22, Chapter 8]; by (41), γ and δ tend to 0 when a tends to 0 (with $b = \frac{1}{k}$ fixed); it follows that W^- tends to 0. This, of course, is also a consequence of Theorem 2.

Remark 6. The geometric situation described in this paper fits well with the general existence theorems of extremal Kähler metrics on blow-ups of extremal Kähler manifolds established by Arezzo–Pacard–Singer in [5], in particular with [5, Corollary 2.2].

Acknowledgment It is a pleasure and an honor for me to dedicate this paper to my colleague and friend, Charles Boyer, for the celebration of his 65th birthay. I also warmly thank Vestislav Apostolov, Liana David, and Andrei Moroianu for helpful discussions.

References

1. M. Abreu, *Kähler metrics on toric orbifolds*, J. Differential Geom. **58** (2001), 151–187.
2. V. Apostolov, D. M. J. Calderbank, P. Gauduchon, *The geometry of weakly self-dual Kähler surfaces*, Compositio Math. **135**, No. 3 (2003), 279–322.
3. V. Apostolov, D. M. J. Calderbank, P. Gauduchon, C. W. Tønnesen-Friedman, *Hamiltonian 2-forms in Kähler geometry, II: Global classification*, J. Differential Geom. **68** (2004), 277–345.
4. V. Apostolov, D. M. J. Calderbank, P. Gauduchon, C. W. Tønnesen-Friedman, *Hamiltonian 2-forms in Kähler geometry, III: Extremal metrics and stability*, Invent. Math. **173** (2008), 547–601.
5. C. Arezzo, F. Pacard and M. Singer, *Extremal metrics on blow-ups*, arXiv:math.DG/0701028v1.
6. S. Bando and T. Mabuchi, *Uniqueness of Einstein Kähler Metrics Modulo Connected Group Actions*, Algebraic Geometry, Sendai (1985), 11–40, Adv. Stud. Pure Math. **10**, North-Holland, Amsterdam and New York, 1987.
7. O. Biquard, *Métriques kählériennes à courbure scalaire constante: unicité, stabilité*, Séminaire Bourbaki, 57 ème année, 2004-2005, n° 938, Novembre 2004.
8. C. Boyer and K. Galicki, *Sasakian Geometry*, Oxford Mathematical Monographs, Oxford University Press, Oxford, 2008.
9. R. Bryant, *Bochner-Kähler metrics*, J. Am. Math. Soc. **14** (2001), 623–715.
10. E. Calabi, *Extremal Kähler metrics*, in *Seminar of Differerential Geometry*, ed. S. T. Yau, Annals of Mathematics Studies **102**, Princeton University Press (1982), 259–290.
11. E. Calabi, *Extremal Kähler metrics, II*, in *Differential Geometry and Complex Analysis*, eds. I. Chavel and H. M. Farkas, Springer-Verlag (1985), 95–114.
12. X. X. Chen and G. Tian, *Uniqueness of extremal Khler metrics*, C. R. Math. Acad. Sci. Paris **340**, No. 4 (2005), 287–290.
13. X. X. Chen and G. Tian, *Partial regularity for homogeneous complex Monge-Ampere equations* C. R. Math. Acad. Sci. Paris **340**, No. 5 (2005), 337–340.
14. X. X. Chen and G. Tian, *Geometry of Kähler metrics and holomorphic foliation by discs*, arXiv:math.DG/0507148 v1.
15. X. X. Chen, C. LeBrun and B. Weber, *On conformally Kähler, Einstein manifolds*, arXiv:0705.0710.
16. X. X. Chen and B. Weber, *Moduli spaces of critical Riemannian metrics with $L^{n/2}$-norm curvature bounds*, arXiv:0705.4440.
17. T. Delzant, *Hamiltoniens périodiques et images convexes de l'application moment*, Bulletin de la S.M.F., tome **116**, No. 3 (1988), 315–339.
18. L. David, P. Gauduchon, *The Bochner-Flat Geometry of Weighted Projective Spaces*, CRM Proceedings and Lecture Notes, Volume **40** (2006), 109–156.
19. I. V. Dolgachev, *Weighted Projective Varieties*, in *Group Actions and Vector Fields* (Vancouver, B.C., 1981), Lecture Notes in Math. **956**, Springer-Verlag (1982), 34–71.
20. S. K. Donaldson, *Scalar curvature and projective embeddings, II*, Q. J. Math. **56**, No. 3 (2005), 345–356.
21. S. K. Donaldson, *Lower bounds on the Calabi functional*, J. Differential Geom. **70** (3) (2005), 453–472.
22. P. Gauduchon, *Calabi extremal Kähler metrics.* (lecture notes in progress).
23. V. Guillemin, *Kähler structures on toric varieties*, J. Differential Geom. **40** (1994), no. 2, 285–309.
24. V. Guillemin, *Moment Maps and Combinatorial Invariants of Hamiltonian $\mathbb{T}^n$-Spaces*, Progress in Mathematics **122**, Birkäuser (1994).
25. F. Hirzebruch, *Über eine Klasse von einfachzusammenhängenden komplexen Mannigfaltigkeiten*, Math. Ann. **124** (1951), 77–86.
26. A. D. Hwang, *On existence of Kähler metrics with constant scalar curvature*, Osaka J. Math. **31** (1994), 561–595.

27. A. D. Hwang and M. A. Singer, *A momentum construction for circle-invariant Kähler metrics*, Trans. Am. Math. Soc. **354** (2002), 2285–2325.
28. E. Lerman, S. Tolman, *Hamiltonian torus actions on symplectic orbifolds and toric varieties*, Trans. AMS, **349** (1997), 4201–4230.
29. T. Mabuchi, *Uniqueness of extremal Kähler metrics for an integral Kähler class*, Int. J. Math. **15**, No. **6** (2004), 531–546.
30. G. Székelyhidi, *Extremal metrics and K-stability* (Ph.D. thesis), math.AG/0410401 v2 (1 Nov 2005).
31. G. Székelyhidi, *The Calabi functional on a ruled surface*, arXiv:math.DG/0703562v1.
32. C. W. Tønnesen-Friedman, *Extremal metrics on minimal ruled surfaces*, J. Reine Angew. Math. **502** (1998), 175–197.
33. S. M. Webster, *On the pseudo-conformal geometry of Kähler manifolds*, Math. Z. **157**, No. 3, (1977), 265–270.

Quaternionic Kähler Moduli Spaces

Nigel Hitchin

Abstract We describe in differential-geometric language a class of naturally occurring quaternionic Kähler moduli spaces due originally to the physicists Ferrara and Sabharwal. This class yields an example in real dimension $4n$ for every projective special Kähler manifold of real dimension $2n - 2$ and can be applied in particular to the case of the moduli space of complex structures on a Calabi–Yau threefold.

1 Introduction

The study of quaternionic Kähler orbifolds of positive scalar curvature is equivalent to the study of 3-Sasakian manifolds, and the procedure of quaternionic Kähler reduction gives many examples of these, starting from a finite-dimensional quaternionic projective space (see [3]). But whereas hyperkähler reduction has been used in an infinite-dimensional context to produce many examples of hyperkähler metrics on gauge-theoretic moduli spaces (monopoles, instantons, and Higgs bundles), there seems to be no such source of examples of quaternionic Kähler geometries. Physics has, however, produced a class of examples of naturally occurring quaternionic Kähler moduli spaces, in the construction of Ferrara and Sabharwal [6], which yields an example in real dimension $4n$ for every projective special Kähler manifold M of real dimension $(2n - 2)$. Particularly interesting examples of projective special Kähler manifolds arise from the moduli spaces of Calabi–Yau threefolds or their generalized versions as in [10]. The $4n$-manifold is a bundle over M whose fiber is a quotient of the unit ball in $\mathbf{C}^{n+1}$, considered as a Siegel domain of the second kind.

The purpose of this paper is to describe in more standard differential-geometric language this construction and to illustrate it (following [9]) in the case where the projective special Kähler manifold is the moduli space $\mathcal{M}$ of Calabi–Yau complex structures on a six-dimensional real manifold Y. We proceed via the construc-

N. Hitchin
University of Oxford, Mathematical Institute, 24-29 St. Giles, Oxford OX1 3LB, U.K.

K. Galicki and S.R. Simanca (eds.), *Riemannian Topology and Geometric Structures on Manifolds,* Progress in Mathematics 271, DOI 10.1007/978-0-8176-4743-8,

tion due to Cecotti, Ferrara, and Girardello [4] of a natural associated indefinite hyperkähler manifold. In the Calabi–Yau case, one starts with the gauged moduli space $\mathcal{U}$ of complex structures together with a choice of holomorphic three-form, a $\mathbf{C}^*$-bundle over $\mathcal{M}$. Then the hyperkähler manifold can be thought of as a flat bundle over $\mathcal{U}$ whose fiber is the torus $H^3(Y, \mathbf{R})/H^3(Y, \mathbf{Z})$. With one of the three hyperkähler complex structures this is a holomorphic fibration whose fiber is the complex torus known as the Griffiths intermediate Jacobian. The corresponding quaternionic Kähler manifold fibers over $\mathcal{M}$ with fiber a punctured disc bundle over the complex torus known as the Weil intermediate Jacobian.

2 Special Kähler Manifolds

The starting point for the construction is special Kähler geometry (see [7, 8]).

Definition. *A special Kähler manifold is a complex manifold with*

- *a Kähler metric g with Kähler form ω*
- *a flat torsion-free connection ∇ such that*
- *$\nabla \omega = 0$ and $d_\nabla I = 0$*

The last property is to be interpreted by thinking of the complex structure, an endomorphism of the tangent bundle T, as $I \in \Omega^1(T)$, a one-form with values in T. The connection ∇ defines a covariant exterior derivative $d_\nabla : \Omega^p(T) \to \Omega^{p+1}(T)$ and we require $d_\nabla I = 0 \in \Omega^2(T)$. We shall consider such structures where the metric has various signatures.

The flat torsion-free connection ∇ gives distinguished flat local coordinates x_i such that

$$\omega = \sum \omega_{ij} dx_i \wedge dx_j$$

has constant coefficients. As shown in [8], the structure is then determined by a locally defined real function ϕ where the metric has Hessian form

$$g_{ij} = \frac{\partial^2 \phi}{\partial x_i \partial x_j}.$$

If X is the Hamiltonian vector field for ϕ, then $d_\nabla X = I$, and hence since ∇ is flat, $d_\nabla I = d_\nabla^2 X = 0$. The condition $I^2 = -1$ and integrability of I impose conditions on ϕ, but ϕ is (locally and noncanonically) determined by an arbitrary holomorphic function.

Example. Consider the moduli space of pairs (Y, Ω) where Y is a compact Kähler threefold and Ω is a nonvanishing holomorphic three-form on Y. The isomorphism $H^1(Y, \mathcal{O}(T)) \cong H^1(Y, \mathcal{O}(\Lambda^2 T^*))$ and the Tian–Todorov theorem gives a smooth moduli space $\mathcal{M}$ of complex dimension $h^{2,1}$, and the choice of Ω defines a $\mathbf{C}^*$-bundle $\mathcal{U}$ over $\mathcal{M}$.

It is known (see [9] for example) that the de Rham cohomology class $[\text{Re}\,\Omega]$ gives local coordinates on $\mathcal{U}$, and the cup product on $H^3(Y,\mathbf{R})$ is a constant-coefficient symplectic form. This defines ω and the flat affine structure. The function ϕ is

$$\phi = \int_Y \text{Re}\,\Omega \wedge \text{Im}\,\Omega.$$

The tangent space to $\mathcal{U}$ is an extension

$$0 \to H^0(Y, \mathcal{O}(\Lambda^3 T^*)) \to TU \to H^1(Y, \mathcal{O}(\Lambda^2 T^*)) \to 0.$$

Now

$$dz_1 \wedge dz_2 \wedge dz_3 \wedge d\bar{z}_1 \wedge d\bar{z}_2 \wedge d\bar{z}_3 = -d\bar{z}_1 \wedge dz_2 \wedge dz_3 \wedge dz_1 \wedge d\bar{z}_2 \wedge d\bar{z}_3,$$

which means that the Kähler metric, which is defined by the symplectic form and complex structure, has opposite sign on the one-dimensional space of multiples of $[\Omega]$ to that on $H^1(Y, \mathcal{O}(\Lambda^2 T^*))$. For convenience we choose an orientation such that the overall metric has Hermitian signature $(h^{2,1}, 1)$.

There is a particular class of structures when the vector field X preserves the metric (or equivalently the complex structure). If X integrates to an action of S^1, then we can take the Kähler quotient and get a Kähler metric on

$$M/\mathbf{C}^* \cong \phi^{-1}(c)/S^1,$$

which is called a projective special Kähler structure.

Example. In the Calabi–Yau case, there is a natural S^1-action $\Omega \mapsto e^{i\theta}\Omega$ on $\mathcal{U}$. Differentiating at $\theta = 0$, we get $\dot{\Omega} = i\Omega$, so the action defines the vector field $X = -[\text{Im}\,\Omega] \in H^3(Y,\mathbf{R})$ which is the Hamiltonian vector field of ϕ. Since X and IX give the holomorphic $\mathbf{C}^*$-action on $\mathcal{U}$, this action preserves the complex structure. Thus $\mathcal{M} = \mathcal{U}/\mathbf{C}^*$ is a projective special Kähler manifold with positive definite signature.

3 The Hyperkähler Manifold

To any special Kähler manifold of real dimension $2n$, Cecotti, Ferrara, and Girardello [4] associate a semiflat hyperkähler manifold on a bundle over it.

Its local geometry is a product,

$$M \times \mathbf{R}^{2n}$$

and the metric is

$$\sum g_{ij}(dx_i dx_j + dy_i dy_j), \tag{1}$$

where

$$g_{ij} = \frac{\partial^2 \phi}{\partial x_i \partial x_j}.$$

The full hyperkähler structure is determined by the three closed symplectic two-forms $\omega_1, \omega_2, \omega_3$ defined by

$$\omega_1 = 2 \sum \frac{\partial^2 \phi}{\partial x_j \partial x_k} dx_j \wedge dy_k, \tag{2}$$

$$\omega_2 + i\omega_3 = \sum \omega_{jk} d(x_j + iy_j) \wedge d(x_k + iy_k). \tag{3}$$

Note that

$$\omega_2 = \sum \omega_{jk} dx_j \wedge dx_k - \sum \omega_{jk} dy_j \wedge dy_k = \omega - F,$$

where ω is the pullback of the Kähler form on the special Kähler manifold M. The translation group $\mathbf{R}^{2n}$ acts freely on $M \times \mathbf{R}^{2n}$, and the action preserves all three symplectic forms and thus acts via triholomorphic isometries.

The three hyperkähler complex structures J_1, J_2, J_3 for which $\omega_1, \omega_2, \omega_3$ are the Kähler forms are given by

$$J_1 = \begin{pmatrix} 0 & -I \\ I & 0 \end{pmatrix}, \quad J_2 = \begin{pmatrix} I & 0 \\ 0 & -I \end{pmatrix}, \quad J_3 = \begin{pmatrix} 0 & I \\ I & 0 \end{pmatrix}.$$

Remark. The local coordinate description above extends to a more global interpretation. In flat coordinates x_i, we can write a tangent vector as $\sum y_i \partial/\partial x_i$ and use the connection ∇ to split the tangent bundle of $\pi : TM \to M$ into $\pi^*(T \oplus T)$. Then in these coordinates, (1) provides a well-defined hyperkähler metric on the total space of TM. Alternatively, the Hermitian metric gives an antilinear identification of T with T^*, and it is in some ways more natural to use this, in which case the structure is defined on the total space of the cotangent bundle T^*M, and then we can write

$$J_1 = \begin{pmatrix} 0 & -g \\ g^{-1} & 0 \end{pmatrix}, \quad J_2 = \begin{pmatrix} I & 0 \\ 0 & I \end{pmatrix}, \quad J_3 = \begin{pmatrix} 0 & -\omega^{-1} \\ \omega & 0 \end{pmatrix}.$$

Then J_2 is the canonical complex structure on T^*M, and ω_1 is the canonical symplectic structure.

The Hamiltonian vector field X of ϕ on M is

$$X = \sum X_j \frac{\partial}{\partial x_j} = \sum \omega^{ij} \frac{\partial \phi}{\partial x_i} \frac{\partial}{\partial x_j},$$

and preserves ω. Its trivial extension to the product preserves F, so the form ω_2 is left fixed by the action. Now consider its effect on the other symplectic forms. We have

$$\mathcal{L}_X \omega_3 = \mathcal{L}_X \left(2 \sum \omega_{jk} dx_j \wedge dy_k \right) = 2 \sum \omega_{jk} dX_j \wedge dy_k.$$

But

$$dX_j = d\sum \omega^{ij} \frac{\partial \phi}{\partial x_i} = \sum \omega^{ij} g_{il} dx_l \,,$$

so

$$\mathcal{L}_X \omega_3 = 2\sum \omega_{jk} \omega^{ij} g_{il} dx_l \wedge dy_k = 2\sum g_{kl} dx_l \wedge dy_k = \omega_1 \,.$$

In the case when X preserves the metric g on M, we have

$$\mathcal{L}_X g = \mathcal{L}_X \left(\sum g_{jk} dx_j \otimes dx_k \right) = 0 \,.$$

If we set

$$\theta_k = \sum g_{jk} dx_j \,,$$

then we have

$$\sum \mathcal{L}_X \theta_k \otimes dx_k + \theta_k \otimes dX_k = 0 \,,$$

which implies

$$\mathcal{L}_X \theta_j = -\sum \theta_k \frac{\partial X_k}{\partial x_j} = -\sum g_{ik} \omega^{lk} \frac{\partial^2 \phi}{\partial x_l \partial x_j} dx_i$$

$$= -\sum g_{ik} \omega^{lk} g_{lj} dx_i = \sum \omega_{ji} dx_i \,.$$

Thus

$$\mathcal{L}_X \omega_1 = \mathcal{L}_X \left(2\sum g_{ij} dx_i \wedge dy_j \right) = \mathcal{L}_X \left(2\sum \theta_j \wedge dy_j \right) = 2\sum \omega_{ji} dx_i \wedge dy_j = -\omega_3 \,.$$

Consequently, X acts as an infinitesimal isometry on $M \times \mathbf{R}^{2n}$, defining an infinitesimal rotation in the three-dimensional space of covariant constant two-forms $\langle \omega_1, \omega_2, \omega_3 \rangle$, fixing ω_2.

Example. In the Calabi–Yau case where $M = \mathcal{U}$, the tangent space is $H^3(Y, \mathbf{R})$, which has an integral lattice, the image of $H^3(Y, \mathbf{Z})$, so we can put the hyperkähler metric on a torus bundle over M. The cup product is unimodular, so we can canonically identify this torus with its dual.

Since $J_2 = (I, I)$, the projection $\pi : T^*M \to M$ is holomorphic in complex structure J_2 and so the fibers are complex tori varying holomorphically (in fact $\omega_3 + i\omega_1$ vanishes on each fiber so they are also complex Lagrangian). Now $H^3(Y, \mathbf{R})/H^3(Y, \mathbf{Z})$ has two natural structures as a complex torus – the *Griffiths* intermediate Jacobian and the *Weil* intermediate Jacobian (see [2]). In our case, I acts as $-i$ on $H^{0,3}$ and $+i$ on $H^{1,2}$, and this is the definition of the Griffiths Jacobian. Its characteristic property, in contrast with the Weil Jacobian, is that it varies holomorphically in families, and this we have just seen for this example. On the other hand, the symplectic form defines an indefinite Hermitian form, so the torus is not naturally an Abelian variety.

This metric was also discussed in [5].

4 The Heisenberg Group

Returning to the hyperkähler manifold, the symplectic forms ω_1, ω_3 vanish on the orbits of the group $\mathbf{R}^{2n}$, but ω_2 restricts to the invariant form $-F$, where

$$F = \sum \omega_{jk} dy_j \wedge dy_k.$$

Using this skew form, we can define a canonical Heisenberg Lie algebra: a central extension of $\mathbf{R}^{2n}$ defined in terms of generators $Y_1, \ldots, Y_{2n}, Z$ by

$$[Y_i, Y_j] = 2\omega_{ij} Z, \qquad [Y_i, Z] = 0.$$

There is a simply connected Heisenberg group H with this Lie algebra.

Remark. If $\mathbf{R}^{2n}$ contains a lattice, making the quotient a torus T^{2n}, and if the de Rham cohomology class $[F] \in H^2(T^{2n}, \mathbf{R})$ is integral, then this Lie algebra is the algebra of a compact group that can be considered as the total space of a principal S^1-bundle $P \to T^{2n}$, defining the quantum line bundle for the symplectic structure F. It is the quotient $H/H_{\mathbf{Z}}$ for a discrete subgroup $H_{\mathbf{Z}} \subset H$.

We extend this structure to a Lie algebra $\mathfrak{g}$ by adjoining a "homogeneity operator" E:

$$[E, Y_i] = Y_i, \qquad [E, Z] = 2Z.$$

A simply connected group G with this Lie algebra is provided by the Type II Siegel domain $D = \{(v, w) \in \mathbf{C} \times \mathbf{C}^n : \operatorname{Im} v > \|w\|^2\}$. The translations in $\mathbf{C}^n$ act as

$$(v, w) \mapsto (v + 2i \operatorname{Re}\langle w, c \rangle + i\|c\|^2, w + c),$$

and Z and E generate the following actions:

$$(v, w) \mapsto (v + t, w), \qquad (v, w) \mapsto (\lambda^2 v, \lambda w).$$

The group acts simply transitively on D. The map $(v, w) \mapsto w$ expresses D as a bundle over $\mathbf{C}^n$ with fiber $\{v \in \mathbf{C} : \operatorname{Im} v > \|w\|^2\}$, which is a copy of the upper half-plane.

As a complex manifold, D is a familiar object: the map

$$z_0 = \frac{v - i}{v + i}, \qquad z = \frac{2w}{v + i}$$

maps D biholomorphically to the unit disc $|z_0|^2 + \|z\|^2 < 1$ in $\mathbf{C}^{n+1}$.

Remark. If we have a complex structure on $\mathbf{R}^{2n}$ that makes ω into a positive definite Hermitian form, then the group G has the complex structure of the domain D. If there is an integral structure, then we can form $G/H_{\mathbf{Z}}$. This is a holomorphic bundle over a torus whose fiber is the upper half-plane modulo the $\mathbf{Z}$-action generated by a real translation, which is the punctured disc $\{z \in \mathbf{C} : 0 < |z| < 1\}$. If ω is integral and unimodular, then the torus is an Abelian variety, and the theta divisor defines a line bundle with a Hermitian form. The complex manifold G/H_Z is then a bundle of punctured discs inside this line bundle.

Let $\eta_i, \varepsilon, \zeta$ be the basis of $\mathfrak{g}$ dual to Y_i, E, Z. Then the Lie algebra relations above for left-invariant vector fields on G define the following relations for one-forms:

$$d\eta_i = -\varepsilon \wedge \eta_i, \tag{4}$$

$$d\varepsilon = 0, \tag{5}$$

$$d\zeta = -2\varepsilon \wedge \zeta - \sum \omega_{kl} \eta_k \wedge \eta_l = -2\varepsilon \wedge \zeta - \tilde{F}, \tag{6}$$

where from the last formula we also obtain

$$d\tilde{F} = -2\varepsilon \wedge \tilde{F}.$$

5 Quaternionic Structures

The standard examples of quaternionic Kähler manifolds (and indeed the only ones with positive scalar curvature to date) are provided by the Wolf spaces, which are symmetric spaces (see [1]). To describe the characteristic features of the general construction, we first consider the well-known negatively curved Wolf space: the $4n$-dimensional space $U(2,n)/U(2) \times U(n)$. This is the Grassmannian of two-dimensional subspaces of $\mathbf{C}^{2,n}$ on which the induced Hermitian form is positive definite.

Let $\mathbf{C}^{2,n} = \mathbf{C}^{1,1} \oplus \mathbf{C}^{1,n-1}$ and suppose $V \subset \mathbf{C}^{2,n}$ is a two-dimensional subspace with positive definite induced Hermitian form. Pull back the indefinite form using the projection $\pi : \mathbf{C}^{2,n} \to \mathbf{C}^{1,1}$, then using the induced Hermitian form this is a self-adjoint transformation B with distinct eigenvalues. Hence over the Grassmannian, the tautological rank two bundle defined by V splits non-holomorphically into a sum of line bundles — the eigenspaces of B. The bundle of trace zero skew-adjoint transformations of V becomes in the quaternionic Kähler structure the bundle of imaginary quaternions, so there is a reduction of the structure group of this bundle from $SU(2)$ to $U(1)$.

If $v \in V$ is an eigenvector of B with nonpositive eigenvalue, then $v = \pi v + w$ where $w \in \mathbf{C}^{1,n-1}$ has $\|w\|^2 > 0$, and so defines a point in the unit ball in $\mathbf{C}^{n-1}$.

The above features — a projection onto a $(2n-2)$-manifold with a Kähler structure, and a reduction from $SU(2)$ to S^1 of the bundle of imaginary quaternions — are present in the general construction.

Thus return to the general case and suppose the special Kähler metric on M is indefinite with Hermitian signature $(n-1,1)$, and that the fundamental vector field X satisfies $g(X,X) < 0$ and generates an isometric S^1 action. Consider the manifold

$$M \times G$$

of dimension $4n + 2$. We shall put a positive definite quaternionic Kähler metric on the $4n$-dimensional manifold $\bar{M} \times G$ where $\bar{M}$ is the Kähler quotient of M by the S^1-action.

First define on $M \times G$ three two-forms $\tilde{\omega}_1, \tilde{\omega}_2, \tilde{\omega}_3$ by replacing dy_j by η_j in the formulas (2), (3) for $\omega_1, \omega_2, \omega_3$. These are all clearly G-invariant, but not closed. From (4) we have

$$d\tilde{\omega}_1 = -\varepsilon \wedge \tilde{\omega}_1, \qquad d\tilde{\omega}_2 = 2\varepsilon \wedge \tilde{F}, \qquad d\tilde{\omega}_3 = -\varepsilon \wedge \tilde{\omega}_3. \tag{7}$$

On the other hand, the algebraic relations they satisfy are the same as those on $M \times \mathbf{R}^{2n}$, and so they give the subbundle of the tangent bundle spanned by TM and $Y_1, \ldots, Y_{2n}$ a quaternionic structure with compatible (indefinite) metric. Thus

$$T(M \times G) = TM \oplus U_1 \oplus U_2,$$

where $U_1 = \langle Y_1, \ldots, Y_{2n} \rangle, U_2 = \langle E, Z \rangle$ are trivial bundles over G spanned by the corresponding invariant vector fields, and $TM \oplus U_1$ has an $Sp(n-1,1)$ structure.

The action by the vector field X is unchanged, so we have as before

$$\mathcal{L}_X \tilde{\omega}_1 = -\tilde{\omega}_3 \qquad \mathcal{L}_X \tilde{\omega}_2 = 0 \qquad \mathcal{L}_X \tilde{\omega}_3 = \tilde{\omega}_1.$$

Now $X, J_1 X, J_2 X, J_3 X$ lie in the $Sp(n-1,1)$ bundle $TM \oplus U_1$. Let φ_i ($0 \le i \le 3$) be the linear forms dual to these four vectors using the metric, and extend them to one-forms on $M \times G$ by using $T^*(M \times G) = (TM \oplus U_1)^* \oplus U_2^*$. Then for $i = 1, 2, 3$, $\varphi_i = i_X \tilde{\omega}_i$, which means that

$$d\varphi_1 = d(i_X \tilde{\omega}_1) = \mathcal{L}_X \tilde{\omega}_1 - i_X d\tilde{\omega}_1. \tag{8}$$

But from (7), $i_X d\tilde{\omega}_1 = \varepsilon \wedge i_X \tilde{\omega}_1 = \varepsilon \wedge \varphi_1$ so together with (8), we obtain

$$d\varphi_1 = \mathcal{L}_X \tilde{\omega}_1 - \varepsilon \wedge \varphi_1 = -\tilde{\omega}_3 - \varepsilon \wedge \varphi_1,$$

and similarly

$$d\varphi_3 = \tilde{\omega}_1 - \varepsilon \wedge \varphi_3$$

As for $d\varphi_0$, we find

$$\varphi_0 = \sum g_{jk} X_j dx_k = \sum g_{jk} \omega^{ij} \frac{\partial \phi}{\partial x_i} dx_k = \sum \frac{\partial \phi}{\partial x_i} I_k^i dx_k.$$

But as $d_\nabla I = 0$ (the special Kähler condition),

$$d\varphi_0 = \sum \frac{\partial^2 \phi}{\partial x_i \partial x_l} dx_l \wedge I_k^i dx_k = \sum g_{il} I_k^i dx_l \wedge dx_k = -\sum \omega_{lk} dx_l \wedge dx_k = -\omega.$$

To summarize:

$$d\varphi_0 = -\omega, \qquad d\varphi_1 = -\tilde{\omega}_3 - \varepsilon \wedge \varphi_1, \qquad d\varphi_3 = \tilde{\omega}_1 - \varepsilon \wedge \varphi_3. \tag{9}$$

Denote by $Q_X \subset TM \oplus U_1$ the subbundle spanned by the vector fields $X, J_1 X, J_2 X, J_3 X$. Since $J_2 = (I, -I)$, it follows that $J_2 X$ is tangent to M, but $J_3 X$ and $J_1 X$ are sections of U_1 and hence tangential to G. The orthogonal projection P onto Q_X is defined by

$$PV = g(X,X)^{-1}(\varphi_0(V)X + \varphi_1(V)J_1 X + \varphi_2(V)J_2 X + \varphi_3(V)J_3 X). \quad (10)$$

Consider now the three two-forms obtained by restricting to $Q_X^{\perp}$:

$$\tilde{\omega}_i(V - PV, V' - PV') = \tilde{\omega}_i(V, V') - \tilde{\omega}_i(PV, PV').$$

Because the hyperkähler metric had quaternionic signature $(n-1, 1)$ with $g(X,X) < 0$, these define a (positive definite) $Sp(n-1)$ structure on the orthogonal complement H_X to Q_X in $TM \oplus U_1$. From (10) we have

$$\tilde{\omega}_1(PV, PV') = g(X,X)^{-1}(\varphi_0 \wedge \varphi_1 - \varphi_2 \wedge \varphi_3)(V, V'),$$

since

$$\tilde{\omega}_1(J_2 X, J_3 X) = -\tilde{\omega}_1(X, J_2 J_3 X) = -\tilde{\omega}_1(X, J_1 X).$$

Similar formulae for $\tilde{\omega}_2$ and $\tilde{\omega}_3$ show that this $Sp(n-1)$ structure is defined by

$$\begin{aligned}
\tilde{\omega}_1 - g(X,X)^{-1}(\varphi_0 \wedge \varphi_1 - \varphi_2 \wedge \varphi_3), \\
\tilde{\omega}_2 - g(X,X)^{-1}(\varphi_0 \wedge \varphi_2 - \varphi_3 \wedge \varphi_1), \\
\tilde{\omega}_3 - g(X,X)^{-1}(\varphi_0 \wedge \varphi_3 - \varphi_1 \wedge \varphi_2).
\end{aligned} \quad (11)$$

In conclusion, we have split the tangent bundle of $M \times G$ as

$$T(M \times G) \cong H_X \oplus Q_X \oplus U_2,$$

and have described the $Sp(n-1)$ structure on H_X by the above two-forms.

6 Taking the Quotient

To construct the Kähler quotient of M by the S^1-action, we use the moment map ϕ of X and a regular value c and take the quotient of the submanifold $\phi^{-1}(c)$ by S^1 — this is just symplectic reduction, but the Kähler metric descends to the quotient $\bar{M}$. Its complex structure is given by the holomorphic quotient $M/\mathbf{C}^*$ (or strictly speaking the quotient of a suitable open set of stable points). We shall work on the S^1-principal bundle $P = \phi^{-1}(c)$ in order to do our computations. The vector field IX is normal to $P \subset M$, and X is tangential.

The pull-back of the tangent bundle of $\bar{M}$ to P is naturally identified with the orthogonal complement of X. Recall that H_X is the orthogonal complement in $TM \oplus U_1$ to the quaternionic space Q_X generated by X; it follows that we can write

$$T(\bar{M} \times G) \cong H_X \oplus \langle J_3 X, J_1 X \rangle \oplus \langle E, Z \rangle.$$

Our strategy is to use the $Sp(n-1)$ structure on H_X defined in the previous section over $P \times G$ and to define a complementary $Sp(1)$ structure on the four-dimensional space spanned by $J_3 X, J_1 X, E, Z$.

Remark. Since the action of the circle rotates the space of two-forms spanned by $\tilde{\omega}_3$ and $\tilde{\omega}_1$, the quaternionic action will not descend from the principal bundle $P \times G$ to $\bar{M} \times G$. But this is in any case a feature of a quaternionic Kähler manifold: in general, the tangent bundle is a module over a nontrivial bundle of quaternion algebras (the imaginary part is called $\mathcal{Q}$ in [3]). In our case, the bundle becomes trivial when pulled back to P.

First note that

$$d\phi = i_X \omega = i_X \tilde{\omega}_2 = \varphi_2,$$

so φ_2 vanishes when restricted to $P = \phi^{-1}(c)$. Secondly observe that

$$d(g(X,X)) = d(i_X \varphi_0) = \mathcal{L}_X \varphi_0 - i_X d\varphi_0.$$

But since the S^1-action preserves the metric and the vector field X, $\mathcal{L}_X \varphi_0 = 0$. Hence

$$d(g(X,X)) = -i_X d\varphi_0 = i_X \omega = d\phi$$

using $d\varphi_0 = -\omega$. Thus

$$d(g(X,X) - \phi) = 0,$$

and $g(X,X)$ differs from ϕ by a constant. Hence on $P = \phi^{-1}(c)$, the function $g(X,X)$ is a negative constant $-k^{-1}$. Using $\varphi_2|_P = 0$, the quaternionic structure on H_X is now defined (using (11) and (9)) by the three two-forms:

$$\begin{aligned}
\mu_1 &= \tilde{\omega}_1 + k\varphi_0 \wedge \varphi_1 = d\varphi_3 + \varepsilon \wedge \varphi_3 + k\varphi_0 \wedge \varphi_1, \\
\mu_2 &= \tilde{\omega}_2 - k\varphi_3 \wedge \varphi_1 = \omega - \tilde{F} - k\varphi_3 \wedge \varphi_1, \\
\mu_3 &= \tilde{\omega}_3 + k\varphi_0 \wedge \varphi_3 = -d\varphi_1 - \varepsilon \wedge \varphi_1 + k\varphi_0 \wedge \varphi_3.
\end{aligned} \qquad (12)$$

Now consider the space $\langle J_3 X, J_1 X, E, Z \rangle$ and the restriction of the three two-forms

$$\begin{aligned}
\nu_1 &= -\varepsilon \wedge \varphi_3 - k\zeta \wedge \varphi_1, \\
\nu_2 &= -2\varepsilon \wedge \zeta + 2k\varphi_3 \wedge \varphi_1, \\
\nu_3 &= \varepsilon \wedge \varphi_1 - k\zeta \wedge \varphi_3.
\end{aligned}$$

We can write

$$\begin{aligned}
\nu_3 + i\nu_1 &= (\varphi_3 + i\varphi_1) \wedge (k\zeta + i\varepsilon), \\
\nu_2 &= \frac{1}{i} \left[k(\varphi_3 + i\varphi_1) \wedge (\varphi_3 - i\varphi_1) + k^{-1}(k\zeta + i\varepsilon) \wedge (k\zeta - i\varepsilon) \right],
\end{aligned}$$

from which it is clear that they define an $Sp(1)$ structure for the positive definite inner product

$$k(\varphi_3^2 + \varphi_1^2 + k^{-2}\varepsilon^2 + \zeta^2)$$

on this space.

Remark. Note that $\varphi_3(J_3 X) = g(X,X) = -k^{-1}$, so the inner product on the space spanned by $J_3 X, J_1 X$ in the hyperkähler picture is $-k(\varphi_3^2 + \varphi_1^2)$. The quaternionic structure thus changes the sign of the metric on this two-dimensional space.

To obtain a metric on $\bar{M} \times G$, we take the forms

$$\rho_i = \mu_i + \nu_i,$$

and these then define an $Sp(n)$ structure on $H_X \oplus \langle J_3 X, J_1 X, E, Z \rangle$. Explicitly, we have

$$
\begin{aligned}
\rho_1 &= d\varphi_3 + k(\varphi_0 - \zeta) \wedge \varphi_1, \\
\rho_2 &= \omega - \tilde{F} + k\varphi_3 \wedge \varphi_1 - 2\varepsilon \wedge \zeta, \\
\rho_3 &= -d\varphi_1 + k(\varphi_0 - \zeta) \wedge \varphi_3.
\end{aligned}
\tag{13}
$$

By construction, the forms ρ_i satisfy $i_X \rho_i = 0$, since they were defined in terms of linear forms on the orthogonal complement of X. The four-form $\Omega = \rho_1^2 + \rho_2^2 + \rho_3^2$ therefore satisfies $i_X \Omega = 0$, but is also S^1-invariant—the action fixes ρ_2 and rotates ρ_1 and ρ_3. Thus Ω is the pull-back of a four-form $\bar{\Omega}$ on $\bar{M} \times G$ and defines an $Sp(n)$ structure on the tangent bundle.

Using $d\varphi_0 = -\omega$ and $d\zeta = -2\varepsilon \wedge \zeta - \tilde{F}$ and the constancy of k, we find from (13) that

$$
\begin{aligned}
d\rho_1 &= -k\varphi_1 \wedge \rho_2 + k(\varphi_0 - \zeta) \wedge \rho_3, \\
d\rho_2 &= k\varphi_1 \wedge \rho_1 + k\varphi_3 \wedge \rho_3, \\
d\rho_3 &= -k\varphi_3 \wedge \rho_2 - k(\varphi_0 - \zeta) \wedge \rho_1,
\end{aligned}
\tag{14}
$$

which shows that the ρ_i generate a differential ideal. But also

$$d(\rho_3^2 + \rho_1^2) = -2k\rho_2 \wedge (\varphi_3 \wedge \rho_3 + \varphi_1 \wedge \rho_1) = -2\rho_2 \wedge d\rho_2,$$

so

$$d(\rho_1^2 + \rho_2^2 + \rho_3^2) = d\Omega = 0.$$

Now the form Ω defines a quaternionic Kähler structure for $n \geq 3$ if and only if it is closed, and for $n = 2$ if it is closed and the rank 3 bundle of two-forms generates a differential ideal (see [11,12]), thus $\bar{M} \times G$ with this structure is quaternionic Kähler. This is the metric written down by Ferrara and Sabharwal [6].

7 Properties of the Metric

By construction, the metric has an isometric action of the $(2n+2)$-dimensional Heisenberg group.

As for the curvature, we know that any quaternionic Kähler manifold is an Einstein manifold, so we find next the Einstein constant.

From (14), the exterior derivatives of ρ_i satisfy

$$d\rho_i = \sum \alpha_{ij} \wedge \rho_j$$

for a skew-symmetric matrix α_{ij} of one-forms, where

$$\alpha_{12} = -k\varphi_1, \qquad \alpha_{23} = k\varphi_3, \qquad \alpha_{31} = -k(\varphi_0 - \zeta).$$

For any quaternionic Kähler manifold, the matrix α_{ij} is uniquely determined and is the connection form for the rank three bundle of two-forms or $\mathcal{Q}$ (see [3], for example). The curvature is then

$$d\alpha_{ij} - \sum \alpha_{ik} \wedge \alpha_{kj} = \lambda \sum \varepsilon_{ijk} \rho_k,$$

where $\pm\lambda$ (depending on the orientation of the three local sections ρ_i) is the Einstein constant Λ. In this case,

$$\Lambda = -k$$

is negative.

Remark. The group generated by the vector fields E and Z has, as described in Section 4, orbits with the structure of the upper half plane. A short calculation shows that these have constant Gaussian curvature $-4k$ in the induced metric.

Although the forms ρ_1, ρ_2, ρ_3 do not descend from the principal S^1-bundle P to the quaternionic Kähler base $\bar{M} \times G$, the form ρ_2 does, and consequently there is a section of the bundle $\mathcal{Q}$ that defines an almost complex structure. We shall see that this is never integrable. Indeed, although the Grassmannian example above is a complex manifold, this complex structure does not belong to the quaternionic family (see [1], 14.53).

This almost complex structure is defined equally by the locally defined complex two-form $\rho_3 + i\rho_1$: the $(0,1)$ vectors V are defined by $i_V(\rho_3 + i\rho_1) = 0$, and then $\rho_3 + i\rho_1$ is of type $(2,0)$. It follows that the Lie bracket $[V, V']$ of any two such vector fields is also of type $(0,1)$ if the three-form $d(\rho_3 + i\rho_1)$ has no component of type $(1,2)$. Now from (14)

$$d(\rho_3 + i\rho_1) = -k(\varphi_3 + i\varphi_1) \wedge \rho_2 - ik(\varphi_0 - \zeta) \wedge (\rho_3 + i\rho_1).$$

The second term is of type $(3,0) + (2,1)$ so the obstruction only involves the first term. Now ρ_2 is a Hermitian form, of type $(1,1)$ and nondegenerate, so the first term has vanishing $(1,2)$ component if and only if $\varphi_3 + i\varphi_1$ is of type $(1,0)$. But the complex structure corresponds to J_2 and $J_2\varphi_3 = \varphi_1, J_2\varphi_1 = -\varphi_3$, which means that $\varphi_3 + i\varphi_1$ is of the opposite type $(0,1)$. Since it is nonvanishing, integrability never holds.

Remark. The Weil intermediate Jacobian for a Calabi–Yau threefold has complex structure defined by the Hodge star operator on harmonic three-forms. The skew pairing on $H^3(Y, \mathbf{R})$ then becomes a positive Hermitian form rather than one of signature $(h^{2,1}, 1)$ and so defines the structure of a principally polarized Abelian variety. The S^1-action is on the one-dimensional $H^{3,0}$ part, and this in the hyperkähler manifold we started from is the space spanned by J_2X, J_3X. As remarked above, the change from hyperkähler to quaternionic Kähler changed the sign of the metric on this space: this is the change from the Griffiths complex structure to the Weil complex structure. In this case, instead of a holomorphic fibration of Griffiths intermediate Jacobians, as in the indefinite hyperkähler space, we get for the quaternionic Kähler manifold $\mathcal{M} \times G/H_{\mathbf{Z}}$ a non-holomorphic fibration by punctured disc bundles over the Weil Jacobian.

Acknowledgments Dedicated to Charles Boyer on the occasion of his 65th birthday.

References

1. A.L. Besse, "Einstein manifolds," Ergebnisse der Mathematik und ihrer Grenzgebiete **10** Springer-Verlag, Berlin (1987).
2. C. Birkenhake & H. Lange, "Complex tori," Progress in Mathematics, **177** Birkhäuser Boston Inc., Boston, MA (1999).
3. C.P. Boyer & K. Galicki, "Sasakian Geometry," Oxford University Press, Oxford, (2008).
4. S. Cecotti, S. Ferrara & L. Girardello, Geometry of type II superstrings and the moduli of superconformal field theories, *Int. J. Mod. Phys. A* **4** (1989) 2475–2529.
5. V. Cortés, On hyper-Kähler manifolds associated to Lagrangian Kähler submanifolds of T^*C^n, *Trans. Am. Math. Soc.* **350** (1998) 3193–3205.
6. S. Ferrara & S. Sabharwal, Quaternionic manifolds for type II superstring vacua of Calabi-Yau spaces, *Nucl. Phys. B* **332** (1990) 317–332.
7. D.S. Freed, Special Kähler manifolds, *Commun. Math. Phys.* **203** (1999) 31–52.
8. N.J. Hitchin, The moduli space of complex Lagrangian submanifolds, *Asian J. Math.* **3** (1999) 77–91.
9. N.J. Hitchin, The geometry of three-forms in six dimensions, *J. Differential Geom.* **55** (2000) 547–576.
10. N.J. Hitchin, Generalized Calabi-Yau manifolds, *Quart. J. Math. Oxford*, **54** (2003) 281–308.
11. S. Salamon, "Riemannian geometry and holonomy groups," Pitman Research Notes in Mathematics **201**, Longman, Harlow (1989).
12. A. Swann, Hyperkähler and quaternionic Kähler geometry, *Math.Ann.* **289** (1991) 421–450.

Homological Mirror Symmetry and Algebraic Cycles

Ludmil Katzarkov

Abstract In this paper we describe a Homological Mirror Symmetry approach to classical problems in Algebric Geometry — rationality questions and the Hodge conjecture. Several examples are studied in detail.

1 Introduction

This paper summarizes a talk given at a conference at UNM in October 2006. It introduces a new approach to two classical problems in algebraic geometry — rationality of algebraic varieties and the Hodge conjecture. Our new approach is based on homological mirror symmetry (HMS). We concentrate mainly on the connection between algebraic cycles on Abelian varieties and four-dimensional cubics and their tropical interpretation on the mirrors. For more information on the classical treatment of these questions, we refer to papers [18,39,41] and the vast literature cited there.

One of the most interesting open questions in mirror symmetry is to understand how HMS is affected by moving in moduli spaces. Some answers of this question can be found in [7,17,30] in the cases of Abelian varieties and Del Pezzo surfaces. In this paper, we suggest a new approach Exploiting the rich geometry of Shimura varieties and of the moduli spaces of four-dimensional cubics—see [18] and [39]. We introduce the notion of an earthquake—the mirror image of the process of jumping from one Noether–Lefschetz locus to another. We consider many examples and formulate a tropical version of the Hodge conjecture. Some of the examples in this paper will appear in full detail in joint papers with Auroux, Abouzaid, Gross [2] and Kontsevich [26]. Here we only give the flavor of the arguments and techniques involved and outline the examples. It is quite possible that different approaches to the above two classical questions might prove more fruitful.

L. Katzarkov
Department of Mathematics, University of Miami, Coral Gables, Florida, USA

K. Galicki and S.R. Simanca (eds.), *Riemannian Topology and Geometric Structures on Manifolds,* Progress in Mathematics 271, DOI 10.1007/978-0-8176-4743-8,

Recall that mirror symmetry was introduced as a special duality between two $N = 2$ super conformal field theories. Traditionally, an $N = 2$ super conformal field theory (SCFT) is constructed as a quantization of a σ-model with target a compact Calabi–Yau manifold equipped with a Ricci flat Kähler metric and a closed 2-form —the so called B-field. We say that two Calabi–Yau manifolds X and Y form a *mirror pair $X|Y$* if the associated $N = 2$ SCFTs are mirror dual to each other [13].

Our approach to questions of rationality and to earthquakes can be summarized as follows:

- On the algebro-geometric side, we consider a degeneration the manifold X, and following Bardelli [8], we record the data of this degeneration in a mixed hodge structure (MHS).
- On the mirror side, these data (and more) are recorded by the MHS associated with the perverse sheaf of vanishing cycles—see Theorem 4.16. We express the nonrationality condition of Bardelli in terms of geometry of the perverse sheaf of vanishing cycles—see Conjecture 4.11.
- Through local and global HMS, we translate and extend Bardelli nonrationality condition to the example of four-dimensional cubic.

We start with a quick introduction of homological mirror symmetry. After that, we introduce our construction of the mirror Landau–Ginzburg model, which differs from constructions in [19]. Connection with birational geometry is discussed in Section 4. Questions related to Hodge conjecture appear in Section 5. We consider many examples and outline the flavor of the arguments—full details will appear in [2] and [26].

2 Definitions

We first recall a definition that belongs to Seidel [36]. Historically, the idea was introduced first by M. Kontsevich and later by K. Hori. We begin by briefly reviewing Seidel's construction of a Fukaya-type A_∞-category associated to a symplectic Lefschetz pencil see [36, 37].

Let (Y, ω) be an open symplectic manifold of dimension $\dim_{\mathbb{R}} Y = 2n$. Let $w : Y \to \mathbb{C}$ be a symplectic Lefschetz fibration, i.e., w is a C^∞ complex-valued function with isolated nondegenerate critical points $p_1, \ldots, p_r$ near which w is given in adapted complex local coordinates by $w(z_1, \ldots, z_n) = f(p_i) + z_1^2 + \cdots + z_n^2$, and whose fibers are symplectic submanifolds of Y. Fix a regular value λ_0 of w, and consider arcs $\gamma_i \subset \mathbb{C}$ joining λ_0 to the critical value $\lambda_i = f(p_i)$. Using the horizontal distribution that is symplectic orthogonal to the fibers of w, we can transport the vanishing cycle at p_i along the arc γ_i to obtain a Lagrangian disc $D_i \subset Y$ fibered above γ_i, whose boundary is an embedded Lagrangian sphere L_i in the fiber $\Sigma_{\lambda_0} = w^{-1}(\lambda_0)$. The Lagrangian disc D_i is called the *Lefschetz thimble* over γ_i, and its boundary L_i is the vanishing cycle associated to the critical point p_i and to the arc γ_i.

After a small perturbation, we may assume that the arcs $\gamma_1, \ldots, \gamma_r$ in $\mathbb{C}$ intersect each other only at λ_0, and are ordered in the clockwise direction around λ_0. Similarly, we may always assume that the Lagrangian spheres $L_i \subset \Sigma_{\lambda_0}$ intersect each other transversely inside Σ_0.

Definition 2.1 (Seidel). *Given a coefficient ring R, the R-linear directed Fukaya category $FS(Y, w; \{\gamma_i\})$ is the following A_∞-category: the objects of $FS(Y, w; \{\gamma_i\})$ are the Lagrangian vanishing cycles $L_1, \ldots, L_r$; the morphisms between the objects are given by*

$$\mathrm{Hom}(L_i, L_j) = \begin{cases} CF^*(L_i, L_j; R) = R^{|L_i \cap L_j|} & \text{if } i < j \\ R \cdot id & \text{if } i = j \\ 0 & \text{if } i > j; \end{cases}$$

and the differential m_1, composition m_2, and higher-order products m_k are defined in terms of Lagrangian Floer homology inside Σ_{λ_0}.

More precisely,

$$m_k : \mathrm{Hom}(L_{i_0}, L_{i_1}) \otimes \cdots \otimes \mathrm{Hom}(L_{i_{k-1}}, L_{i_k}) \to \mathrm{Hom}(L_{i_0}, L_{i_k})[2-k]$$

is trivial when the inequality $i_0 < i_1 < \cdots < i_k$ fails to hold (i.e., it is always zero in this case, except for m_2 where composition with an identity morphism is given by the obvious formula).

When $i_0 < \cdots < i_k$, m_k is defined by fixing a generic ω-compatible almost-complex structure on Σ_{λ_0} and counting pseudo-holomorphic maps from a disc with $k+1$ cyclically ordered marked points on its boundary to Σ_{λ_0}, mapping the marked points to the given intersection points between vanishing cycles, and the portions of boundary between them to $L_{i_0}, \ldots, L_{i_k}$, respectively (see [37]).

Remark 2.2. Clearly, the directed Fukaya category $FS(Y, w; \{\gamma_i\})$ depends on the choice of paths. As shown in [37], the change of paths leads to an autoequivalence of $FS(Y, w; \{\gamma_i\})$—a mutation.

We will denote the category defined above by $FS(Y, w)$ and will refer to it as the *Fukaya-Seidel* category of w.

Let Y be a complex algebraic variety (or a complex manifold) and let $w : Y \to \mathbb{C}$ be a holomorphic function. Following [34], we define:

Definition 2.3 (Orlov). *The derived category $D^b(Y, w)$ of a holomorphic potential $w : Y \to \mathbb{C}$ is defined as the disjoint union*

$$D^b(Y, w) := \prod_t D^b_{\mathrm{sing}}(Y_t)$$

of the derived categories of singularities $D^b_{\mathrm{sing}}(Y_t)$ of all fibers $Y_t := w^{-1}(t)$ of w.

The category $D^b_{\mathrm{sing}}(Y_t) = D^b(Y_t)/\mathrm{Perf}(Y_t)$ is defined as the quotient category of derived category of coherent sheaves on Y_t modulo the full triangulated subcategory of perfect complexes on Y_t. Note that $D^b_{\mathrm{sing}}(Y_t)$ is nontrivial only for singular fibers Y_t.

In what follows, we will use notions defined in [19], section 7.3. Let X be a manifold of general type, i.e., a sufficiently high power of the canonical linear system on X defines a birational map. We will use the above definitions to formulate and motivate an analogue of Kontsevich's homological mirror symmetry (HMS) conjecture for such manifolds as well as for Fano manifolds—manifolds whose anticanonical linear system defines a birational map.

A quantum sigma-model with target X is free in the infrared limit, whereas in the ultraviolet limit it is strongly coupled. In order to make sense of this theory at arbitrarily high energy scales, one has to embed it into some asymptotically free $N = 2$ field theory, for example into a gauged linear sigma-model (GLSM). Here "embedding" means finding a GLSM such that the low-energy physics of one of its vacua is described by the sigma-model with target X. In mathematical terms, this means that X has to be realized as a complete intersection in a toric variety.

The GLSM usually has additional vacua, whose physics is not related to X. Typically, these extra vacua have a mass gap. To learn about X by studying the GLSM, it is important to be able to recognize the extra vacua. Let Z be a toric variety defined as a symplectic quotient of $\mathbb{C}^N$ by a linear action of the gauge group $G \simeq U(1)^k$. The weights of this action will be denoted Q_{ia}, where $i = 1, \ldots, N$ and $a = 1, \ldots, k$. Let X be a complete intersection in Z given by homogeneous equations $G_\alpha(X) = 0$, $\alpha = 1, \ldots, m$. The weights of G_α under the G-action will be denoted $d_{\alpha a}$. The GLSM corresponding to X involves chiral fields Φ_i, $i = 1, \ldots, N$ and Ψ_α, $\alpha = 1, \ldots, m$. These are equivariant functions on $\mathbb{C}^N$ whose charges under the gauge group G are given by matrices Q_{ia} and $d_{\alpha a}$, respectively. The Lagrangian of the GLSM depends also on complex parameters t_a, $a = 1, \ldots, k$. On the classical level, the vector t_a is the level of the symplectic quotient, and thus parameterizes the complexified Kähler form on Z. The Kähler form on X is the induced one. On the quantum level, the parameters t_a are renormalized and satisfy linear renormalization semigroup equations, which, in the Calabi–Yau case, imply the vanishing of $\sum_i Q_{ia} - \sum_\alpha d_{\alpha a}$.

According to [20], the Landau–Ginzburg model (Y, w) that is mirror to X will have (twisted) chiral fields Λ_a, $a = 1, \ldots, k$, Y_i, $i = 1, \ldots, N$ and Υ_α, $\alpha = 1, \ldots, m$ and a superpotential is given by

$$w = \sum_a \Lambda_a \left(\sum_i Q_{ia} Y_i - \sum_\alpha d_{\alpha a} \Upsilon_\alpha - t_a \right) + \sum_i e^{-Y_i} + \sum_\alpha e^{-\Upsilon_\alpha}.$$

The vacua are in one-to-one correspondence with the critical points of w. By definition, massive vacua are those corresponding to nondegenerate critical points. An additional complication is that before computing the critical points, one has to partially compactify the target space of the LG model—this partial compactification is not always well defined except in the case of Fano manifolds when generic fiber of Landau–Ginzburg model (Y, w) is the mirror of the an anticanonical section—a Calabi–Yau manifold on its own (see [4]).

One can determine which vacua are "extra" (i.e., unrelated to X) as follows. The infrared limit is the limit $\mu \to 0$. Since t_a depend on μ, so do the critical points of w. A critical point is relevant for X (i.e., is not an extra vacuum) if and only if the critical values of e^{-Y_i} all go to zero as μ goes to zero. In terms of the original variables Φ_i, this means that vacuum expectation values of $|\Phi_i|^2$ go to $+\infty$ in the infrared limit. This is precisely the condition that justifies the classical treatment of vacua in the GLSM. Recall also that the classical space of vacua in the GLSM is precisely X.

Now let us state the analogue of the HMS for complete intersections X of general type. We will write $D^b(X)$ for the standard derived category of bounded complexes of coherent sheaves on X and by $\text{Fuk}(X, \omega)$ the standard Fukaya category of a symplectic manifold X with a symplectic form ω.

Conjecture 2.4. *Let X be a variety of general type that is realized as a complete intersection in a toric variety, and let (Y, w) be the mirror LG model. The derived Fukaya category $\text{Fuk}(X, \omega)$ of X embeds as a direct summand into the category $D^b(Y, \text{w})$ (the category of B-branes for the mirror LG model). If the extra vacua are all massive, the complement of the Fukaya category of X in $D^b(Y, \text{w})$ is very simple: each extra vacuum contributes a direct summand that is equivalent to the derived category of graded modules over a Clifford algebra.*

There is also a mirror version of this conjecture:

Conjecture 2.5. *The derived category $D^b(X)$ of coherent sheaves on X embeds as a full subcategory into the derived Fukaya–Seidel category $FS(Y, \text{w})$ of the potential* w.

With an appropriate generalization of Fukaya–Seidel category to the case of non-isolated singularities, we arrive at Table 1 summarizing our previous considerations. The categories $D^b_{\lambda_i, r_i}(Y, \text{w})$ and $FS_{(\lambda_i, r_i)}(Y, \text{w}, \omega)$ appearing in the last row of Table 1 are modifications that contain information only about singular fibers with base points contained in a disc with a radius—a real number λ, are introduced in order to deal with problems arising from massless vacua.

In this formalism, we have to take a Karoubi closure of the categories on both sides.

3 The Construction

The standard construction of Landau–Ginzburg models was done for smooth complete intersections in toric varieties in [19]. We describe a new procedure that is a natural continuation of our previous results [5] and of the works of Moishezon and Teicher. Our procedure underlines the geometric nature of HMS, and several categorical correspondences come naturally. We will restrict ourselves to the case of three-dimensional Fano manifolds X with the anticanonical linear system on it. For simplicity, we will assume that $|-K_X|$ contains a pencil.

Table 1 Kontsevich's HMS conjecture.

A side	B side
X — Compact manifold, ω — Symplectic form on X.	X — Smooth projective variety over $\mathbb{C}$
$\mathrm{Fuk}(X,\omega) = \begin{cases} \mathrm{Obj}: (L_i, \mathcal{E}) \\ \mathrm{Mor}: HF(L_i, L_j) \end{cases}$	$D^b(X) = \begin{cases} \mathrm{Obj}: C_i^{\bullet} \\ \mathrm{Mor}: Ext(C_j^{\bullet}, C_j^{\bullet}) \end{cases}$
L_i — Lagrangian submanifold of X, $\mathcal{E}$ — Flat $U(1)$-bundle on L_i	$C_i^{\bullet}$ — Complex of coherent sheaves on X

(Y, ω) — Open symplectic manifold, $\mathrm{w}: Y \to \mathbb{C}$ — A proper C^{∞} map with symplectic fibers.	Y — Smooth quasi-projective variety over $\mathbb{C}$, $\mathrm{w}: Y \to \mathbb{C}$ — Proper algebraic map.
$FS(Y, \mathrm{w}, \omega) = \begin{cases} \mathrm{Obj}: (L_i, \mathcal{E}) \\ \mathrm{Mor}: HF(L_i, L_j) \end{cases}$	$D^b(Y, \mathrm{w}) = \bigsqcup_t D^b(Y_t) \big/ \mathrm{Perf}(Y_t)$
L_i — Lagrangian submanifold of Y_{λ_0}, $\mathcal{E}$ — Flat $U(1)$-bundle on L_i.	
$FS_{\lambda_i, r_i}(Y, \mathrm{w}, \omega)$—The Fukaya–Seidel category of $(Y_{|t-\lambda_i|<r_i}, \mathrm{w}, \omega)$.	$D^b_{\lambda_i, r_i}(Y, \mathrm{w}) = \bigsqcup_{|t-\lambda_i| \leq r_i} D^b(Y_t) \big/ \mathrm{Perf}(Y_t)$

Step 1. Choose a Moishezon degeneration [5] $\mathcal{X} \to \Delta$ of X corresponding to a projective embedding of X given by a multiple of the ample line bundle $-K_X$. This means that we choose a generic Noether normalization projection $X \to \mathbb{P}^3$ in the given projective embedding, and then degenerate the covering map to a totally split cover. The central fiber $\mathcal{X}_0$ of this degeneration is a configuration of spaces intersecting over rational surfaces and curves. Next we choose a generic pencil : $\mathcal{X} \dashrightarrow \mathbb{P}^1$ in the linear system $|-K_{\mathcal{X}/\Delta}|$.

Step 2. The degeneration $\mathcal{X} \to \Delta$ can be seen as a logarithmic structure (see [17]) $X_0^{\dagger}$ on X_0—roughly a union of toric varieties, intersecting over toric varieties plus monodromy data.

Step 3. We apply a Legendre transform [17] to $(X_0^{\dagger}, f^{\dagger})$ and obtain a new logarithmic structure $(Y_0^{\dagger}, \mathrm{w}^{\dagger})$ with a pencil on it. An explanation on the construction of the potential $\mathrm{w}^{\dagger}$ can be found in [4] and [16]. To get $Y_0^{\dagger}$, we take the dual intersection complex and then for vertices we specify the normal fan structure and corresponding monodromy.

Step 4. We use the logarithmic structure $Y_0^{\dagger}$ to smooth Y_0 and get a mirror degenerating family $\mathcal{Y} \to \Delta$ and a mirror Landau–Ginzburg potential $\mathrm{w}: \mathcal{Y} \to \mathbb{P}^1$. Basically, on the generic fiber Y of $\mathcal{Y} \to \Delta$, the potential $\mathrm{w}: Y \dashrightarrow \mathbb{P}^1$ is an anticanonical pencil of mirror manifolds to the anticanonical $K3$ surfaces in X. As before, we move the $\mathbb{P}^1 \subset |-K_Y|$ corresponding to the pencil $\mathrm{w}: Y \dashrightarrow \mathbb{P}^1$ so that it intersects the discriminant locus Δ_Y in a minimal number of points (0 and ∞ among them—in physics language this is the point of maximal degeneration and the Gepner point).

Step 5. The construction above allow us to work not only with X and its Landau–Ginzburg model Y but also with the algebraic cycles in X and the mirror Lagrangians in Y. We briefly describe how this works—for more see [25]. Let us restrict ourselves to the case of a curve C in three-dimensional Fano manifold X. The construction from Section 2 allows us to follow what happens with the mirror image of C. In the example of an algebraic curve C and its tropical realization T [32], we get that the image of T under the Legendrian transform is the conormal tori fibration—see the diagram below. The new Lagrangian cycle L is the mirror the curve C. Observe that the construction works on the nose only in the case when C is a linear cycle in a toric variety, where existing toric fibrations can be used. In general, we need to work with generalized toric varieties [3] and then smooth. The symplectic counterpart is the construction of pair of pants decomposition introduced by Abouzaid and Seidel—so-called local HMS. Gluing local Fukaya categories should correspond to gluing categories of singularities on the B side via smoothing procedure.

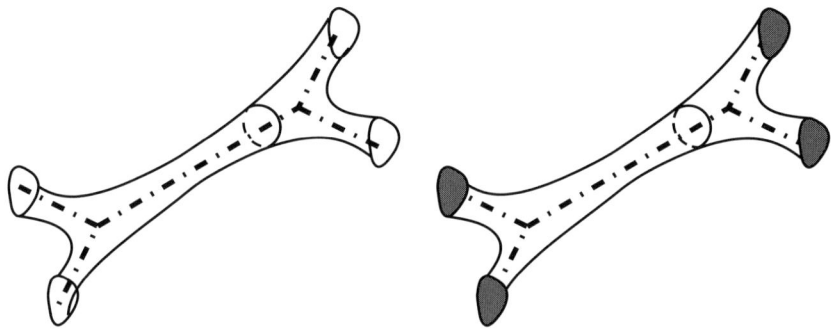

Remark 3.1. The above construction suggest that we can approach HMS in the following way:

1. Degenerate;
2. Do HMS locally for each component;
3. Finally put everything together.

This procedure allow computations in K^0 of the Fukaya category. Some of it is demonstrated in the examples that follow.

Example 3.2. We describe our procedure in the case of $X = \mathbb{CP}^2$ in Fig. 1.

Example 3.3. We also describe the above procedure in the case $X = \mathbb{CP}^1 \times \mathbb{CP}^1$ and $L = (2,2)$. Moishezon degeneration can be summarized by Fig. 2.

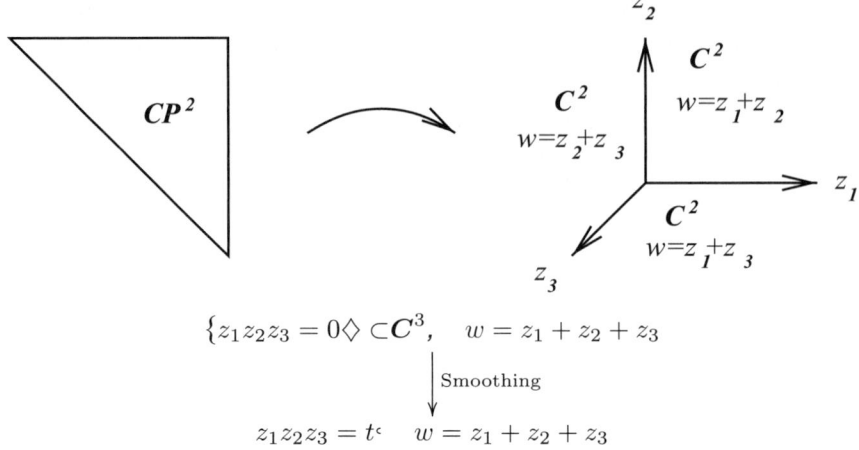

$$\{z_1 z_2 z_3 = 0\} \subset \pmb{C}^3, \quad w = z_1 + z_2 + z_3$$

Smoothing

$$z_1 z_2 z_3 = t^c \quad w = z_1 + z_2 + z_3$$

Fig. 1 The mirror of $\mathbb{C}\mathbb{P}^2$.

Fig. 2 Moisheson's degeneration.

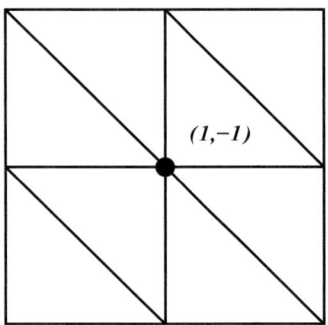

We apply Legendre transform to the result of Moishezon's degeneration replacing two affine structures (see [17]) at the ends by $\mathbb{C}^2$, the middle six by $\mathbb{C}\mathbb{P}^1 \times \mathbb{C}^1$ and the central point by Del Pezzo surface of degree three in order to get the following degeneration of the Landau–Ginzburg model (cf. Fig. 3).

Fig. 4 illustrates the singular fiber over 0 in the Landau–Ginzburg model.

The fiber over infinity is a degenerate elliptic curve consisting of eight rational curves (cf. Fig. 5).

4 Birational Transformations, Earthquakes, and HMS

Let X be a smooth projective variety and Z be a smooth subvariety. As a consequence of the weak factorization theorem [40], it is enough to study the Landau–Ginzburg mirrors of blow-ups and blow-downs with smooth centers. Recall the following result.

Fig. 3 The intersection
complex.

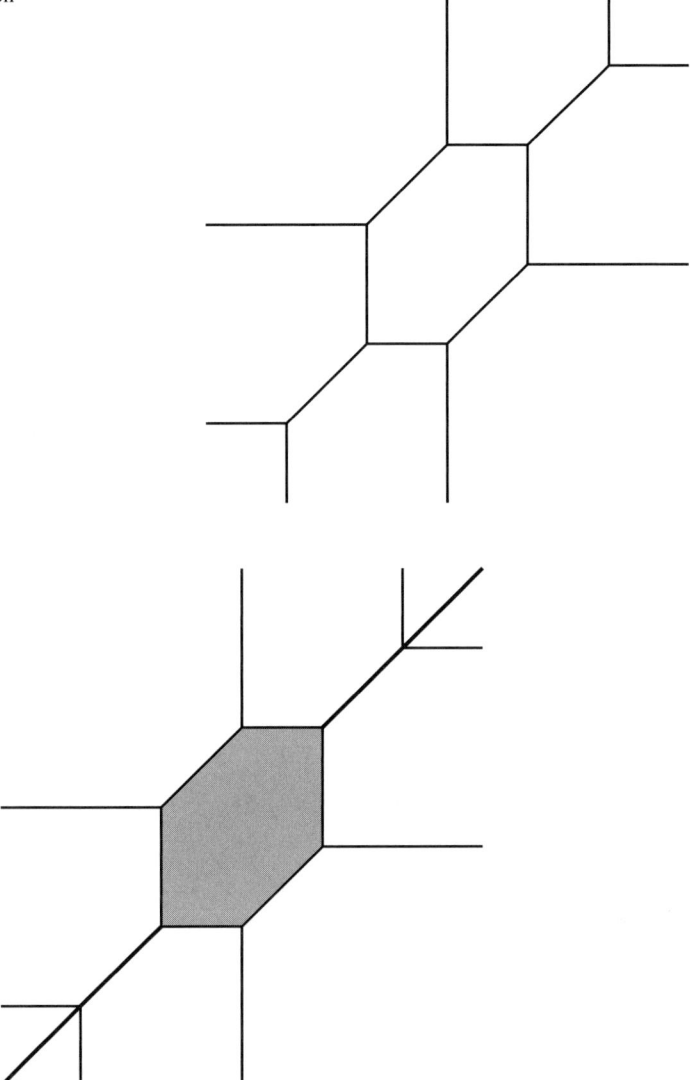

Fig. 4 The fiber over 0 of the LG potential.

Theorem 4.1. (*Orlov [33]*) *Let X be a smooth projective variety and X_Z be a blow-up of X in a smooth subvariety Z of codimension k. Then $D^b(X_Z)$ has a semiorthogonal decomposition $(D^b(X), D^b(Z)_{k-1}, \ldots, D^b(Z)_0)$. Here $D^b(Z)_i$ are corresponding twists by $\mathcal{O}(i)$ (see [33]).*

This *B*-side statement has an *A*-model counterpart. In a joint work in progress with D. Auroux, M. Abouzaid, and M. Gross [2], we work on the following:

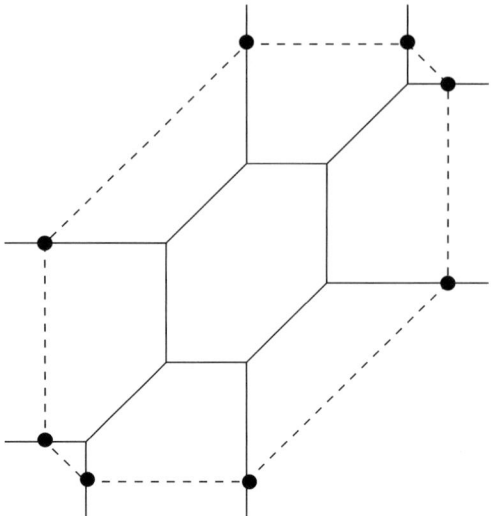

Fig. 5 The fiber over ∞.

Theorem 4.2. *(2]) The Landau–Ginzburg mirror* $(T, \mathbf{g}, \omega_T)$ *of* X_Z *is the fiber sum of the Landau–Ginzburg mirror* $(Y, \mathrm{w}, \omega_Y)$ *of* X *and the Landau–Ginzburg mirror of* $Z \times (\mathbb{C}^*)^k$. *On the level of categories,* $FS(T, \mathbf{g}, \omega_T)$ *has a semiorthogonal decomposition*

$$(FS(T, \mathbf{g}, \omega_T), FS(S, \mathbf{f}, \omega_S)_{k-1}, \dots, FS(S, \mathbf{f}, \omega_S)_0),$$

Here $S, \mathbf{f}, \omega_S$ *is the mirror of* Z, *and* $FS(S, \mathbf{f}, \omega_S)_i$ *are the categories of vanishing cycles located at* $k - 1$ *roots of unity around infinity.*

The idea behind the proof of this theorem is based on local studies of the behavior of LG mirror under birational transformations. We first start with looking at the LG mirror of $\mathbb{C}^2$ blown up at the point $(0, 1)$. It is given by the equations

$$(u, v, y) \in \mathbb{C}^* \times \mathbb{C}^2, vy = 1 + u.e^{-\varepsilon}.$$

Here, ε is the volume of the exceptional divisor, and the potential is $f = u$. Without including instanton corrections, the potential has no singular point $f = u$ but if we include instanton corrections, a singular point appears—for more details see [2]). The next step is to find an explicit family of Lagrangian tori on $\mathbb{C}^3$ blown up at a curve in $\mathbb{C}^2 \times 0$. Then we glue these local pictures around singular points.

We discuss some simple examples in order to illustrate Theorem 4.2.

Example 4.3. We discuss the LG mirror to $\mathbb{CP}^3$ blown up in a point. In this case $k = 3$. The LG mirror of the blown-up manifold is a family of K3 surfaces with 6 singular fibers. Four of these fibers correspond to the LG mirror of $\mathbb{CP}^3$ and they are situated near zero. The two remaining fibers are sitting over second roots of unity in the local chart around ∞—see Fig. 6.

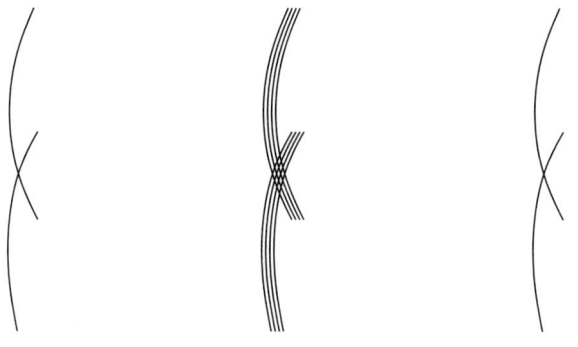

Fig. 6 LG model of $\mathbb{CP}^3$ blown up at a point.

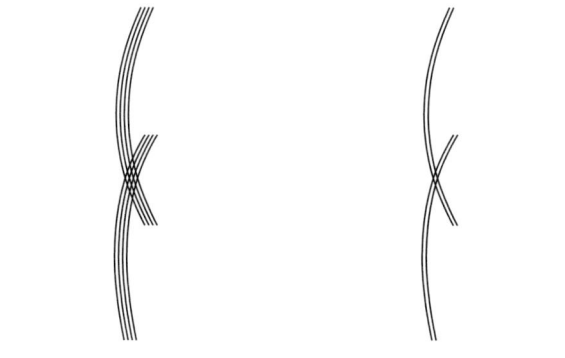

Fig. 7 LG model of $\mathbb{CP}^3$ blown up at a line.

Example 4.4. We discuss LG mirror to $\mathbb{CP}^3$ blown up in a line. In this case $k = 2$. We get as LG model of the blown-up manifold a family of K3 surfaces with 6 singular fibers. Four of these fibers correspond to the LG model of $\mathbb{CP}^3$, and they are situated near zero. The two other fibers are very close to each other. We briefly describe the procedure in this case. We start with the LG model for $\mathbb{CP}^3$

$$\mathsf{w} = x + y + z + \frac{1}{xyz}.$$

We add the additional term μxy, with μ corresponding to the volume of the blown-up $\mathbb{CP}^1$. The critical points of

$$\mathsf{w} = x + y + z + \frac{1}{xyz} + \mu xy$$

split in two groups described in Fig. 7.

Remark 4.5. The example above is rather instructive. Indeed the line in $\mathbb{CP}^3$ defined by $x = y = 1$ has its Landau–Ginzburg mirror defined by

$$w = 2 + z + \frac{1}{z} + 1.$$

This suggests that the Landau–Ginzburg mirror of $\mathbb{CP}^1$ is contained in the Landau–Ginzburg mirror of $\mathbb{CP}^3$.

Example 4.6. We discuss the LG model mirror to $\mathbb{CP}^n$ blown up in $\mathbb{CP}^k$. As in the previous examples, it can be seen as the gluing of several Landau–Ginzburg models mirroring to $\mathbb{CP}^n$ and $\mathbb{CP}^k \times (\mathbb{C}^*)^{n-k}$. The result is a LG model with $n + 1$ singular fibers located around zero (corresponding to $\mathbb{CP}^n$) glued to $n - k - 1$ other LG models of $\mathbb{CP}^k$ with singular fibers located around $k - 1$ roots of unity far away from zero—see also [28].

Example 4.7. Let us look at the Landau–Ginzburg mirror to the manifold X, which is the blow-up of $\mathbb{CP}^3$ in a genus two curve embedded as a $(2, 3)$ curve on a quadric surface. We describe the mirror of X. It is a deformation of the image of the compactification D of the map

$$(t : u_1 : u_2 : u_3) \mapsto (t : u_1 : u_2 : u_3 : u_1 \cdot u_2 : u_2 \cdot u_3 : u_1 \cdot u_3)$$

of $\mathbb{CP}^3$ in $\mathbb{CP}^7$.

    The function w gives a pencil of degree 8 on the compactified D. We can interpret this pencil as obtained by first taking Landau–Ginzburg mirror of $\mathbb{CP}^3$ and then adding to it a new singular fiber consisting of three rational surfaces intersecting over degenerated genus two curve (Fig. 8). This process is the A-model counterpart

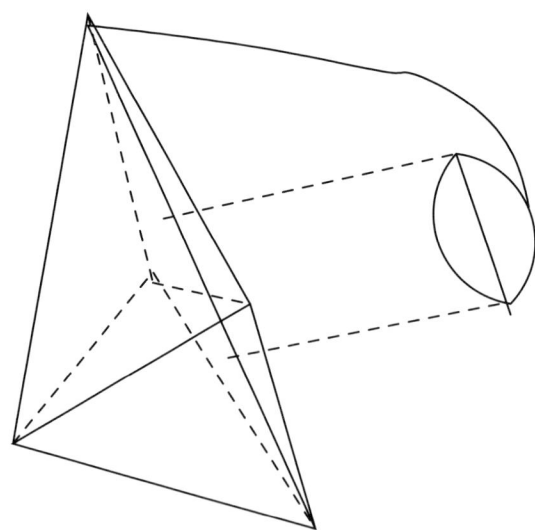

Fig. 8 LG of blow-up of $\mathbb{CP}^3$ in a genus 2 curve.

of blowing up of the genus two curve in $\mathbb{CP}^3$ and according to Orlov's theorem, stated above, can be used to define HMS for manifolds of general type.

The potential in the equation above is

$$w = t + u_1 + u_2 + u_3 + \frac{1}{u_1 \cdot u_2} + \frac{1}{u_2 \cdot u_3} + \frac{1}{u_1 \cdot u_3}.$$

Clearing denominators and substituting $t = u_3 = 0$ and $u_1^2 = u_2^3$, we get a singularity. Its resolution produces $D^b(Y, w)$ a quiver category with relations equivalent to the Fukaya category of genus two curve. But observe that this singularity is very different from the one of the $\mathbb{CP}^1$. This is precisely the point we would like to explore. The rational varieties produce simple singularities.

Let us work out the example of LG model of $\mathbb{CP}^4$ blown up in the intersection of a smooth cubic and quadric surfaces. In the case of the cubic surface, we have the compactification D of the map

$$(t : u_1 : u_2 : u_3 : u_4) \mapsto (t : u_1 : u_2 : u_3 : u_4 : u_1 \cdot u_2 \cdot u_4 : u_2 \cdot u_3 \cdot u_4 : u_1 \cdot u_2 \cdot u_3).$$

This procedure when restricted to dimension 2 produces LG model from [6]. Indeed after change of coordinates and restricting to dimension 2, we get

$$wXYZ = c(X + Y + Z)^3,$$

where c is a constant.

We notice here that the procedure described above allows us to look at the Landau–Ginzburg models in a different way. Namely, we can see that a blow-up of subvariety X in V is nothing else but moving the line in the dual space and creating a new pencil K_Y with one more singular fiber.

We summarize our findings:

Birational Transformations	Homological Mirror Symmetry
X	$w : Y \to \mathbb{CP}^1$
Blowing up of X	Moving to a less singular point in Δ_Y
Blowing up of X	Creating a new singular fiber in K_Y

4.1 Studying Nonrationality

In this section, we introduce a new tool: the perverse sheaf of vanishing cycles on the LG mirror.

We suggest a procedure to geometrize the use of categories in the nonrationality questions—perverse sheaf of vanishing cycle. We start by recalling some results from [16]. Suppose M is a complex manifold equipped with a proper holomorphic map $f : M \to \Delta$ onto the unit disc, which is submersive outside of $0 \in \Delta$. For simplicity, we will assume that f has connected fibers. In this situation, there is a natural deformation retraction $r : M \to M_0$ of M onto the singular fiber $M_0 := f^{-1}(0)$ of f. The restriction $r_t : M_t \to M_0$ of the retraction r to a smooth fiber $M_t := f^{-1}(t)$ is the "specialization to 0" map in topology. The complex of *nearby cocycles* associated with $f : M \to \Delta$ is by definition the complex of sheaves $Rr_{t*}\mathbb{Z}_{M_t} \in D^-(M_0, \mathbb{Z})$. Since for the constant sheaf we have $\mathbb{Z}_{M_t} = r_t^*\mathbb{Z}_{M_0}$, we get (by adjunction) a natural map of complexes of sheaves

$$\mathrm{sp} : \mathbb{Z}_{M_0} \to Rr_{t*}r_t^*\mathbb{Z}_{M_0} = Rr_{t*}\mathbb{Z}_{M_t}.$$

The complex of *vanishing cocycles* for f is by definition the complex **cone**(sp) and thus fits in an exact triangle

$$\mathbb{Z}_{M_0} \xrightarrow{\mathrm{sp}} Rr_{t*}\mathbb{Z}_{M_t} \to \mathbf{cone}(\mathrm{sp}) \to \mathbb{Z}_{M_0}[1],$$

of complexes in in $D^-(M_0, \mathbb{Z})$). Note that $\mathbb{H}^i(M_0, Rr_{t*}\mathbb{Z}_{M_t}) \cong H^i(M_t, \mathbb{Z})$, and so if we pass to hypercohomology, the exact triangle above induces a long exact sequence

$$\ldots H^i(M_0, \mathbb{Z}) \xrightarrow{r_t^*} H^i(M_t, \mathbb{Z}) \to \mathbb{H}^i(M_0, \mathbf{cone}(\mathrm{sp})) \to H^{i+1}(M_0, \mathbb{Z}) \to \ldots \quad (1)$$

Since M_0 is a projective variety, the cohomology spaces $H^i(M_0, \mathbb{C})$ carry the canonical mixed Hodge structure defined by Deligne. Also, the cohomology spaces $H^i(M_t, \mathbb{C})$ of the smooth fiber of f can be equipped with the Schmid–Steenbrink limiting mixed Hodge structure, which captures essential geometric information about the degeneration $M_t \rightsquigarrow M_0$. With these choices of Hodge structures, it is known from the works of Scherk and Steenbrink that the map r_t^* in (1) is a morphism of mixed Hodge structures and that $\mathbb{H}^i(M_0, \mathbf{cone}(\mathrm{sp}))$ can be equipped with a mixed Hodge structure so that (1) is a long exact sequence of Hodge structures.

Now given a proper holomorphic function $w : Y \to \mathbb{C}$, we can perform, above construction near each singular fiber of w in order to obtain a complex of vanishing cocycles supported on the union of singular fibers of w. We will write $\Sigma \subset \mathbb{C}$ for the discriminant of w, $Y_\Sigma := Y \times_\mathbb{C} \Sigma$ for the union of all singular fibers of w, and $\mathscr{F}^\bullet \in D^-(Y_\Sigma, \mathbb{Z})$ for the complex of vanishing cocycles. Slightly more generally, for any subset $\Phi \subset \Sigma$ we can look at the union Y_Φ of singular fibers of w sitting over points of Φ and at the corresponding complex

$$\mathscr{F}_\Phi^\bullet = \bigoplus_{\sigma \in \Phi} \mathscr{F}_{|Y_\sigma}^\bullet \in D^-(Y_\Phi, \mathbb{Z})$$

of cocycles vanishing at those fibers. In the following, we take the hypercohomology $\mathbb{H}^i(Y_\Sigma, \mathscr{F}^\bullet)$ and $\mathbb{H}^i(Y_\Phi, \mathscr{F}_\Phi^\bullet)$ with their natural Scherk–Steenbrink mixed Hodge structure. For varieties with anti-ample canonical class, we have:

Theorem 4.8. *Let X be a d-dimensional Fano manifold realized as a complete intersection in some toric variety. Consider the mirror Landau–Ginzburg model* w : Y → ℂ. *Suppose that Y is smooth, and that all singular fibers of* w *are either normal crossing divisors or have isolated singularities.*

Then the Deligne $i^{p,q}$ numbers for the mixed Hodge structure on $\mathbb{H}^{\bullet}(Y_{\Sigma}, \mathscr{F}^{\bullet})$ satisfy the identity

$$i^{p,q}(\mathbb{H}^{\bullet}(Y_{\Sigma}, \mathscr{F}^{\bullet})) = h^{d-p,q-1}(X).$$

When combined with Theorem 4.2, the above theorem suggests that in the case of three- and four-dimensional Fano manifolds, we have.

Conjecture 4.9. *If X is rational Fano threefold or fourfold, the monodromy of $\mathscr{F}$ restricted to all components of the singular set is trivial.*

The reason behind this conjecture is as follows. In Theorem 4.2, what we get as a new singular fiber after blowing up Z in the corresponding Landau–Ginzburg mirror of X_Z is a new singular fiber with singularities the mirror S of Z. Then the monodromy of $\mathscr{F}$ is trivial. According to a theorem of [40], we only need to care about blow-ups—blow-downs just remove singular fibers in the Landau–Ginzburg mirror.

Similarly we have:

Theorem 4.10. *Suppose that X is a variety with an ample canonical class realized as a complete intersection in a toric variety. Let (Y, w) be the mirror Landau–Ginzburg model. Suppose that all singular fibers of* w *are either normal crossing divisors or have isolated singularities.*

Then there exists a Zariski open set $U \subset \mathbb{C}$ so that Deligne's $i^{p,q}$ numbers of the mixed Hodge structure on $\mathbb{H}^{\bullet}(Y_{\Sigma \cap U}, \mathscr{F}_U^{\bullet})$ satisfy the identity

$$i^{p,q}(\mathbb{H}^i(Y_{\Sigma \cap U}, \mathscr{F}_{\Sigma \cap U}^{\bullet})) = h^{d-p,q-i+1}(X).$$

We consider the example of three-dimensional cubic. Applying the procedure described in the previous section (and in this case the procedure from [20]), we get, after smoothing, the following Landau–Ginzburg mirror:

$$xyuvw = (u+v+w)^3 . t$$

with potential $x + y$. Here u, v, w are in $\mathbb{CP}^2$, and x, y are in $\mathbb{C}^2$. The singular set W of this Landau–Ginzburg model looks as in Fig. 9.

These singularities are produced as intersection of six surfaces —see Fig. 10.

Topologically the sheaf of vanishing cycles is a fibration of tori over the singular set described above. We desingularize the singular set of $\mathscr{F}$ of rational curves with three points taken out. Then $\mathscr{F}$ restricted on such a rational curve produces an S^1 local system with nontrivial monodromy. Recall (see [12]) that three-dimensional cubic is a conic bundle over $\mathbb{CP}^2$ with a degeneration curve a smooth curve of degree five. The two sheeted covering corresponds to an odd theta characteristic, and

Fig. 9 The singular set for the
LG of a 3-D cubic.

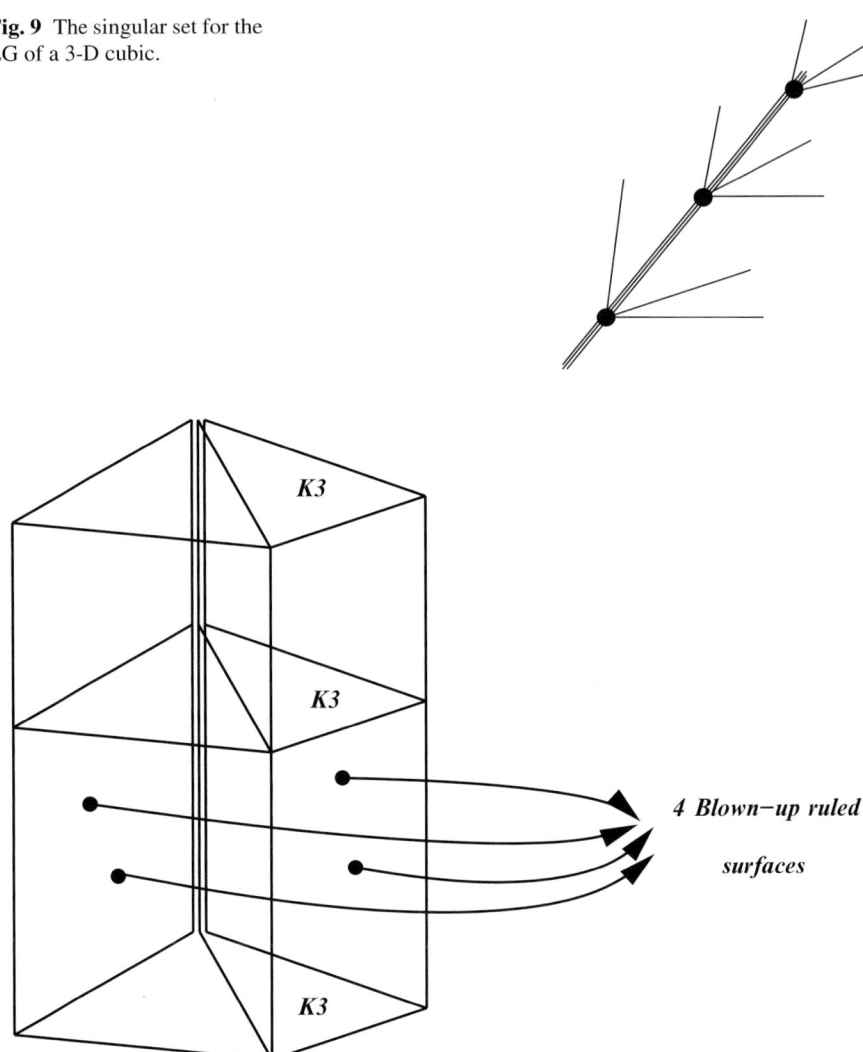

Fig. 10 Fiber over zero for the LG mirror of a 3-D cubic.

this is exactly the reason why $\mathscr{F}$ restricted on such rational curve produces a S^1 local system with nontrivial monodromy (see also [8]). We will briefly work out the case of a conic bundle over $\mathbb{CP}^2$ with a degeneration curve—a curve of degree five corresponding to an even theta characteristic.

Following the procedure from the previous sections, we get the Landau–Ginzburg mirror for such conic bundles—see Fig. 11. In this case, $\mathscr{F}$ restricted on rational curves with 3 points taken out produces a S^1 local system with trivial monodromy only. We arrive at:

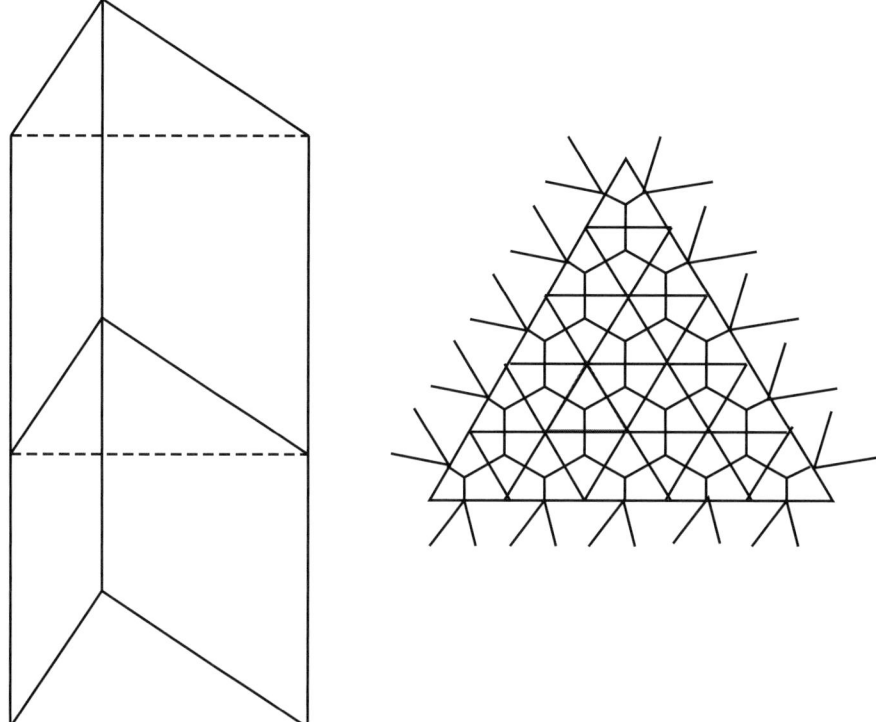

Fig. 11 The mirror of the rational conic bundle.

Conjecture 4.11. *Let* X *be a three-dimensional conic bundle and* $Y \to \mathbb{C}$ *is its Landau–Ginzburg mirror. If there exists a rational curve in the singular set such that* $\mathscr{F}$ *restricted on it produces a local system with nontrivial monodromy, then* X *is nonrational.*

4.2 Homological Mirror Symmetry and Earthquakes

In this section, we introduce a new approach to studying the mirror map—the technique of earthquakes. Our main goal is to demonstrate how the mirror image changes when we move in the moduli space.

We start by degenerating four-dimensional cubic X to three $\mathbb{CP}^4$ and then applying our standard procedure. After smoothing, we get the following Landau–Ginzburg mirror (see also [20]):

$$xyzuvw = (u+v+w)^3 \cdot t$$

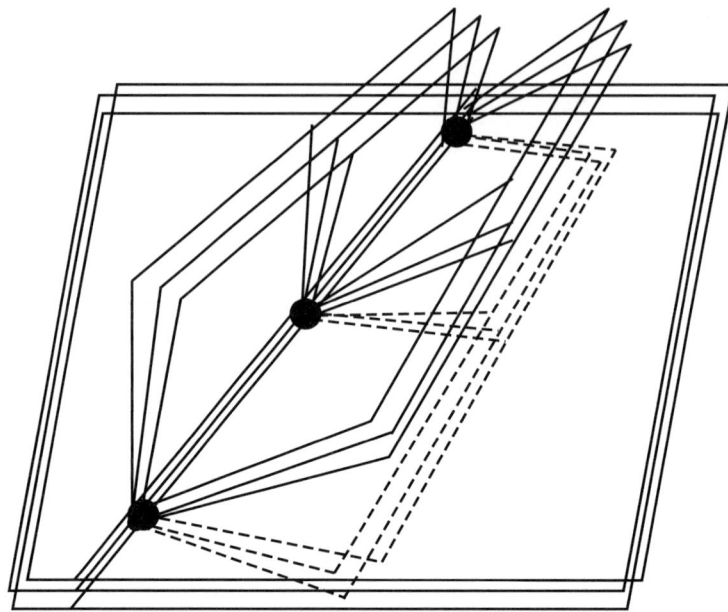

Fig. 12 Singular set for the LG model of the 4-D cubic.

with a potential $x + y + z$. Here u, v, w are in $\mathbb{CP}^2$, and x, y, z are in $\mathbb{C}^3$. The singular set W of this Landau–Ginzburg model can be seen on Fig. 12—it consists of twelve rational surfaces intersecting as shown.

Conjecture 4.12. *The generic four-dimensional cubic X is not rational.*

The idea for this conjecture is again based on the analysis of perverse sheaf of vanishing cycles $\mathscr{F}$. For generic four-dimensional cubic, there exists an open rational surface in the singular set such that $\mathscr{F}$ restricted on it produces a nontrivial local system with nontrivial monodromy. By the argument above, such singular fiber cannot correspond to a blowing up of an algebraic surface. As it follows from the work of Kulikov, there is only one fiber whose vanishing cycles contribute to the primitive cohomologies of X, and our analysis of $\mathscr{F}$ identifies this fiber. (For more details, see [2]).

Remark 4.13. The condition of triviality of the monodromy of $\mathscr{F}$ should correspond to the condition of stable degeneration of MHS associated to curves and surfaces on the B side. It will be interesting to investigate this connection further using ideas of [3].

Many examples of four-dimensional cubics have been studied by Beauville and Donagi [9], Voisin [38], Hasset [18], Iliev [23], Kuznetsov [31]. We will analyze some of these examples now.

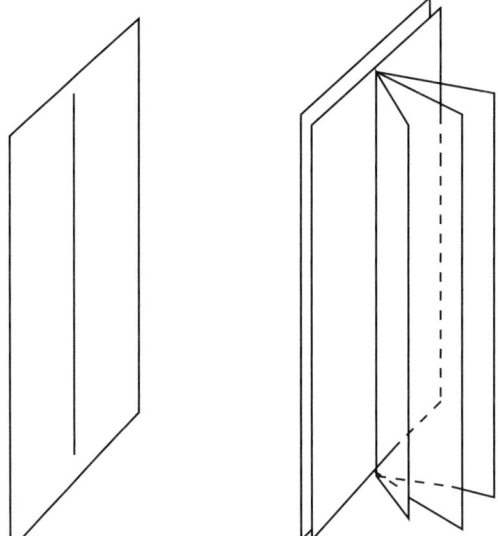

Fig. 13 Singular set for cubics containing a plane.

Example 4.14. Example of cubics containing a plane P—see Fig. 13. In this case, the singular set looks as shown. As step 5 of the construction suggests, the appearance of only one plane implies that $\mathscr{F}$ has nontrivial monodromy. Cubics with one plane contain another Del Pezzo surface of degree 4, whose mirrors can be seen in the figure. They project to a curve of degree 4 in P—compare with [18].

Example 4.15. Example of cubic containing two planes P_1 and P_2. In this case, the fiber Y_0 of the LG model Y looks as on Fig. 14. One can check that the appearance of two planes makes the monodromy of $\mathscr{F}$ trivial.

Theorem 4.16. *Let X be a four-dimensional cubic containing a plane P. Then the perverse sheaf of vanishing cycles $\mathscr{F}$ on the LG mirror of X has trivial monodromy iff the projection of X from P has a section. In this case, the singular set of Y is a degenerated K3 surface and X is rational.*

Example 4.17. Example of cubic containing a plane P_1 and being Pfaffian—Fig. 15. In this particular case, we get a section in the projection from P that is a singular Del Pezzo surface consisting of elliptic quintic curves—see the Fig. 15. Recall that Pfaffian cubics are obtained by from $\mathbb{CP}^4$ by blowing a K3 surface and blowing down a scroll over a Del Pezzo surface of degree 5—compare with [18].

Example 4.18. Hassett's examples containing a K3 surface—they are obtained [18] from a singular four-dimensional cubic by changing vanishing cycles (Fig. 16). On

Fig. 14 Singular set for cubics containing two planes.

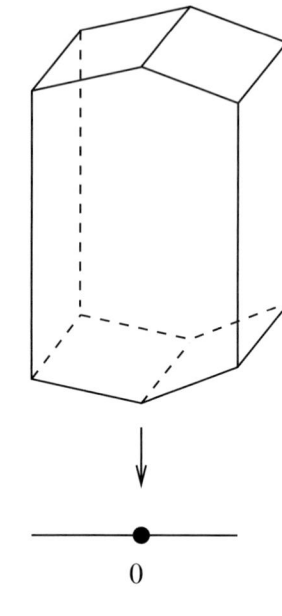

Fig. 15 Singular set for a Pfaffian cubic containing a plane.

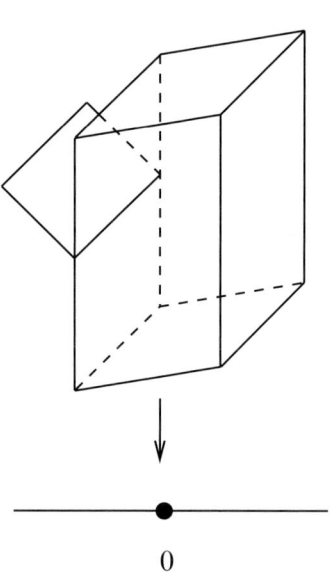

the LG mirror, this results in modifying the Lefschetz thimbles. Let, restrict ourselves to the case of C_{26}—cubics with Fano varieties of lines isomorphic to symmetric power of K3 surface of degree 26.

In this case, we start with a LG model of small resolution of a singular cubic. We modify it by creating (see the diagram below) additional thimbles with intersection form in H^4.

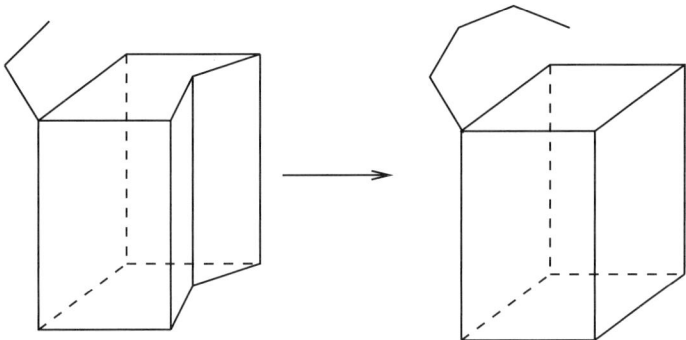

Fig. 16 Singular set for cubics from C_{26}.

	h^2	T
h^2	3	7
T	7	$19 + 2e$

We obtain a family of three-dimensional thimbles corresponding under mirror symmetry with a ruled surface T of degree 7 with e the number of double points on T. In case of C_{26}, $e = 3$. Using Torelli theorem for four-dimensional cubics, we get:

Theorem 4.19. *All cubics in Hassett's divisor C_{26} are rational.*

Example 4.20. Modifications of the singular set of the LG model allows us to construct rational cubics with Fano varieties of lines nonisomorphic to symmetric power of K3 surfaces. In a similar way as explained in a previous example, we modify the thimbles so that Fukaya–Seidel category corresponds to a noncommuative Poisson deformation of a K3 surface. Under this deformation, hypercohomology of $\mathscr{F}$ stays unchanged.

In the discussion above, we have used HMS in full —we will briefly discuss how one can try proving HMS for cubics, and we will extend Seidel's ideas from the case of the quintic. We first have:

Conjecture 4.21. *HMS holds for the total space of $\mathcal{O}_{\mathbb{CP}^4}(3)$.*

The next step is

Conjecture 4.22. *The Hochschild cohomology $HH^\bullet(D_0^b(\mathrm{tot}(\mathcal{O}_{\mathbb{CP}^4}(3)))$ is isomorphic to the space all homogeneous polynomials of degree three in five variables. Here $(D_0^b$ stands for category with support at the zero section.*

The Hochschild cohomology $HH^\bullet(D_0^b(\mathrm{tot}(\mathcal{O}_{\mathbb{CP}^4}(3))))$ produces all B side deformations of categories of the three-dimensional cubic, and we have similar phenomenon on the A side as well. For the opposite statement of HMS and more general case, see [25].

Table 2 Earthquake for Fano: $F_0 \longrightarrow F_2$.

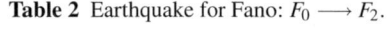

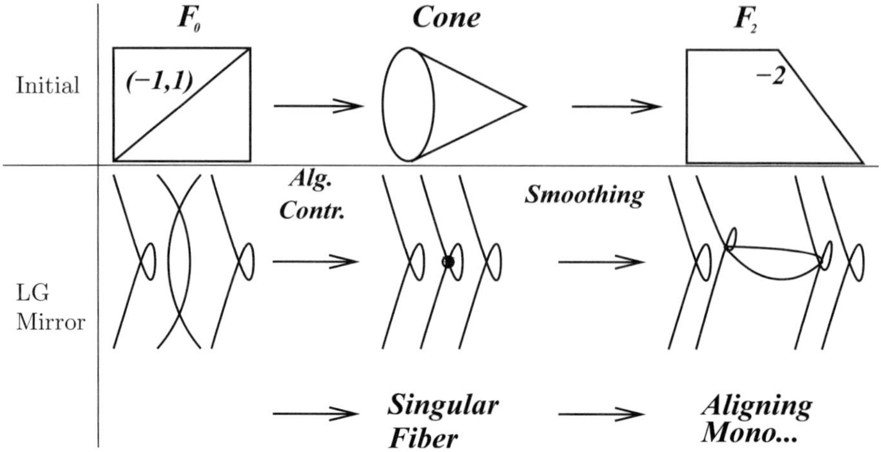

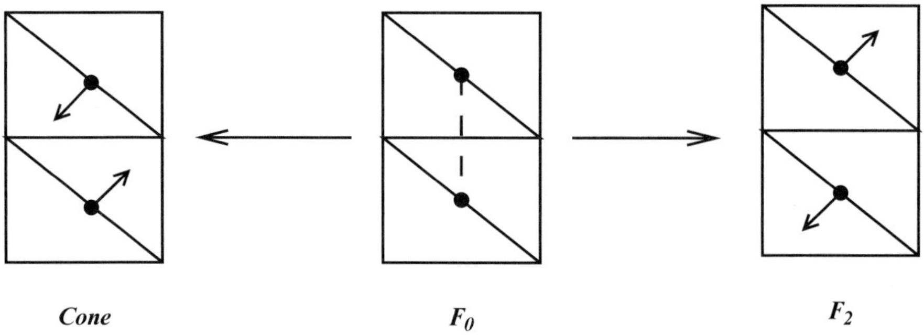

Fig. 17 Monodromy with common eigenvalues.

Remark 4.23. The above argument should work for any degree and for any dimension toric hypersurfaces.

In what follows, we put examples we have studied in a new prospective—how the mirror map changes when we move in the moduli space. We introduce a definition:

Definition 4.24. *We call the mirror image of the moving from one Noether–Lefschetz loci to another an earthquake.*

We start with the already considered example of F_0 and its deformation to F_2.

Already in this example, we see the phenomenon of creating a new Lagrangian which in LG mirror corresponds to the fact that monodromies in two singular fibers have a common eigenvalue—see [17]. In the language of affine structures, this could be explained in Table 2 and Fig. 17.

We move now to the example of the three-dimensional cubic—see Table 3.

In this case, the vanishing S^3 Lagrangian sphere as an object of Fukaya category corresponds to O_Y as an element of the derived category of singularities, Here Y is

Table 3 Earthquake for Fano 3-D cubic.

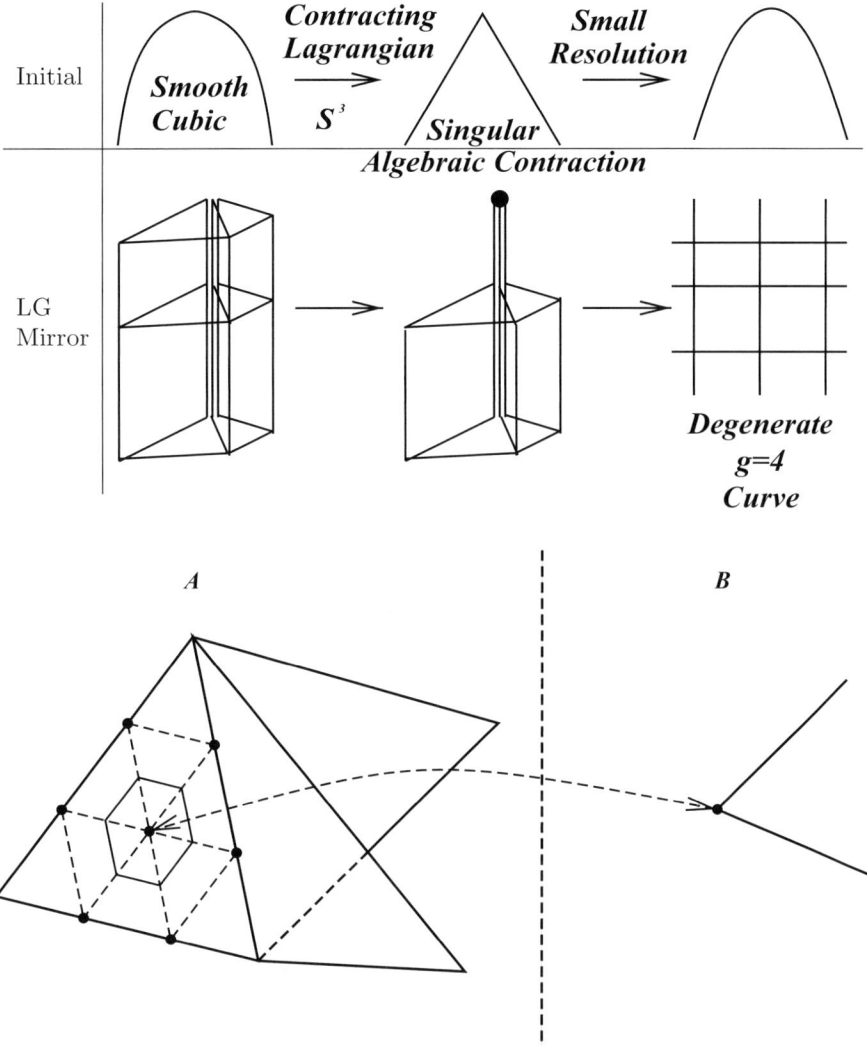

Fig. 18 S^3 –Del Pezzo correspondence in LG model of 3-D cubic.

one of the Del Pezzo surfaces of degree 6 participating in the singular fiber over 0 of the LG mirror of the three-dimensional cubic. An easy calculation shows that mod 2

$$Ext(O_Y, O_Y) = HF(S^3, S^3).$$

This correspondence is described in Fig. 18.

We also pictorially work out an example of four-dimensional cubic in Table 4.

Table 4 Earthquake for Fano's 4-D cubic.

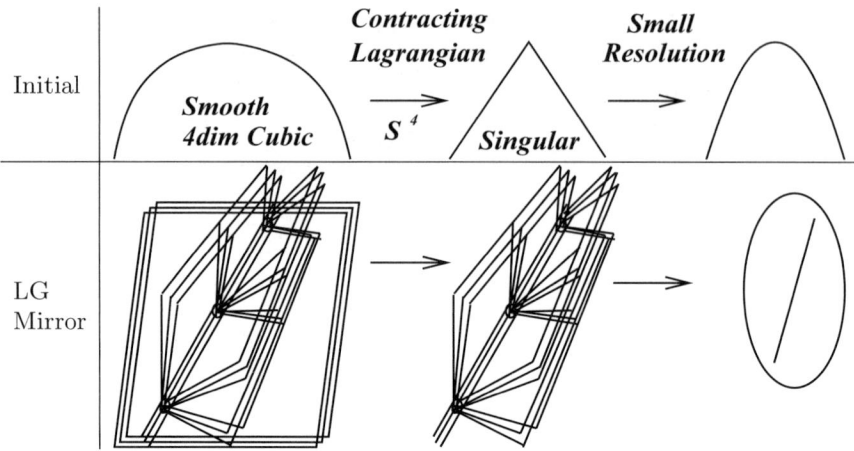

The earthquake in Table 4 explains why appearance of more algebraic cycle makes the cubic containing them rational. As in the case of F^0, we have new directions, new singular sets over which the monodromy of $\mathscr{F}$ is trivial—see [17]. Observe that the appearance of new Lagrangians makes the monodromy jump and move in a noncontinuous way. Example 4.20 is a very good illustration of this phenomenon.

We would like to outline a connection of the notion of earthquakes with local HMS and vanishing cycles. We consider the case of classical Horikawa surface, a two-sheeted cover of $\mathbb{CP}^1 \times \mathbb{CP}^1$ ramified at 6, 12 curve. We first apply Moishezon's degeneration —see Fig. 19. Then following our procedure, we can do local HMS piece by piece, then glue together and finally smooth. Under this procedure, we can follow the images of the -2 spheres and compute K^0 of the category of singularities.

Remark 4.25. It will be interesting to investigate if there is connection between the MHS of the above degeneration (compare with [8]) and K^0 of the category of singularities and therefore with K^0 of the Fukaya category.

5 Homological Mirror Symmetry and Tropical Hodge Conjecture

Algebraic cycles define objects of the category $D^b(X)$. According to homological mirror symmetry, they define objects in $FS(Y, \mathrm{w}, \omega)$.

In this section, we will try to see how a cycle being algebraic can be encoded in the Lagrangian side and then translate Hodge conjecture on the language of tropical geometry. We start with two examples:

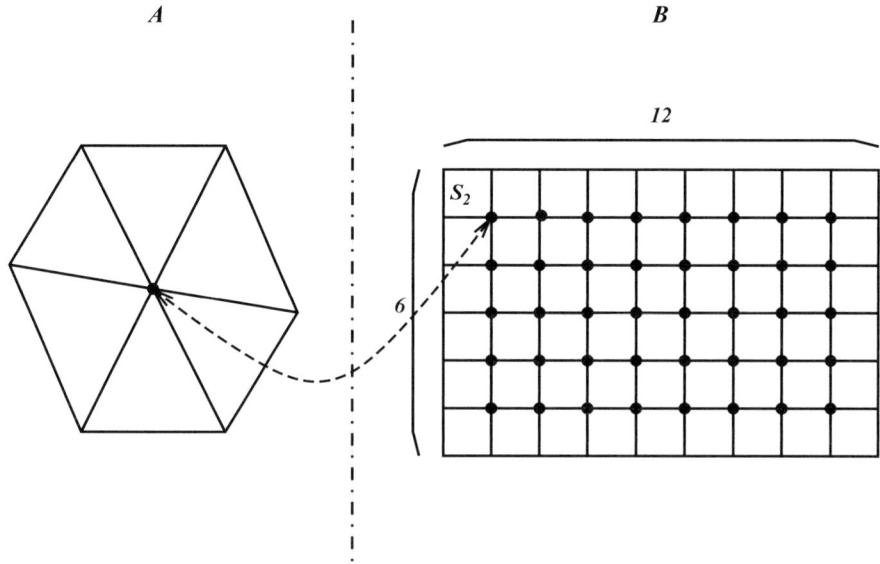

Fig. 19 Degeneration of Horikawa surface.

Example 5.1. Let us consider a smooth plane cubic C in $\mathbb{CP}^3$. We have that $dimExt^1(O_C, O_C) = 13$. Now we can associate with C a tropical curve T and let us modify it extending the horns h_1, h_2, h_3 (see Fig. 20). Using step 5 of our construction, we get a Lagrangian L and $dimH^1(L) = 14$. This leads to $m_0(L) \neq 0$, and therefore L does not represent an object in Fukaya–Seidel category.

We extend the above considerations to a counterexample to the Hodge conjecture for X—a complex two-dimensional torus. We have:

Large complex structure limit: Going to a cusp at the boundary of the moduli of complex tori leads to a Gromov–Hausdorff colapse of X to a real two-dimensional torus: $X \rightsquigarrow T := \mathbb{R}^2/\Gamma$, where Γ is rank 2 lattice.

Mirror symplectic torus (Y, ω):   Obtained as the quotient of T^*T by the fiber-wise action of the lattice $\widehat{\Gamma} \subset (\mathbb{R}^2)^*$ of all linear functionals on $\mathbb{R}^2$ that take integral values on Γ; ω is induced from the standard form on T^*T.

Note:

- T can be viewed as a tropical degeneration of X.
- Objects in $Fuk(Y, \omega)$ can be constructed as images of conormal bundles to piecewise linear submanifolds in T.

The mirror map between the period matrices for X and the complexified symplectic forms $B + \sqrt{-1}\omega$ on Y can be given explicitly:

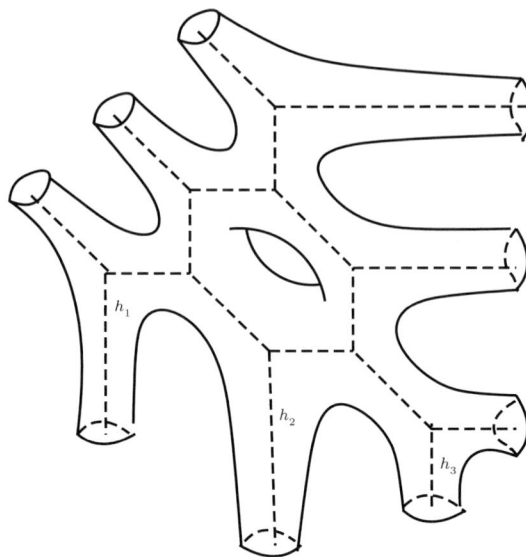

Fig. 20 Nonrealizable cubic.

$$\begin{pmatrix} a & b \\ c & d \end{pmatrix} \leftrightarrow a \cdot dx_1 \wedge dx_2 + b \cdot dx_3 \wedge dx_2 + c \cdot dx_1 \wedge dx_4 + d \cdot dx_3 \wedge dx_4$$

We can use this mirror map to run some consistency checks for HMS in this case. The main idea here is that the mirror map should transform Noether–Lefschetz loci of periods (= loci of periods with extra objects in $D^b(X)$) into loci of complexified symplectic forms with additional objects in the Fukaya category, and the move from one to another constitutes an earthquake. We will use conormal bundles to tropical subvarieties in T to construct the additional objects in $Fuk(Y, B + \sqrt{-1})$. We have the following matches:

- $\left\{ \begin{matrix} \text{period matrices} \\ \text{with } b = -c \end{matrix} \right\} \leftrightarrow \left\{ \begin{matrix} \text{images of} \\ \text{conormal} \\ \text{bundles of} \end{matrix} \quad \begin{matrix} -1 \\ -1 \end{matrix} \right\}.$

  Note that in this case, the Lagrangian in Y corresponding to the above tropical scheme has $m_0 \neq 0$. It does correspond to a Hodge but nonalgebraic cycle—well-known counterexample to the Hodge conjecture for complex nonprojective tori in dimension 2 [41].

- $\left\{ \begin{matrix} \text{period matrices} \\ \text{with } b = c \end{matrix} \right\} \leftrightarrow \left\{ \begin{matrix} \text{images of} \\ \text{conormal} \\ \text{bundles of} \end{matrix} \right\}.$

The corresponding Lagrangian in Y is the mirror image for the theta divisor of the genus two Jacobian.

- $\left\{\begin{array}{l}\text{period matrices with} \\ b = c = 0\end{array}\right\} \leftrightarrow \left\{\begin{array}{l}\text{conormal bundles of tropical} \\ \text{elliptic curves}\end{array}\right\}.$

The corresponding Lagrangians in Y are the mirrors of genus one surfaces.

- $\left\{\begin{array}{l}\text{period matrices} \\ \text{with } b = c = 0, \\ a = d = 1\end{array}\right\} \leftrightarrow \left\{\begin{array}{l}\text{image of the} \\ \text{conormal bun-} \\ \text{dle of}\end{array}\right\}$

In this case, a new A brane appears —a coisotropic brane. The last one is lost in the tropical interpretation. In general, we indeed loose some information by going to a tropical limit. As we show in [26], enough information survives. We combine the ideas of the previous two subsections in [26] to study the Hodge conjecture for some Weil type of Abelian varieties in dimension 4,6,8.

We briefly introduce our main tool—an analogue of Bloch's cycle complex in tropical geometry. Let us consider a degeneration of an Abelian variety A where half of the cycles vanish and form a lattice Γ_1. We denote by Γ_2 the remaining lattice—the dual with respect to the polarization. Consider the tropical Bloch complex characterizing all tropical algebraic cycles in A.

$$B^{\bullet, k} = \bigoplus_{A_{i_0}, \dots, A_{i_k}} \mathbb{Z}\left[A_{i_0}, \dots, A_{i_k}\right] \otimes \left(\bigwedge{}^{\bullet}\Gamma_2 \wedge \det\left(T_{A_{i_0}, \dots, A_{i_k}} \cap \Gamma_2\right)\right), \qquad (2)$$

with differential defined as follows:

- On the chains $\bigoplus_{A_{i_0}, \dots, A_{i_k}} \mathbb{Z}\left[A_{i_0}, \dots, A_{i_k}\right]$ for $k > 1$

$$\partial \left[A_{i_0}, \dots, A_{i_k}\right] = \sum_{r=0}^{k} (-1)^r \left[A_{i_0}, \dots, \widehat{A}_{i_r}, \dots, A_{i_k}\right] \qquad (3)$$

- On $\bigwedge^{\bullet}\Gamma_2$, ∂ is trivial.

Here $\left[A_{i_0}, \dots, A_{i_k}\right]$ are flags of lines with Γ_2—rational slopes associated with the vertexes of tropical varieties—see the example that follows. The Γ_1 equivariant part of H^0 in the above complex characterizes the tropical algebraic cycles.

We characterize tropical Hodge cycles:

$$\ker\left[(\Lambda^k\Gamma_1 \otimes \Lambda^k\Gamma_2) \to (\Lambda^{k+1}V \otimes \Lambda^{k-1}\Gamma_2)\right]$$

We formulate the tropical Hodge conjecture as surjectivity of the map.

$$H^0(B^{(k)}) \xrightarrow{\ s\ } \ker\left[(\Lambda^k\Gamma_1 \otimes \Lambda^k\Gamma_2) \to (\Lambda^{k+1}V \otimes \Lambda^{k-1}\Gamma_2)\right]$$

The tropical algebraic deformations of Y as a cycle in A are governed by the following complex.

$$0 \longrightarrow \bigoplus_{v \in V} N_v \longrightarrow \bigoplus_{e \in E} N_e / T_e \longrightarrow \bigoplus_{\sigma \in T_l} \Theta_l \qquad (4)$$

Here E and V are the sets of vertexes and edges of the tropical realization of the cycle Y, and Θ is the restriction of the tangent bundle to A on Y. N and T are the corresponding tropical normal and tangent bundles.

To illustrate the above definitions, we compute briefly the example of a differential applied to the tropical Hodge cycle R with $m_0 \neq 0$. R consists of two flags and the differential applied to each produces three terms recorded in the sum below. The overall summation of the differentiation applied to R results to zero.

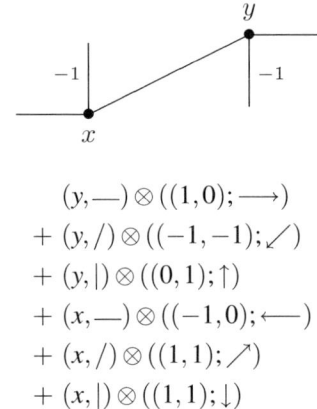

$$(y, —) \otimes ((1,0); \longrightarrow)$$
$$+ (y, /) \otimes ((-1,-1); \diagup)$$
$$+ (y, |) \otimes ((0,1); \uparrow)$$
$$+ (x, —) \otimes ((-1,0); \longleftarrow)$$
$$+ (x, /) \otimes ((1,1); \nearrow)$$
$$+ (x, |) \otimes ((1,1); \downarrow)$$

There is a certain duality between (4) and (2), which allows us to build an intersection theory functional F defined over all tropical Hodge cycles. This functional measures realizability of any Hodge tropical cycle by an algebraic variety.

Conjecture 5.2. *F is trivial on all tropical algebraic cycles.*

To make a connection with the previous section, we formulate:

Conjecture 5.3. *The zeros of F characterize all earthquakes.*

In [26], we investigate the following.

Conjecture 5.4. *There exists a six-dimensional complex Abelian variety A and tropical Hodge cycle Y in it such that F is nontrivial on Y.*

Acknowledgments This work is dedicated to Charles Boyer with admination. We are grateful to D. Auroux, M. Gross, T. Pantev, B. Hassett, A. Iliev, A. Kuznetsov, P. Seidel, D. Orlov, and M. Kontsevich for many useful conversations. Many thanks go to V. Boutchaktchiev without whom this paper would not have been written. While working on this paper, I learned about the tragic death of a close friend, Kris Galicki. I dedicate this paper to his memory as well.

This work was partially supported by NSF grants DMS0600800 and DMS0652633, FWF grant P20778, ERC grant, and by the Clay Math Institute.

References

1. M. Abouzaid, *On the Fukaya category of higher genus surfaces*, preprint arXiv:math/0606598.
2. M. Abouzaid, D. Auroux, M. Gross & L. Katzarkov, *Homological mirror symmetry and Birational Geometry*, in preparation.
3. V. Alexev, *Complete moduli in the presence of semiabelian group action*, Ann. Math. (2) 155 (2002), 3, pp. 611–708.
4. D. Auroux, *Mirror symmetry and T-duality in the complement of an anticanonical divisor*, preprint arXiv:0706.3207.
5. D. Auroux, L. Katzarkov, S. Donaldson & M. Yotov, *Fundamental groups of complements of plane curves and symplectic invariants*, Topology 43 (2004), 6, pp. 1285–1318.
6. D. Auroux, L. Katzarkov & D. Orlov, *Mirror symmetry for Del Pezzo surfaces: Vanishing cycles and coherent sheaves*, Invent. Math. (3) 166 (2006), pp. 537–582.
7. D. Auroux, L. Katzarkov, T. Pantev & D. Orlov, *Mirror symmetry for Del Pezzo surfaces II*, in preparation.
8. F. Bardelli, *Polarized mixed Hodge structures: on irrationality of threefolds via degeneration*, Ann. Mat. Pura Appl. (4) 137 (1984), pp. 287–369.
9. A. Beauville & R. Donagi, *La varité des droites d'une hypersurface cubique de dimension 4*, C. R. Acad. Sci. Paris Sér. I Math. 301 (1985), 14, pp. 703–706.
10. F. Bogomolov, *The Brauer group of quotient spaces of linear representations*, Izv. Akad. Nauk SSSR Ser. Mat. 51 (1987), 3, pp. 485–516.
11. H. Clemens, *Cohomology and obstructions II*, AG 0206219.
12. H. Clemens & P. Griffiths, *The intermediate Jacobian of the cubic threefold*, Ann. Math. (2) 95 (1972), pp. 281–356.
13. D. Cox & S. Katz, *Mirror symmetry and algebraic geometry*, Mathematical Surveys and Monographs, vol. 68, American Mathematical Society, Providence, RI, 1999.
14. T. deFernex, *Adjunction beyond thresholds and birationally rigid hypersurfaces*, AG/0604213.
15. M. Green & P. Griffiths, Algebraic Cycles I, II, preprints 2006.
16. M. Gross & L. Katzarkov, *Mirror Symmetry and vanishing cycles*, in preparation.
17. M. Gross & B. Siebert, *Mirror Symmetry via logarithmic degeneration data I*, J. Diff. Geom. (2) 72 (2006), pp. 169–338.
18. B. Hassett, *Some rational cubic fourfolds*. J. Algebraic Geom. 8 (1999), 1, pp. 103–114.
19. K. Hori, S. Katz, A. Klemm, R. Pandharipande, R. Thomas, C. Vafa, R. Vakil & E. Zaslow, *Mirror symmetry*, Clay Mathematics Monographs, vol. 1, American Mathematical Society, Providence, RI, 2003.
20. K. Hori & C. Vafa, *Mirror symmetry*, 2000, hep-th/0002222.
21. V.A. Iskovskikh, *On Rationality Criteria for Connic Bundles*, Mat. Sb. (1996), 7, pp. 75–92.
22. V.A. Iskovskikh & Y.I. Manin, *Three-dimensional quartics and counterexamples to the Lüroth problem*, Mat. Sb. (N.S.) 86 (1971), pp. 140–166.
23. A. Iliev & L. Manivel, *Cubic hypersurfaces and integrable systems*, preprint 0606211.
24. A. Kapustin, L. Katzarkov, D. Orlov & M. Yotov, *Homological mirror symmetry for manifolds of general type*, 2004, preprint.
25. L. Katzarkov, *Mirror symmetry and nonrationality*, Rationality, Tschinckel and Bogomolov Eds., Birkhauser 2008.
26. L. Katzarkov & M. Kontsevich, *Homological Mirror Symmetry and the Hodge conjecture*, in preparation.
27. L. Katzarkov, M. Kontsevich & T. Pantev, *Hodge Theoretic Aspects of Mirror Symmetry*, arXiv:0806.0107, Proc. Amer. Math. Soc. (to appear).
28. G. Kerr, *Weighted Blow-ups and HMS for toric surfaces*, preprint, to appear in Advances in Mathematics.
29. J. Kollár, *Singularities of pairs*, in Algebraic geometry, Santa Cruz 1995, pp. 221–287, Proc. Sympos. Pure Math., 62, Part 1, Am. Math. Soc., Providence, RI, 1997.
30. M. Kontsevich & Y. Soibelman, *Homological mirror symmetry and torus fibrations*, Symplectic geometry and mirror symmetry (Seoul, 2000), pp. 203–263, World Sci. Publ., River Edge, NJ, 2001.

31. A. Kuznetsov, *Derived Categories of the Cubic and V_{14} Fano*, preprint, arXiv:math/0303037
32. G. Mikhalkin, *Tropical geometry and applications*, preprint arXiv:math/0601041.
33. D. Orlov, *Derived categories of coherent sheaves and equivalences between them*, Uspekhi Mat. Nauk 58 (2003), no. 3 (351), pp. 89–172; translation in Russian Math. Surveys 58 (2003), no. 3, pp. 511–591.
34. D. Orlov, *Triangulated categories of singularities and D-branes in Landau–Ginzburg models*, Tr. Mat. Inst. Steklova, 246 (Algebr. Geom. Metody, Svyazi i Prilozh., pp. 240–262, 2004.
35. A. Pukhlikov, *Birationally rigid Fano varieties*, The Fano Conference, pp. 659–681, Univ. Torino, Turin, 2004.
36. P. Seidel, *Vanishing cycles and mutation*, European Congress of Mathematics, vol. II (Barcelona, 2000), pp. 65–85, Progr. Math., 202, Birkhäuser, Basel, 2001.
37. P. Seidel, *Fukaya categories and deformations*, Proceedings of the International Congress of Mathematicians, Vol. II (Beijing, 2002), pp. 351–360, Higher Ed. Press, Beijing, 2002.
38. C. Voisin, *Torelli theorems for cubics in $\mathbb{CP}^5$*, Inv. Math. 86, 1986, 3, pp. 577–601.
39. A. Weil, *On algebraic cycles in Abelian varieties*, Scientific works. Collected papers. vol. III, Springer-Verlag, New York and Heidelberg, 1979.
40. J. Wlodarczyk, *Birational cobordisms and factorization of birational maps*, J. Algebraic Geom. 9 (2000), 3, pp. 425–449.
41. S. Zucker, *The Hodge conjecture for four-dimensional cubic*, Comp. Math. 34 (1977), pp. 199–209.

Positive Sasakian Structures on 5-Manifolds

János Kollár

Abstract The aim of this paper is to study 5-manifolds that carry a positive Sasakian structure. Strong restrictions are derived for the integral hemology groups. In some cases, all positive sasakian structures are classified. A key step is the study of log Del Pezzo surfaces whose boundary divisor contains positive genus curves.

A quasi-regular Sasakian structure on a manifold L is equivalent to writing L as the unit circle subbundle of a holomorphic Seifert $\mathbb{C}^*$-bundle over a complex algebraic orbifold $(X, \Delta = \sum(1 - \frac{1}{m_i})D_i)$. The Sasakian structure is called positive if the orbifold first Chern class $c_1(X) - \Delta = -(K_X + \Delta)$ is positive. We are especially interested in the case when the Riemannian metric part of the Sasakian structure is Einstein. By a result of [2,6], this happens iff

1. $-(K_X + \Delta)$ is positive,
2. the first Chern class of the Seifert bundle $c_1(L/X)$ is a rational multiple of $-(K_X + \Delta)$, and
3. there is an orbifold Kähler–Einstein metric on (X, Δ).

If the first two conditions hold, we say that $f : L \to (X, \Delta)$ is a *pre-SE* (or pre–Sasakian–Einstein) Seifert bundle.

Note that if $H_2(L, \mathbb{Q}) = 0$, then $H_2(S, \mathbb{Q}) \cong \mathbb{Q}$, and so the second condition is automatic.

Whereas it is not true that all pre-SE Seifert bundles carry a Sasakian–Einstein structure, all known topological obstructions to the existence of a Sasakian–Einstein structure in dimension 5 are consequences of the pre-SE condition.

A 2-dimensional orbifold (S, Δ) such that $-(K_S + \Delta)$ is positive is also called a *log Del Pezzo surface*. For any such, S is a rational surface with quotient singularities. By [7, 2.4], Seifert $\mathbb{C}^*$-bundles over $(S, \sum(1 - \frac{1}{m_i})D_i)$ are uniquely classified by a homology class $B \in H_2(S, \mathbb{Z})$ and integers $0 < b_i < m_i$ with $(b_i, m_i) = 1$. We denote the corresponding Seifert $\mathbb{C}^*$-bundle (resp. S^1-bundle) by

J. Kollár
Princeton University, Princeton, New Jersey, USA

K. Galicki and S.R. Simanca (eds.), *Riemannian Topology and Geometric Structures on Manifolds,* Progress in Mathematics 271, DOI 10.1007/978-0-8176-4743-8,
© Springer Science+Business Media LLC 2009

$$Y\left(S,B,\textstyle\sum \frac{b_i}{m_i}D_i\right) \quad \text{resp.} \quad L\left(S,B,\textstyle\sum \frac{b_i}{m_i}D_i\right).$$

Problem 1 (Classification problems). Following [7], we want to address 3 problems.

1. Given (S,Δ) and B, determine the 5-manifold $L(S,B,\sum \frac{b_i}{m_i}D_i)$.
2. Given a 5-manifold L, decide if it can be written as $L(S,B,\sum \frac{b_i}{m_i}D_i)$ for some (S,Δ) and B.
3. Given a 5-manifold L, describe all of its representations as $L(S,B,\sum \frac{b_i}{m_i}D_i)$ for some (S,Δ) and B.

The case when L is simply connected was studied in [7]. Problem (1) was solved in [7, 5.7]. As to Problem (2), note that by a result of [14], such manifolds are determined by the second homology group $H_2(L,\mathbb{Z})$ if $w_2 = 0$. (See [1] for the $w_2 \neq 0$ case.) The torsion part of $H_2(L,\mathbb{Z})$ can be completely described as follows. (There are partial results about the rank of the free part.)

Theorem 2. [7, 1.4] *Let L be a compact 5-manifold such that $H_1(L,\mathbb{Z}) = 0$. Assume that L has a positive Sasakian structure $f : L \to (S,\Delta)$. Then the torsion subgroup of the second homology, $\operatorname{tors} H_2(L,\mathbb{Z})$, is determined by (S,Δ), and it is one of the following.*

1. *$(\mathbb{Z}/m)^2$ for some $m \geq 1$,*
2. *$(\mathbb{Z}/5)^4$ or $(\mathbb{Z}/4)^4$,*
3. *$(\mathbb{Z}/3)^4, (\mathbb{Z}/3)^6$ or $(\mathbb{Z}/3)^8$,*
4. *$(\mathbb{Z}/2)^{2n}$ for some $n \geq 1$.*

Note that the theorem gives restrictions on L but does not say anything about (S,Δ). Whereas it seems hopelessly complicated to describe all Seifert bundle structures on S^5, there are few positive Sasakian structures on complicated 5-manifolds. In particular, Problem (3) was settled for $\operatorname{tors} H_2(L,\mathbb{Z}) \cong (\mathbb{Z}/5)^4$ [7, 1.8.2].

The aim of this note is fourfold. First, we extend several of the results to 5-manifolds that are not simply connected. Second, I would like to correct the mistake in the classification of the case (2.1), which was brought to my attention by K. Galicki. Third, Problem (3) is solved in the $(\mathbb{Z}/4)^4$ and $(\mathbb{Z}/3)^8$ cases. Finally, we show that in many cases, all positive Sasakian structures are given by hypersurfaces in $\mathbb{C}^4$ and also give explicit equations for these.

Toward the first goal, we show that the groups listed in (2.1–4) give large subgroups of all possible $\operatorname{tors} H_2(L,\mathbb{Z})$.

Theorem 3. *Let L be a compact 5-manifold with a positive Sasakian structure. Then $\operatorname{tors} H_2(L,\mathbb{Z})$ has a subgroup G as in (2.1–4) such that the quotient group $\operatorname{tors} H_2(L,\mathbb{Z})/G$ is $\mathbb{Z}/r$, $\mathbb{Z}/r + \mathbb{Z}/2$ or $\mathbb{Z}/r + \mathbb{Z}/3$ for some r.*

A recent result of [16, Cor. 1.4] implies that the list of possible quotients is even smaller: $\mathbb{Z}/r$, $\mathbb{Z}/4 + \mathbb{Z}/2$, $\mathbb{Z}/(4r+2) + \mathbb{Z}/2$, $(\mathbb{Z}/3)^2$ or $\mathbb{Z}/6 + \mathbb{Z}/3$.

Our next aim is to prove that in most cases, the existence of the subgroup G controls L and the Sasakian structure very tightly.

Theorem 4. *Let* $f : L \to (S,\Delta)$ *be a compact 5-manifold with a positive Sasakian structure. Assume that* $H_2(L,\mathbb{Z}) \supset (\mathbb{Z}/m)^2$ *where* $m \geq 28$ *or* $m \geq 12$ *and* $(m,6) = 1$. *Then*

1. *S is a Del Pezzo surface with cyclic Du Val singularities only and* $\Delta = (1 - \frac{1}{m'})C$ *where* $C \in |-K_S|$ *is a smooth elliptic curve and* m' *is a multiple of m.*
2. *There are 132 families of such* (S,C) *up to deformations that preserve the singularities,*
3. *For every such S,* $\pi_1(S^0)$ *is Abelian of order* ≤ 9, *where* $S^0 := S \setminus \mathrm{Sing}\, S$ *denotes the set of smooth points.* $\pi_1^{orb}(S, (1 - \frac{1}{m'})C)$ *is also Abelian of order* ≤ 9, *save 3 cases that are non-Abelian of order* $2^3, 2^4$, *or* 3^3. *(The precise values are given later in Table 2.)*
4. *The fundamental group of L is given by a central extension*

$$0 \to \mathbb{Z}/d \to \pi_1(L) \to \pi_1^{orb}(S, (1 - \frac{1}{m'})C) \to 1,$$

   for some d.
5. *The torsion of the second homology group of L sits in an exact sequence*

$$0 \to (\mathbb{Z}/m')^2 \to \mathrm{tors}\, H_2(L,\mathbb{Z}) \to \pi_1(S^0).$$

It is quite likely that the above theorem holds for all $m \geq 12$. However, for $m < 12$, there are many counter examples, see [7, 6.6–7].

Remark 5 (Sasakian–Einstein structures). In each of the 132 cases, the existence of positive Sasakian and of pre–Sasakian–Einstein structures over (S,Δ) is effectively decidable, but some of the computations may be lengthy.

There are 93 cases of (S,C) when $\pi_1(S^0) = 1$. In [7, 1.8.1] I claimed, incorrectly, that they all admit a Seifert bundle with Sasakian–Einstein structure. We see in (20) that these all admit Seifert bundles with a positive Sasakian structure, but only 19 of them admit a Seifert bundle with pre-SE structure.

For the remaining 39 cases with $\pi_1(S^0) \neq 1$, my computations are not yet complete. There are some cases that do not have any smooth Seifert bundles over them (26), and in many other cases Sasakian–Einstein structures exist.

Example 6. Concrete examples of Sasakian structures can be obtained using $\mathbb{C}^*$-actions on algebraic hypersurfaces. A linear $\mathbb{C}^*$-action on $\mathbb{C}^n$ can be diagonalized and so given by weights $(w_1, \ldots, w_n)$. We consider only actions that are effective (thus the weights are relatively prime) and the origin is an attracting fixed point (thus $w_i > 0$ for every i). Let $Y \subset \mathbb{C}^n$ be an algebraic variety, smooth away from the origin, which is invariant under the $\mathbb{C}^*$-action. Then its link $L := Y \cap S^{2n-1}(1)$ inherits a natural quasi-regular Sasakian structure, and every quasi-regular Sasakian structure arises this way.

This description is especially simple when Y is a hypersurface. In this case, everything is described by the weights $(w_1, \ldots, w_n)$ and a weighted homogeneous polynomial $f(x_1, \ldots, x_n)$ that defines Y.

Thus, up to deformations, we need to specify the weights $(w_1, \ldots, w_n)$ and the weighted degree d of f. I write $L^*(w_1, \ldots, w_n; d)$ for any such quasi-regular Sasakian manifold. (The $*$ is to remind one that using weights is in some sense dual to the notation in [3].)

The simplest examples where $\pi_1(L) \neq 1$ are created by taking a quotient by a subgroup of $\mathbb{C}^*$. These are all cyclic. Thus the symbol

$$L^*(w_1, \ldots, w_n; d)/(\mathbb{Z}/m) \quad \text{or} \quad L^*(w_1, \ldots, w_n; d)/\tfrac{1}{m}(w_1, \ldots, w_n)$$

stands for any Sasakian manifold obtained as $L/(\mathbb{Z}/m) \subset Y/(\mathbb{Z}/m)$ where $Y := (f = 0)$ is the zero set of a degree d weighted homogeneous polynomial such that Y is smooth outside the origin, and we take the quotient by the $\mathbb{Z}/m$-action generated by

$$(x_1, \ldots, x_n) \mapsto (\varepsilon^{w_1} x_1, \ldots, \varepsilon^{w_n} x_n) \quad \text{where} \quad \varepsilon = e^{2\pi i/m}.$$

We call these the *obvious quotients*.

One can also take quotients by groups that are not contained in $\mathbb{C}^*$. For instance, one can take quotients

$$L^*(w_1, \ldots, w_n; d)/\left(\tfrac{1}{r}(a_1, \ldots, a_n) \times \tfrac{1}{m}(w_1, \ldots, w_n)\right)$$

where the first factor of $\mathbb{Z}/r \times \mathbb{Z}/m$ acts via

$$(x_1, \ldots, x_n) \mapsto (\eta^{a_1} x_1, \ldots, \eta^{a_n} x_n) \quad \text{where} \quad \eta = e^{2\pi i/r},$$

and the second factor acts as above. One should keep in mind that not all hypersurfaces $L^*(w_1, \ldots, w_n; d)$ admit such a $\mathbb{Z}/r$-action. In using this notation, we assume that we consider a case when the action exists.

We are now ready to state that classification theorem for those cases when $H_2(L, \mathbb{Z})$ contains a large torsion subgroup.

Theorem 7. *Let L be a compact 5-manifold with a positive Sasakian structure.*

1. *If $H_2(L, \mathbb{Z}) \supset (\mathbb{Z}/5)^4$, then $H_2(L, \mathbb{Z}) = (\mathbb{Z}/5)^4$ and $L = L^*(5, 5, 15, 6; 30)$ or an obvious quotient by $\mathbb{Z}/m$ for $(m, 15) = 1$. The simplest equation is*

$$(x^6 + y^6 + z^2 + t^5 = 0)/\tfrac{1}{m}(5, 5, 15, 6).$$

The moduli space of Sasakian–Einstein structures is naturally parameterized by the moduli space of genus 2 curves.

2. *If $H_2(L, \mathbb{Z}) \supset (\mathbb{Z}/3)^8$ then $H_2(L, \mathbb{Z}) = (\mathbb{Z}/3)^8$ and $L = L^*(3, 3, 15, 10; 30)$ or an obvious quotient by $\mathbb{Z}/m$ for $(m, 15) = 1$. The simplest equation is*

$$(x^{10} + y^{10} + z^2 + t^3 = 0)/\tfrac{1}{m}(3, 3, 15, 10).$$

The moduli space of Sasakian–Einstein structures is naturally parameterized by the moduli space of hyperelliptic curves of genus 4.

3. *If $H_2(L,\mathbb{Z}) \supset (\mathbb{Z}/4)^4$ then either $H_2(L,\mathbb{Z}) = (\mathbb{Z}/4)^4$ or $H_2(L,\mathbb{Z}) = \mathbb{Z} + (\mathbb{Z}/4)^4$. In the first case, there are 5 families:*

a. *$L = L^*(4,4,8,5;20)$ or an obvious quotient by $\mathbb{Z}/m$ for $(m,2) = 1$, with sample equation*

$$(x^5 + y^5 + yz^2 + t^4 = 0)/\tfrac{1}{m}(4,4,8,5),$$

b. *$Y = L^*(2,2,6,3;12)/\tfrac{1}{2}(1,1,1,e)$ where $e \in \{0,1\}$ or an obvious quotient by $\mathbb{Z}/m$ for $(m,6) = 1$. The simplest equation is*

$$(x^6 + y^6 + z^2 + t^4 = 0)/\left(\tfrac{1}{2}(1,1,1,e) \times \tfrac{1}{m}(2,2,6,3)\right).$$

c. *$L = L^*(4,8,20,9;40)$ or an obvious quotient by $\mathbb{Z}/m$ for $(m,6) = 1$. The simplest equation is*

$$(x^{10} + y^5 + z^2 + xt^4 = 0)/\tfrac{1}{m}(4,8,20,9).$$

d. *$L = L^*(2,4,10,5;20)/\tfrac{1}{2}(1,0,1,1)$ or an obvious quotient by $\mathbb{Z}/m$ for $(m,10) = 1$. The simplest equation is*

$$(x^{10} + y^5 + z^2 + t^4 = 0)/\left(\tfrac{1}{2}(1,0,1,1) \times \tfrac{1}{m}(2,4,10,5)\right).$$

In the second case, there are infinitely many positive Sasakian structures with the same π_1 but only one that is pre-SE. These are

e. *$L = L^*(4,4,12,5;24)$ or an obvious quotient by $\mathbb{Z}/m$ for $(m,10) = 1$. The simplest equation is*

$$(x^6 + y^6 + z^2 + xt^4 = 0)/\tfrac{1}{m}(4,4,12,5).$$

The moduli space of Sasakian–Einstein structures is naturally parameterized by the moduli space of pairs (C,p) where $\pi : C \to \mathbb{P}^1$ is a genus 2 curve and $p \in C$ is not a branch point in cases (a),(b),(e) and $p \in C$ is a branch point in cases (c),(d).

The above parameterizations of the moduli spaces of Sasakian–Einstein structures are one-to-one. However, if we consider the moduli spaces of the resulting Einstein metrics, we get a two-to-one parameterization since a curve and its complex conjugate give the same Einstein metric, see [3, 18–21].

I see no a priori reason why all these cases should be realizable as hypersurface quotients. K. Galicki told me that he found the example $(x^6 + y^6 + z^2 + t^5 = 0)$ with $H_2(L,\mathbb{Z}) = (\mathbb{Z}/5)^4$. This led me to realize that in fact all examples with $H_2(L,\mathbb{Z}) = (\mathbb{Z}/5)^4$ are hypersurfaces. The equations are also easy to see in the $(\mathbb{Z}/3)^8$ case but are rather mysterious from the point of view of my proof for several of the $(\mathbb{Z}/4)^4$ cases.

1 Reduction to Algebraic Geometry

Remark 8. As in [3, 7], to $L = Y \cap S^{2n-1}(1)$ we associate a projective algebraic variety $X = Y \setminus \{0\}/\mathbb{C}^*$ with cyclic quotient singularities. There is also a natural $\mathbb{Q}$-divisor $\Delta = \sum (1 - \frac{1}{m_i})D_i$ on X where $D_i \subset X$ is a divisor such that the stabilizer of the $\mathbb{C}^*$-action has order m_i over the points of D_i. The positivity of the Sasakian structure is equivalent to the log Fano condition:

$$-\left(K_X + \sum (1 - \tfrac{1}{m_i})D_i\right) \quad \text{is ample.}$$

Thus if $\dim L = 5$, we are looking for pairs (S, Δ) where

1. S is a projective surface with cyclic quotient singularities,
2. $\Delta = \sum (1 - \frac{1}{m_i})D_i$ is a $\mathbb{Q}$-divisor, and
3. $-(K_S + \Delta)$ is ample.

There are a few more conditions, which we do not need for now.

A rather easy result [7, 6.3] on log Del Pezzo surfaces gives the following restriction on the D_i:

4. there is at most one index i such that $g(D_i) > 0$.

Because of the special role of D_i, we frequently set $D := D_i$, $r := m_i$ and use the notation $(S, (1 - \frac{1}{r})D + \Delta)$.

It turns out that the ampleness of $-(K_S + (1 - \frac{1}{r})D + \Delta)$ forces $g(D)$ or r to be quite small, and this accounts for the restrictions in Theorem 2.

This is relevant to our purposes since [7, 5.7] relates the torsion of $H_2(L, \mathbb{Z})$ to the divisor $\sum (1 - \frac{1}{m_i})D_i$:

5. If $H_1(L, \mathbb{Z}) = 0$ then $\operatorname{tors} H_2(L, \mathbb{Z}) = \sum (\mathbb{Z}/m_i)^{2g(D_i)}$. Thus, in the log Del Pezzo case, $\operatorname{tors} H_2(L, \mathbb{Z}) \cong (\mathbb{Z}/r)^{2g(D)}$.

If $H_1(L, \mathbb{Z}) \neq 0$, the spectral sequence in [7, 5.10] becomes quite messy, but in the log Del Pezzo case, only one nonzero map involves $H^3(L, \mathbb{Z})$. The same proof gives the following weaker result when the first integral homology of the base is trivial.

Proposition 9. Let $f : L^5 \to (S, \Delta = \sum (1 - \frac{1}{m_i})D_i)$ be a smooth Seifert bundle over a projective surface with quotient singularities. Assume that $H_1(S, \mathbb{Z}) = 0$, and let $s = \operatorname{rank} H^2(S, \mathbb{Q})$. Then there is an exact sequence

$$H_1(S^0, \mathbb{Z}) \cong H^3(S, \mathbb{Z}) \to H^3(L, \mathbb{Z}) \to \mathbb{Z}^{s-1} + \sum (\mathbb{Z}/m_i)^{2g(D_i)} \to 0. \qquad (9.1)$$

Proof. The argument very closely follows [7, 5.9–10]. As there, we have exact sequences

$$0 \to R^1 f_* \mathbb{Z}_L \xrightarrow{\tau} \mathbb{Z}_S \to Q \to 0 \qquad (9.2)$$

and

$$0 \to \sum_i \mathbb{Z}_{P_j}/n_j \to Q \to \sum_i \mathbb{Z}_{D_i}/m_i \to 0, \qquad (9.3)$$

where $P_j \in S$ are the singular points. This implies that $H^i(S,Q) = \sum_i H^i(D_i, \mathbb{Z}/m_i)$ for $i \geq 1$. The key piece of the long cohomology sequence of (9.2) is

$$H^1(S,\mathbb{Z}) \to \sum H^1(D_i, \mathbb{Z}/m_i) \to H^2(S, R^1 f_* \mathbb{Z}_L) \to H^2(S,\mathbb{Z}) \to \sum H^2(D_i, \mathbb{Z}/m_i)$$

Here $H^1(S,\mathbb{Z}) = 0$ and $H^2(S,\mathbb{Z}) \cong \mathbb{Z}^s$ by assumption. The right-hand group is torsion, hence there is a noncanonical isomorphism

$$H^2(S, R^1 f_* \mathbb{Z}_L) \cong \mathbb{Z}^s + \sum_i H^1(D_i, \mathbb{Z}/m_i).$$

Therefore, in the Leray spectral sequence $H^i(S, R^j f_* \mathbb{Z}_L) \Rightarrow H^{i+j}(L, \mathbb{Z})$, the E_2 term is

$$\begin{array}{ccccc}
\mathbb{Z} & \text{(torsion)} & \mathbb{Z}^s + \sum_i(\mathbb{Z}/m_i)^{2g(D_i)} & \text{(torsion)} & \mathbb{Z} \\
\mathbb{Z} & 0 & \mathbb{Z}^s & H^3(S,\mathbb{Z}) & \mathbb{Z}.
\end{array} \qquad (9.4)$$

Since $H^3(S,\mathbb{Z}) \cong H_1(S^0, \mathbb{Z})$ by [7, 4.2], we get the exact sequence

$$H_1(S^0, \mathbb{Z}) \to \text{tors} H^3(L, \mathbb{Z}) \to \sum_i H^1(D_i, \mathbb{Z}/m_i) \to 0.$$

$\square$

Corollary 10. *Notation and assumptions as in (9). Let $p_j \in S$ be the singular points and n_j the order of the local fundamental group.*

1. *If any two of the numbers m_i, n_j are relatively prime, then the sequence (9.1) is left exact.*
2. *If, in addition $|H_1(S^0, \mathbb{Z})|$ is relatively prime to $\prod m_i$, then the sequence (9.1) splits.*

Proof. If any two of the numbers m_i, n_j are relatively prime, the sequence in (9.2) splits and $H^0(S,Q) \cong \mathbb{Z}/(\prod n_j \cdot \prod m_i)$. This implies that $H^1(S, R^1 f_* \mathbb{Z}_L) = 0$. The E_2 term of the Leray spectral sequence (9.4) is

$$\begin{array}{ccccc}
\mathbb{Z} & 0 & \mathbb{Z}^s + \sum_i(\mathbb{Z}/m_i)^{2g(D_i)} & \text{(torsion)} & \mathbb{Z} \\
\mathbb{Z} & 0 & \mathbb{Z}^s & H^3(S,\mathbb{Z}) & \mathbb{Z}.
\end{array}$$

Thus $H^3(L,\mathbb{Z})$ sits in an exact sequence

$$0 \to H^3(S,\mathbb{Z}) \to H^3(L,\mathbb{Z}) \to \mathbb{Z}^s + \sum_i(\mathbb{Z}/m_i)^{2g(D_i)} \to \mathbb{Z}$$

As before, $H^3(S,\mathbb{Z}) \cong H_1(S^0, \mathbb{Z})$, and the extension of the torsion part splits if $|H_1(S^0, \mathbb{Z})|$ is relatively prime to $\prod m_i$. $\square$

This result turns out to be quite useful, since the groups $H_1(S^0, \mathbb{Z})$ are rather special when (S, Δ) is a log Del Pezzo surface. Thus S is a rational surface and so $H_1(S, \mathbb{Z}) = 0$. The recent preprint [16] gives an almost complete list of the possible fundamental groups $\pi_1(S^0)$ where S is a rational surface with quotient singularities. We need only an easier result on $H_1(S^0, \mathbb{Z})$.

The proof of (3) is obtained by combining (9) and (11).

Lemma 11. *Let S be a log Del Pezzo surface with quotient singularities. Then $H_1(S^0, \mathbb{Z})$ is either $\mathbb{Z}/m$, $\mathbb{Z}/m + \mathbb{Z}/2$ or $\mathbb{Z}/m + \mathbb{Z}/3$ for some $m \geq 1$.*

Proof. Let $S' \to S$ be the corresponding Galois cover with Galois group G. The stabilizers of points are subgroups of the local fundamental groups at the singularities, and hence cyclic. Moreover, no $1 \neq g \in G$ fixes a curve pointwise. Thus it is enough to prove the following.

Lemma 12. *Let G be an Abelian group acting on a rational surface with cyclic stabilizers. Assume also that no element fixes a curve of genus ≥ 1 pointwise. Then G is either $\mathbb{Z}/m$, $\mathbb{Z}/m + \mathbb{Z}/2$ or $\mathbb{Z}/m + \mathbb{Z}/3$.*

Proof. We can take a G-equivariant resolution and then pass to the G-minimal model T. (Note that if a smooth point $p \in T$ is fixed by an Abelian group H, then H also has a fixed point on the blow-up $B_p T$, so a fixed point on one model gives a fixed point on any other model.)

Our aim is to find a G-invariant subset $Z \subset T$ such that either $Z \cong \mathbb{P}^1$ or Z is at most 3 points. In the first case, G acts on $\mathbb{P}^1$, hence it either has a fixed point or it acts through $(\mathbb{Z}/2)^2$ and an index 2 subgroup has a fixed point. In the second case, a subgroup of index ≤ 3 has a fixed point. Such a subgroup is cyclic and so G is $\mathbb{Z}/m$, $\mathbb{Z}/m + \mathbb{Z}/2$ or $\mathbb{Z}/m + \mathbb{Z}/3$ as required.

Consider first the case when there is a G-equivariant ruling $f : T \to \mathbb{P}^1$. We are done if $f_* : G \to \mathrm{Aut}(\mathbb{P}^1)$ is injective. Otherwise, any $g \in \ker f_*$ fixes 2 points in each fiber of f, thus g fixes either 1 or 2 irreducible curves pointwise and their union is G-invariant. We are done if there is a G-invariant curve.

Otherwise, every $g \in \ker f_*$ fixes 2 disjoint curves E_1, E_2 pointwise and G permutes these curves. By the Hodge index theorem $E_i^2 \leq 0$, thus they generate the "other" extremal ray of the cone of curves. In particular, either $T = \mathbb{P}^1 \times \mathbb{P}^1$ or these curves are unique and every element of $\ker f_*$ fixes the same curves. If G has a fixed point $p \in \mathbb{P}^1$, then $Z := f^{-1}(p) \cap (E_1 \cup E_2)$ is a 2-element set fixed by G. If there is no such p, then G acts on $\mathbb{P}^1$ and also on $E_1 \cup E_2$ through $(\mathbb{Z}/2)^2$. Furthermore, there is an index 2 subgroup $H \subset G$ that fixes each E_i and has a fixed point on each E_i. Thus H is cyclic and so $G \cong \mathbb{Z}/m$ or $G \cong \mathbb{Z}/m + \mathbb{Z}/2$.

The $\mathbb{P}^1 \times \mathbb{P}^1$ case is left to the end.

Otherwise T is a Del Pezzo surface of G-Picard number 1.

Assume that some $g \in G$ pointwise fixes some curves C_i. The C_i are smooth, disjoint, rational curves. By the adjunction formula, $C_i^2 \in \{0, -1\}$ or $T \cong \mathbb{P}^2$. Their sum $\sum C_i$ is G-invariant and not ample, but this contradicts G-minimality. Thus either $T \cong \mathbb{P}^2$ or every element of G acts with isolated fixed points.

Assume that there is $H < G$ with $H \cong (\mathbb{Z}/2)^2$. Then T/H is a Del Pezzo surface with A_1-singularities only. From Table 2 in (24), we see that $\deg T/H = 2$ and $T \cong \mathbb{P}^1 \times \mathbb{P}^1$. Thus, aside from this case, the 2-part of G is cyclic.

If $T = \mathbb{P}^2$, take any $g \in G$. It has either 3 fixed points or a fixed point and a fixed line (and the second case must happen if $g^2 = 1$). In the second case, the fixed point is G-fixed (and so G is cyclic), and in the former case either all 3 points are G-fixed or G permutes them cyclically.

If $\deg T \in \{1, 2, 3, 4, 5\}$ then $\mathrm{Aut}(T)$ acts faithfully on $H^2(T, \mathbb{Z})$. Indeed, if g acts trivially on $H^2(T, \mathbb{Z})$, then it descends to an automorphism of $\mathbb{P}^2$ with ≥ 4 fixed points in general position, hence $g = 1$. Thus, G is a subgroup of the Weyl group of E_8, E_7, E_6, D_5, A_4 (cf. [11, Sec. 25]).

If $\deg T = 1$, then $|K_T|$ has a unique base point that is fixed by $\mathrm{Aut}(T)$, so G is cyclic.

If $\deg T = 2$, then there is a unique degree two morphism $T \to \mathbb{P}^2$. If $|G|$ is odd and $H < G$, then any H-fixed point on $\mathbb{P}^2$ is dominated by (one or two) H-fixed point(s) on T. If G is even, then a G-fixed point on $\mathbb{P}^2$ is dominated by two points, and each is fixed by an index 2 subgroup of G.

If $\deg T \in \{3, 4, 5\}$, then by looking at the order of the Weyl groups, we see that the odd part $G_{odd} \subset \mathbb{Z}/15$, except possibly when $\deg T = 3$ where the 3-part could be bigger. Any action of a group on a cubic surface T induces an action on $\mathbb{P}^3 \cong |K_T|$. For odd order groups, this action lifts to $\mathbb{C}^4$ since the kernel of $SL_4 \to PSL_4$ is $\mathbb{Z}/4$. Thus we get an eigenvector on $\mathbb{C}^4$ and a fixed point on the cubic T.

If $\deg T = 6$, then only $\mathbb{Z}/3$ can act on the (-1)-curves nontrivially, and even for the $\mathbb{Z}/3$-action there is a $\mathbb{Z}/3$-invariant set of 3 disjoint (-1)-curves. Thus the action descends to $\mathbb{P}^2$.

There are no G-minimal Del Pezzo surfaces of degree 7, and in degree 8 we get $\mathbb{P}^1 \times \mathbb{P}^1$.

Thus we are left with G acting on $\mathbb{P}^1 \times \mathbb{P}^1$. If any $g \in G$ interchanges the two factors, then g has 2 fixed points or a fixed rational curve from the Lefschetz fixed point formula. In both cases, an index 2 subgroup of G has a fixed point.

The last case is when G preserves the coordinate projections. The image of G in each $\mathrm{Aut}(\mathbb{P}^1)$ is either cyclic or $(\mathbb{Z}/2)^2$. If the first case happens at least once, then an index 2 subgroup has a fixed point. Finally we have to deal with subgroups of $G \subset (\mathbb{Z}/2)^4$. If the order is 8 or 16, then there are $g_1, g_2 \in G$ that act trivially on the first (resp. second) factor. Thus $\langle g_1, g_2 \rangle$ is a noncyclic subgroup with a fixed point, a contradiction. $\square$

Putting these together, we obtain the following. Note that the torsion in H_2 is dual to the torsion in H^3, thus a quotient of H^3 becomes a subgroup of H_2.

Corollary 13. *Let L be a compact 5-manifold with a positive Sasakian structure. Then the torsion subgroup of the second homology, $\mathrm{tors}\, H_2(L, \mathbb{Z})$ has a subgroup G such that*

1. G is $(\mathbb{Z}/m)^2$, $(\mathbb{Z}/5)^4$, $(\mathbb{Z}/4)^4$, $(\mathbb{Z}/3)^{2n}$ for $n \in \{2, 3, 4\}$ or $(\mathbb{Z}/2)^{2n}$ and
2. $\mathrm{tors}\, H_2(L, \mathbb{Z})/G$ is $\mathbb{Z}/r$, $\mathbb{Z}/r + \mathbb{Z}/2$ or $\mathbb{Z}/r + \mathbb{Z}/3$ for some r.

Remark 14. Most Abelian groups cannot be written in the above form. If G exists, then it is almost always unique up to isomorphism. The only ambiguity is with the 6-torsion part.

To see this, we can consider the p-parts separately. The main cases are

1. $A_p = (\mathbb{Z}/p^a)^2 + \mathbb{Z}/p^b$ and $A_p/G \cong \mathbb{Z}/p^b$,
2. $A_p = \mathbb{Z}/p^a + \mathbb{Z}/p^b$ and $A_p/G \cong \mathbb{Z}/p^{b-a}$.

These are the only possibilities for $p \geq 7$, with a few more cases for $p = 2,3,5$.

Plan 15 (The proofs of (4) and (7)). We follow the approach in [7, 6.8]. Start with $(S, (1 - \frac{1}{r})D + \Delta)$ satisfying the conditions (8.1–3). Let $g : S' \to S$ be the minimal resolution of S and $h : S' \to S^m$ a minimal model of S'. In the sequence of blow-ups leading from S to S' and the subsequent blow-downs leading from S' to S^m, every intermediate surface T satisfies the following condition:

(∗) We can write $-K_T \equiv (1 - \frac{1}{r})D_T + \Delta_T + H_T$ where D_T denotes the birational transform of D on T, Δ_T is an effective linear combination of rational curves (coming from Δ and the exceptional curves of g), and H_T is nef and big (this is a general divisor numerically equivalent to the pull-back/push-forward of $-(K_S + (1 - \frac{1}{r})D + \Delta)$).

This turns out to be very restrictive in many cases.

It is easy to see that S is a rational surface, so S^m is either $\mathbb{P}^2, \mathbb{P}^1 \times \mathbb{P}^1$ or a minimal ruled surface $\mathbb{F}_n$ for some $n \geq 2$. For the latter, let $E \subset \mathbb{F}_n$ denote the negative section and F a fiber. By an easy case analysis (cf. [7, 6.8]) we get the following.

1. If $r \geq 6$ and $g \geq 1$ then $S^m = \mathbb{P}^2, \mathbb{P}^1 \times \mathbb{P}^1$ or $\mathbb{F}_2$ and $D^m \in |-K_{S^m}|$ is smooth and elliptic.
2. If $r \geq 5, g \geq 2$ then $r = 5$, $S^m = \mathbb{F}_3$ and $D^m \in |2E + 6F|$ has genus 2.
3. If $r \geq 3, g \geq 4$ then $r = 3$, $S^m = \mathbb{F}_5$ and $D^m \in |2E + 10F|$ has genus 4.
4. If $r = 4, g \geq 2$ then either $S^m = \mathbb{F}_3$ and $D^m \in |2E + 6F|$ or $S^m = \mathbb{F}_2$ and $D^m \in |2E + 5F|$. In both cases, D^m has genus 2.

The plan is, in each case, to start with $(S^m, (1 - \frac{1}{r})D^m + \Delta^m)$ and to write down all possible $(S', (1 - \frac{1}{r})D' + \Delta')$. It is then easy to get a complete list of all $(S, (1 - \frac{1}{r})D + \Delta)$. Finally, we need to check the additional conditions, especially [7, 3.6 and 4.8].

2 The Exceptional Cases

Case 16 (The $(\mathbb{Z}/5)^4$ and $(\mathbb{Z}/3)^8$ cases).
Assume that $H_2(L.\mathbb{Z})$ contains a subgroup isomorphic to $(\mathbb{Z}/5)^4$ or $(\mathbb{Z}/3)^8$. Let $G \subset H_2(L.\mathbb{Z})$ be the subgroup given in (13). Since the odd torsion in $H_2(L,\mathbb{Z})/G$ is either cyclic or $(\mathbb{Z}/3)^2$, we conclude that either

1. $G = (\mathbb{Z}/5)^4$,
2. $G = (\mathbb{Z}/3)^8$, or

3. $G = (\mathbb{Z}/3)^6$ and $H_2(L,\mathbb{Z})/G = (\mathbb{Z}/3)^2$.

Thus $g(D) = 2$ in the first case and $g(D) \geq 3$ in the second and third cases.

This implies that S^0 is simply connected. Indeed, any nontrivial cover $\pi : S' \to S$ would give another log Del Pezzo surface where $D' := \pi^{-1}(D)$ has genus ≥ 3 in the first case and genus ≥ 5 in the other cases. By the list in (15), there are no such surfaces. In particular, the case $G = (\mathbb{Z}/3)^6$ and $H_2(L,\mathbb{Z})/G = (\mathbb{Z}/3)^2$ does not happen.

Next, as in [7, 9.9], we show that $S' = S^m$ and S is obtained by contracting the negative section E. Indeed, any one point blow-up of $\mathbb{F}_m$ maps either to $\mathbb{F}_{m-1}$ (if we blow up a point not on E) or $\mathbb{F}_{m+1}$ (if we blow-up a point on E). Since S^m is unique in each case by the list in (15), $S' \to S^m$ is an isomorphism.

If $S = S^m$, then

$$\Delta^m + H^m \equiv \sum (1 - \tfrac{1}{r_i})D_i + H$$

where the D_i are rational and H is ample. In the $(\mathbb{Z}/5)^4$ case, this would give

$$\tfrac{2}{5}E + \tfrac{1}{5}F \equiv \sum (1 - \tfrac{1}{r_i})D_i + H.$$

The intersection number of the left-hand side with F is $\tfrac{2}{5} < \tfrac{1}{2}$, so any D_i is a fiber. Now intersecting with E shows that there cannot be any D_i. This is impossible since the left-hand side is not nef and big.

Similarly, in the $(\mathbb{Z}/3)^8$ case we would need to solve

$$\tfrac{2}{3}E + \tfrac{1}{3}F \equiv \sum (1 - \tfrac{1}{r_i})D_i + H,$$

which is also impossible.

This proves (7), except for the equations.

There is a systematic way to obtain Y and the $\mathbb{C}^*$-action from (S,Δ) (cf. [7, 2.5]), but I found it much easier to do this by guessing. Note that by contracting the negative section in $\mathbb{F}_n$, we get the weighted projective plane $\mathbb{P}(1,1,n)$, and $D \in |2E + 2nF|$ is a curve given by a degree $2n$ equation. After completing the square, these are of the form $z^2 + f_{2n}(x,y)$. The explicit relationship between Δ and the weights exhibited in [3, 6] now leads to $f = t^5 + z^2 + f_6(x,y)$ in case $r = 5, S^m = \mathbb{F}_3$ and to $f = t^3 + z^2 + f_{10}(x,y)$ in case $r = 3, S^m = \mathbb{F}_5$.

The case analysis is harder for $(\mathbb{Z}/4)^4$, and we need some way to see how to get S' from S^m.

Remark 17 (Blow-up criterion). Assume that $\pi : T_1 \to T$ is the inverse of the blowing up of $p \in T$ with exceptional curve $E_1 \subset T_1$ and that $(T_1, (1 - \tfrac{1}{r})D_1 + \Delta_1 + H_1)$ satisfies (15.∗). That is, $-K_{T_1} \equiv (1 - \tfrac{1}{m})D_1 + \Delta_1 + H_1$, Δ_1 is effective and H_1 is nef and big.

Set $D := \pi_* D_1, \Delta := \pi_* \Delta_1$ and $H := \pi_* H_1$. Then $(T, (1 - \tfrac{1}{r})D + \Delta + H)$ also satisfies (15.∗). Since $(K_{T_1} \cdot E_1) = -1$, we conclude that

$$\left((1 - \tfrac{1}{r})D_1 + \Delta_1 + H_1 \right) \cdot E_1 = 1,$$

hence

$$\text{mult}_p\left((1-\tfrac{1}{r})D+\Delta+H\right) \geq 1.$$

In practice, we know $(T,(1-\tfrac{1}{r})D)$ and the numerical class of $\Delta+H$, but not $\Delta+H$ itself. If the numerical class of $\Delta+H$ is small, then $\Delta+H$ is very much under control, but we get many possibilities if the numerical class of $\Delta+H$ is big.

Case 18 (The $(\mathbb{Z}/4)^4$ case). By (15.1–4) there are two possibilities for (S^m,D^m) and $S' \to S^m$ need not be an isomorphism.

Let us start with $S^m = \mathbb{F}_2$ and $D^m \in |2E+5F|$.

Case 1. If $S' = S_m$ then, as in (16), from $(E \cdot (\Delta^m+H^m)) < 0$ we see that E must be contracted. Thus S is the weighted projective plane $\mathbb{P}(1,1,2)$, and D is given by a weighted degree 5 equation $(f_5(x,y,z) = 0)$ where $w(z) = 2$. (Alternatively, S is the quadric cone in $\mathbb{P}^3$, and D is a curve of degree 5 passing through its vertex.) Thus we get the equation $(t^4 + f_5(x,y,z) = 0)$ for Y. This gives us (7.3.a).

Case 2. If we perform at least one blow-up in going from S^m to S', then as before, we could have ended up with $S^m = \mathbb{F}_1$ or $S^m = \mathbb{F}_3$ instead. The first case is impossible from the list of (15), and the second one we consider next.

Case 3. Thus assume that $S^m = \mathbb{F}_3$ and $D \in |2E+6F|$ is a smooth curve. Thus $\Delta^m + H^m \equiv \tfrac{1}{2}E + \tfrac{1}{2}F$.

If $S' = S^m$ then, as before, we see that E must be contracted. This leads to the surface $S = \mathbb{P}(1,1,3)$ and $D \in |\mathcal{O}_S(6)|$ a smooth curve.

This never leads to simply connected 5-manifolds by [7, 4.8]. All Seifert bundles are of the form $Y(S,B,\tfrac{b}{4}D)$ where $B = aF$ for some $a \in \mathbb{Z}$ and $b \in \{1,3\}$. The corresponding Chern class is $(a+\tfrac{3b}{2})F$, always a half integer. Thus we get the basic cases with $c_1(Y/S) = \tfrac{1}{2}$ and their obvious quotients.

Since $D \in |\mathcal{O}_S(6)|$ and 6 is even, $H_1^{orb}(S,\tfrac{3}{4}D) = \mathbb{Z}/2$, and there are two basic cases: $a = -1, b = 1$ and $a = -4, b = 3$.

The double cover of $(S,\tfrac{3}{4}D)$ is given by

$$S_6 = (f_6(x,y,z)+T^2 = 0) \subset \mathbb{P}(1,1,3,3),$$

and the involution is $(x:y:z:T) \mapsto (x:y:z:-T)$. Keep in mind that these are projective coordinates, so the same involution can be given as $(x:y:z:T) \mapsto (-x:-y:-z:T)$. The basic Seifert bundle pulls back to $Y = L^*(2,2,6,3;12)$ with typical equation

$$x^6 + y^6 + z^2 + t^4 = 0.$$

There are 2 ways to lift the involution to a fixed point free involution on Y. These are $(x,y,z,t) \mapsto (-x,-y,-z,-t)$ and $(x,y,z,t) \mapsto (-x,-y,-z,t)$. These give the 2 families listed in (7.3.b).

Case 4. Next we blow up at least 1 point on S^m. This point cannot be on E because the resulting surface would also dominate $\mathbb{F}_4$. Write

$$\tfrac{1}{2}E + \tfrac{1}{2}F \equiv \Delta^m + H^m = \sum a_i F_i + bE + R^m,$$

where the F_i are distinct fibers, and R^m has no irreducible component that is E or a fiber. By intersecting with F, we see that $b \leq \frac{1}{2}$. Intersecting with E gives

$$0 \leq (E \cdot R^m) = 3b - 1 - \sum a_i,$$

in particular $b \geq \frac{1}{3}$ and $\frac{1}{2} - b \leq \frac{1}{3}(\frac{1}{2} - \sum a_i)$, thus $\sum a_i \leq \frac{1}{2}$. If p lies on F_1, then

$$\mathrm{mult}_p(\Delta^m + H^m) \leq a_1 + (F \cdot R^m) = a_1 + (\tfrac{1}{2} - b) \leq \tfrac{1}{6} + \tfrac{2}{3}a_1 - \tfrac{1}{3}\sum_{i \neq 1} a_i \leq \tfrac{1}{2}.$$

The condition (15.∗) is

$$\mathrm{mult}_p(\tfrac{3}{4}D + \Delta^m + H^m) \geq 1,$$

which is only possible if $p \in D$ and $a_1 \geq 1/8$. Furthermore, if $a_1 \leq a_2$, then we get $\frac{1}{4} \leq a_1 \leq a_2$. But $\sum a_i \leq \frac{1}{2}$, and this would lead to $R^m = (\frac{1}{2} - b)E$, which is impossible. Thus we conclude:

Claim. There is a unique fiber F such that S' is obtained from S^m by blowing up points in or above $F \cap D$.

The general case is when $F \cap D$ consists of 2 points and the special case is when $F \cap D$ is a single point where F and D are tangent.

Case 5. Let us deal first with the general case and blow up $p \in F \cap D$. We get $S_1 \to S^m$ with exceptional curve G_1. Let F_1 and D_1 be the birational transforms of F and D. Write

$$\Delta_1 + H_1 = aF_1 + bE_1 + cG_1 + R_1.$$

From (17) we see that

$$c \leq \mathrm{mult}_p(\tfrac{3}{4}D + \Delta^m + H^m) - 1 \leq \tfrac{3}{4} + a + \tfrac{1}{3}(\tfrac{1}{2} - a) - 1 = \tfrac{2}{3}a - \tfrac{1}{12}.$$

Thus at every point $q \in G_1 \setminus (D_1 + F_1)$, the multiplicity of $\Delta_1 + H_1$ is at most

$$c + \mathrm{mult}_q R_1 \leq c + \mathrm{mult}_p R^m \leq \tfrac{2}{3}a - \tfrac{1}{12} + \tfrac{1}{6} - \tfrac{1}{3}a = \tfrac{1}{12} + \tfrac{1}{3}a < 1,$$

thus we cannot blow these up. At $G_1 \cap F_1$, the multiplicity is at most

$$a + c + \mathrm{mult}_q R_1 \leq a + \tfrac{1}{12} + \tfrac{1}{3}a < 1.$$

Finally, at $G_1 \cap D_1$, the multiplicity is at most

$$\tfrac{3}{4} + \tfrac{1}{12} + \tfrac{1}{3}a = \tfrac{5}{6} + \tfrac{1}{3}a < 1,$$

except when $a = \frac{1}{2}$. But this again would lead to $R^m = (\frac{1}{2} - b)E$, which is impossible.

Thus we conclude that we can blow up one or both of the points $F \cap D$ and then we get S'.

If we blow up only one point, we have to contract F_1, and we are in the already considered case when $S' = \mathbb{F}_2$.

Case 6. If we blow up both points of $F \cap D$, we have to contract $E_1 \cup F_1$. We get a surface with Picard number 2 and a single cyclic quotient singularity of the form $\mathbb{C}^2 / \frac{1}{5}(1,2)$.

Claim. The surface S is isomorphic to a quasi-smooth hypersurface $S_6 \subset \mathbb{P}(1,1,3,5)$, and D is the complete intersection of S_6 with $(t = 0)$.

I found this isomorphism by computing the quotient by the hyperelliptic involution of D, which acts on S^m, S', and also on S. Once the isomorphism is guessed, it is easier to verify by working backwards from the surface $S_6 \subset \mathbb{P}(1,1,3,5)$. Its equation can be written, after coordinate changes, as

$$S_6 = (z^2 + f_6(x,y) + \ell_1(x,y)t = 0) \subset \mathbb{P}(1,1,3,5).$$

Notice that its intersection with $(\ell_1 = 0)$ is the reducible curve $(z^2 + f_6(x, \alpha x) = 0) \subset \mathbb{P}(1,3,5)$ for some α. The two irreducible components correspond to the two exceptional curves of $S' \to S^m$. The rest is a straightforward computation. The surface is specified by choosing 6 points in $\mathbb{P}^1$ (given by $f_6 = 0$) plus one more corresponding to the choice of ℓ_1.

This leads to the case (7.3.e).

Case 7. Next we deal with the special case when $F \cap D$ is a single point where F and D are tangent. Computations as above yield that in this case, we can blow up the point p on D at most 3-times.

Case 8. If S' is obtained by 1 blow-up, we factor through $\mathbb{F}_2$ as before. If we do 2 blow-ups, we get $S = \mathbb{P}(1,2,5)$ and $D \in |\mathcal{O}(10)|$. As in Case 3, $H_1^{orb}(S, \frac{3}{4}D) = \mathbb{Z}/2$. Set $L := (x = 0)$. There are two basic cases, corresponding to $Y(S, -2L, \frac{1}{4}D)$ and $Y(S, -7L, \frac{3}{4}D)$. The index 2 point $(0 : 1 : 0) \in S$ adds a further complication since the first of these does not satisfy the smoothness condition [7, 4.8.1].

The orbifold double cover of $(S, \frac{3}{4}D)$ is

$$S_{10} := (f_{10}(x,y,z) + T^2 = 0) \subset \mathbb{P}(1,2,5,5)$$

and the involution is $(x : y : z : T) \mapsto (x : y : z : -T)$. Since these are weighted projective coordinates, this is the same as $(x : y : z : T) \mapsto (-x : y : -z : T)$. (Note that -1 acts by sending a coordinate u to $(-1)^{wt(u)}u$, hence the $+$ sign in front of y.)

This leads to the basic examples $L = L^*(2,4,10,5;20)$ with sample equation $(x^{10} + y^5 + z^2 + t^4 = 0)$, The lifting of the involution is either $(x,y,z,t) \mapsto (-x,y,-z,-t)$ or $(x,y,z,t) \mapsto (-x,y,-z,t)$. Here the second action has fixed point; this is consistent with our earlier considerations. Thus we get only one family, as in (7.3.d).

Case 9. Finally, if we blow up 3-times, we can contract the birational transforms of E, F and of the first 2 exceptional curves. This gives a surface with a single singular point of the form $\mathbb{C}^2 / \frac{1}{9}(2,5)$.

Claim. The surface S is isomorphic to a quasi-smooth hypersurface $S_{10} \subset \mathbb{P}(1,2,5,9)$, and D is the complete intersection of S_{10} with $(t = 0)$.

Once again, it is easier to verify this by working backwards. The equation of S_{10} is
$$(f_5(x^2, y) + z^2 + xt = 0) \subset \mathbb{P}(1, 2, 5, 9).$$
Its intersection with $(x = 0)$ is the curve $(ay^5 + bz^2 = 0) \subset \mathbb{P}(2, 5, 9)$. This is a smooth (but not quasi-smooth) rational curve, and it corresponds to the last exceptional curve of $S' \to S^m$. The surface is specified by choosing 6 points in $\mathbb{P}^1$ (given by $xf_5 = 0$), that is, a genus 3 hyperelliptic curve plus a specified branch point. The rest is a straightforward computation.

Thus we obtain (7.3.c).

Example 19. It is easy to write down infinitely many positive Sasakian structures on certain simply connected 5-manifolds L either by hand or by consulting the (partially unpublished) lists of Boyer and Galicki. Below I write the orbifolds (S, Δ) and the simplest equation. (Here C_k denotes a general curve of degree k in the weighted projective plane $\mathbb{P}(a, b, c)$ with coordinates x, y, z and $\ell := (x = 0)$.)

1. $H_2(L, \mathbb{Z}) = (\mathbb{Z}/3)^2$. For any $(k, 6) = 1$, take
$$\left(\mathbb{P}(1, 1, 2), (1 - \tfrac{1}{3})C_4 + (1 - \tfrac{1}{k})\ell \right) \quad \text{and} \quad x^{4k} + y^4 + z^2 + t^3.$$

2. $H_2(L, \mathbb{Z}) = (\mathbb{Z}/5)^2$. For any $(k, 30) = 1$, take
$$\left(\mathbb{P}(1, 2, 3), (1 - \tfrac{1}{5})C_6 + (1 - \tfrac{1}{k})\ell \right) \quad \text{and} \quad x^{6k} + y^3 + z^2 + t^5.$$

3. $H_2(L, \mathbb{Z}) = (\mathbb{Z}/3)^4$. For any $(k, 30) = 1$, take
$$\left(\mathbb{P}(1, 2, 5), (1 - \tfrac{1}{3})C_{10} + (1 - \tfrac{1}{k})\ell \right) \quad \text{and} \quad x^{10k} + y^5 + z^2 + t^3.$$

4. $H_2(L, \mathbb{Z}) = (\mathbb{Z}/2)^{2n}$. For $n \geq 0$ and $(k, 2n(2n+1)) = 1$ take
$$\left(\mathbb{P}(1, 1, n), (1 - \tfrac{1}{2})C_{2n+1} + (1 - \tfrac{1}{k})\ell \right) \quad \text{and} \quad x^{k(2n+1)} + y^{2n+1} + yz^2 + t^2.$$

Sasakian–Einstein structures are known to exist in the first 3 cases, but for the last one the criterion of [4] fails and existence is not known.

The situation is more complicated in the next two cases:

5. $H_2(L, \mathbb{Z}) = (\mathbb{Z}/4)^2$ or $(\mathbb{Z}/6)^2$. For any k, take
$$\left(\mathbb{P}(1, 2, 3), (1 - \tfrac{1}{4})C_6 + (1 - \tfrac{1}{k})\ell \right) \quad \text{resp.} \quad \left(\mathbb{P}(1, 2, 3), (1 - \tfrac{1}{6})C_6 + (1 - \tfrac{1}{k})\ell \right).$$

These are the right examples as log Del Pezzo surfaces, but the condition [7, 4.8.2] fails, so the corresponding Seifert bundles are not simply connected. Since my approach is to rule out everything at the surface level, these cases could be very hard to settle using my methods.

This leaves open the finiteness question for

6. $H_2(L, \mathbb{Z}) = (\mathbb{Z}/m)^2$ with $7 \leq m \leq 11$, and
7. $H_2(L, \mathbb{Z}) = (\mathbb{Z}/3)^6$.

The computations in Section 4 suggest that in these cases, there should be only finitely many families of pre-SE structures.

However, as the examples with $H_2(L,\mathbb{Z}) = (\mathbb{Z}/4)^4$ suggest, the case analysis can be rather tricky, and unexpected special configurations may arise.

3 The Main Series

Let us start by fixing an error in [7].

Correction 20 (Error in [7, Thm. 1.8.1(b)]). The theorem asserts that for each $m \geq 12$, there are exactly 93 pre-SE Seifert bundles $f : L \to (S,\Delta)$ on compact 5-manifolds L satisfying $H_1(L,\mathbb{Z}) = 0$ and tors $H_2(L,\mathbb{Z}) \cong (\mathbb{Z}/m)^2$.

The construction of the 93 families of pairs (S,Δ) is correct. Going from (S,Δ) to the Seifert bundle is, however, done incorrectly because the conditions for a Sasakian structure and for a pre–Sasakian–Einstein structure have been thoroughly mixed up.

The construction of the 93 families given in [7, 7.6] starts with a surface T, which is one of $\mathbb{P}^1 \times \mathbb{P}^1, \mathbb{P}^2, Q, S_5, \mathbb{P}(1,2,3)$. For each of these write $-K_T \sim d(T)H$ where $H \in \mathrm{Weil}(T)$ is a positive generator. We have

$$d(\mathbb{P}^1 \times \mathbb{P}^1) = 2, d(\mathbb{P}^2) = 3, d(Q) = 4, d(S_5) = 5, d(\mathbb{P}(1,2,3)) = 6.$$

Next we perform some weighted blow-ups [7, 7.3] to get $S = B_{m_1,\dots,m_k}T$. There are k exceptional curves $E_1,\dots,E_k$. Each E_i passes through a unique singular point $p_i \in S$, and E_i generates the local class group, which is $\mathbb{Z}/m_i$. Set $d(S) := \gcd(m_1,\dots,m_k,d(T))$.

The divisor class group $\mathrm{Weil}(S)$ is freely generated by

$$\pi^*H, E_1,\dots,E_k \quad \text{and} \quad K_S \sim -d(T)\pi^*H + \sum m_i E_i.$$

The main condition that was overlooked is the smoothness criterion for Seifert bundles [7, 3.6].

In our case, there is only one curve D, and S is smooth along D. Thus, by [7, 3.6], the corresponding Seifert bundle $Y(S, B, \frac{b}{m}D)$ is smooth iff

$$B \text{ generates the local class group at every point.} \tag{20.1}$$

Claim 20.2. Notation as above. Then

$$Y(B_{m_1,\dots,m_k}T, aH + c_1E_1 + \dots + c_kE_k, \frac{b}{m}D) \quad \text{is smooth}$$

iff $(m_i, c_i) = 1$ for $i = 1,\dots,k$ and $(a, d(T)) = 1$ if T is singular.

As a consequence, we see that all 93 cases correspond to positive Sasakian structures on smooth 5-manifolds.

A Seifert bundle is pre-SE iff its Chern class $c_1(Y/S) = B + \frac{b}{m}D$ is a rational multiple of $-(K_S + (1 - \frac{1}{m})D)$. In our case $D \sim -K_S$, hence B itself is a rational multiple of $-K_S$. Thus

$$B = r\left(\frac{d(T)}{d(S)}\pi^*H - \sum \frac{m_i}{d(S)}E_i\right)$$

for some positive integer r. Thus if (20.1) holds, then $-c_i = rm_i/d(S)$ is an integer relatively prime to m_i. Thus $m_i | d(S)$. Since $d(S) | m_i$ by definition, we conclude that $d(S) = m_i$ for every i.

Furthermore, in the cases when T is singular, B generates the local class group at each singular point of T iff $d(T) = d(S)$. Thus we obtain the following.

Claim 20.3. Notation as above. Then

$$Y(B_{m_1,\ldots,m_k}T, aH + c_1E_1 + \cdots + c_kE_k, \frac{b}{m}D) \quad \text{is smooth and pre-SE iff}$$

1. $m_1 = \cdots = m_k = d(S)$,
2. $aH + c_1E_1 + \cdots + c_kE_k = r(-\frac{d(T)}{d(S)}H + E_1 + \cdots + E_k)$ for some $r \in \mathbb{Z}$, and
3. if T is singular, then $(r, d(T)) = 1$ and $d(S) = d(T)$.

These conditions cut down considerably the list given in [7, 7.6], and we get the 19 cases of Table 1.

The condition [7, 4.8.2] says that the resulting Seifert bundle L satisfies $H_1(L, \mathbb{Z}) = 0$ iff $H^2(S, \mathbb{Z}) \to H^2(D, \mathbb{Z}) \to \mathbb{Z}/m$ is surjective. If $S = T$, that is, we do no blow-ups at all, then by [7, 9.8]

$$\text{Im}\left[H^2(T, \mathbb{Z}) \to H^2(D, \mathbb{Z})\right] = d(T)H^2(D, \mathbb{Z}).$$

After blow-ups, the new curves that come in are the exceptional curves E_i. Here $(E_i \cdot D) = 1$, but the E_i pass through the singular points and they are only homology classes. To get a cohomology class (or Cartier divisor), we need to take m_iE_i since m_i is also the index of the singular point. Thus we conclude:

Claim 20.4. For $S = B_{m_1,\ldots,m_k}T$, $\text{Im}\left[H^2(S, \mathbb{Z}) \to H^2(D, \mathbb{Z})\right] = d(S) \cdot H^2(D, \mathbb{Z})$.

Table 1 Del Pezzo surfaces with pre-SE structure.

Surfaces	$d(S)$
$B_1\mathbb{P}^2, B_{11}\mathbb{P}^2, \ldots, B_{11111111}\mathbb{P}^2$	1
$\mathbb{P}^1 \times \mathbb{P}^1, B_2\mathbb{P}^1 \times \mathbb{P}^1, B_{22}\mathbb{P}^1 \times \mathbb{P}^1, B_{222}\mathbb{P}^1 \times \mathbb{P}^1$	2
$\mathbb{P}^2, B_3\mathbb{P}^2, B_{33}\mathbb{P}^2$	3
Q, B_4Q	4
S_5	5
$\mathbb{P}(1,2,3)$	6

Corollary 21. *Let S be a projective surface with Du Val singularities such that* $\pi_1(S^0) = 1$. *Let* $D \in |-K_S|$ *be a smooth elliptic curve. There is a simply connected Seifert bundle* $f : L \to (S, (1 - \frac{1}{m})D)$ *with a pre-SE structure iff*

1. *S is one of the surfaces in Table 1, and*
2. *m is relatively prime to* $d(S)$.

In these cases, $f : L \to (S, (1 - \frac{1}{m})D)$ *is uniquely determined by* $(S, (1 - \frac{1}{m})D)$ *(up to reversing the orientation of the fibers) and L carries a Sasakian–Einstein metric for* $m \geq 7$.

Proof. We have proved eveything, except the claims about the existence of Sasakian–Einstein metrics. This will be established in (33).   □

Example 22. Quite surprisingly, all the singular surfaces on the list can be realized in weighted projective 3-spaces. All of these examples are on the Boyer–Galicki lists, but we also claim the converse: every singular Del Pezzo surface in Table 1 is isomorphic to a corresponding surface below. In all cases $(t = 0)$ is the equation of D.

1. $B_2\mathbb{P}^1 \times \mathbb{P}^1$: $S_3 \subset \mathbb{P}(1,1,1,2)$, e.g. $x^3 + y^3 + z^3 + xt = 0$.
2. $B_{22}\mathbb{P}^1 \times \mathbb{P}^1$: $S_4 \subset \mathbb{P}(1,1,2,2)$, e.g. $x^4 + y^4 + z^2 + zt = 0$.
3. $B_{222}\mathbb{P}^1 \times \mathbb{P}^1$: $S_6 \subset \mathbb{P}(1,2,3,2)$, e.g. $x^6 + y^3 + z^2 + t^3 = 0$.
4. $B_3\mathbb{P}^2$: $S_4 \subset \mathbb{P}(1,1,2,3)$, e.g. $x^4 + y^4 + z^2 + xt = 0$.
5. $B_{33}\mathbb{P}^2$: $S_6 \subset \mathbb{P}(1,2,3,3)$, e.g. $x^6 + y^3 + z^2 + zt = 0$.
6. Q: $S_4 \subset \mathbb{P}(1,1,2,4)$, e.g. $x^4 + y^4 + z^2t = 0$.
7. B_4Q: $S_6 \subset \mathbb{P}(1,2,3,4)$, e.g. $x^6 + y^3 + z^2 + yt = 0$.
8. S_5: (sorry for the notation) $S_6 \subset \mathbb{P}(1,2,3,5)$, e.g. $x^6 + y^3 + z^2 + zt = 0$.
9. $\mathbb{P}(1,2,3)$: $S_6 \subset \mathbb{P}(1,2,3,6)$, e.g. $x^6 + y^3 + z^2 + t = 0$.

Remark 23. There are no hypersurface links on the Boyer–Galicki lists giving infinitely many (S, Δ) where $S \in \{B_1\mathbb{P}^2, \dots, B_{11111}\mathbb{P}^2\}$.

I claim that these cannot be realized as the link L of a hypersurface with a $\mathbb{C}^*$-action, at least when $H_2(L, \mathbb{Z}) \supset (\mathbb{Z}/m)^2$ for $m \geq 12$ and (S, Δ) is log Del Pezzo. Assume the contrary. Then we get that S is a hypersurface in a weighted projective space $\mathbb{P}(a, b, c, d)$ such that $K_S + (1 - \frac{1}{m})D$ is proportional to $H|_S$ where H is the hyperplane class of the weighted projective space. Since $D \in |-K_S|$, we conclude that K_S is proportional to H. By the Grothendieck–Lefschetz theorem, this implies that $-K_S = dH$ for some $d \in \mathbb{Z}$. In our cases, $-K_S$ is not divisible, so $d = 1$. This implies that

$$h^0(S, \mathcal{O}_S(-K_S)) = h^0(\mathbb{P}(a, b, c, d), \mathcal{O}_{\mathbb{P}}(1)) \leq 4,$$

as this dimension is the number of times that 1 occurs among a, b, c, d. In the above cases, however, $h^0(S, \mathcal{O}_S(-K_S)) \geq 5$.

Galicki told me that the link $L^*(2, 3, 4, 7; 14)$ realizes $S = B_{1111}\mathbb{P}^2$, but the corresponding Δ is not a rational multiple of $-K_S$.

Table 2 Minimal Del Pezzo surfaces with $\pi_1(S^0) > 1$.

Degree	Singularities	$\pi_1(S^0)$	Weil / Pic	Univ. cover	$\pi_1(S^0 \setminus D)$
1	A_8	$\mathbb{Z}/3$	$\mathbb{Z}/3$	$B_{3111}\mathbb{P}^2$	$\mathbb{Z}/3$
1	$A_7 + A_1$	$\mathbb{Z}/4$	$\mathbb{Z}/4$	$B_{2111}\mathbb{P}^2$	$\mathbb{Z}/4$
1	$A_5 + A_2 + A_1$	$\mathbb{Z}/6$	$\mathbb{Z}/6$	$B_{111}\mathbb{P}^2$	$\mathbb{Z}/6$
1	$4A_2$	$(\mathbb{Z}/3)^2$	$(\mathbb{Z}/3)^2$	$\mathbb{P}^2$	G_{27}
1	$2A_3 + 2A_1$	$\mathbb{Z}/2 + \mathbb{Z}/4$	$\mathbb{Z}/2 + \mathbb{Z}/4$	$\mathbb{P}^1 \times \mathbb{P}^1$	G_{16}
1	$2A_4$	$\mathbb{Z}/5$	$\mathbb{Z}/5$	$B_{1111}\mathbb{P}^2$	$\mathbb{Z}/5$
1	$A_3 + 4A_1$	$(\mathbb{Z}/2)^2$	$(\mathbb{Z}/2)^3 + \mathbb{Z}/4$	$B_{22}\mathbb{P}^1 \times \mathbb{P}^1$	$\mathbb{Z}/2 + \mathbb{Z}/4$
2	A_7	$\mathbb{Z}/2$	$\mathbb{Z}/4$	$B_{41}\mathbb{P}^2$	$\mathbb{Z}/2$
2	$A_5 + A_2$	$\mathbb{Z}/3$	$\mathbb{Z}/6$	$B_{11}Q$	$\mathbb{Z}/3$
2	$2A_3 + A_1$	$\mathbb{Z}/4$	$\mathbb{Z}/2 + \mathbb{Z}/4$	$\mathbb{P}^1 \times \mathbb{P}^1$	$\mathbb{Z}/2 + \mathbb{Z}/4$
2	$6A_1$	$(\mathbb{Z}/2)^2$	$(\mathbb{Z}/2)^4$	$\mathbb{P}^1 \times \mathbb{P}^1$	G_8
2	$2A_3$	$\mathbb{Z}/2$	$\mathbb{Z}/2 + \mathbb{Z}/4$	$B_{22}\mathbb{P}^1 \times \mathbb{P}^1$	$\mathbb{Z}/4$
3	$A_5 + A_1$	$\mathbb{Z}/2$	$\mathbb{Z}/6$	$B_3\mathbb{P}^2$	$\mathbb{Z}/6$
3	$3A_2$	$\mathbb{Z}/3$	$(\mathbb{Z}/3)^2$	$\mathbb{P}^2$	$(\mathbb{Z}/3)^2$
4	$A_3 + 2A_1$	$\mathbb{Z}/2$	$\mathbb{Z}/2 + \mathbb{Z}/4$	Q	$(\mathbb{Z}/2)^2$
4	$4A_1$	$\mathbb{Z}/2$	$(\mathbb{Z}/2)^3$	$\mathbb{P}^1 \times \mathbb{P}^1$	$(\mathbb{Z}/2)^2$

Remark 24 (The cases with nontrivial fundamental group). The classification of minimal Del Pezzo surfaces with Du Val singularities is completed in [5, 12, 13, 17]. We are interested only in those that have cyclic quotient singularities. The 5 cases where $\pi_1(S^0) = 1$ were considered in [7]. Table 2 lists the remaining ones.

Here G_n is a non-Abelian group of order n.

1. G_8 is the quaternion group,
2. $G_{16} \subset GL_4$ is generated by $(x, y, z, t) \mapsto (z, t, -x, y)$ and $(x, y, z, t) \mapsto (y, -x, t, -z)$.
3. $G_{27} \subset GL_3$ is generated by $(x, y, z) \mapsto (y, z, x)$ and $(x, y, z) \mapsto (x, \varepsilon y, \varepsilon^2 z)$ where $\varepsilon^3 = 1$.

The computation of the table: The papers [5, p. 13–15], [12, p. 71], [13, p. 193], and [17] contain tables for the first 4 columns, except Weil / Pic in the 4 cases with Picard number 2. (The fundamental group of $A_7 + A_1$ is listed in [12, p. 71] as $(\mathbb{Z}/2)^2$, but it is $\mathbb{Z}/4$.) The Picard number and the singularities of the universal cover are listed in [12, p. 71]; from this it is easy to work out where the surface is on the list [7, 7.6].

By [17, 1.2 and 1.6], for each singularity type there is a unique surface, except for $2A_3$ for which there is a 1-parameter family.

Once we have $\bar{S} \to S$ as the universal cover and $\bar{D} \subset \bar{S}$ is the preimage of D, then we have an exact sequence

$$\pi_1(\bar{S}^0 \setminus \bar{D}) \to \pi_1(S^0 \setminus D) \to \pi_1(S^0) \to 1.$$

Each time $\bar{S}$ is obtained by a blow-up of weight 1, the resulting $\mathbb{P}^1 \subset \bar{S}$ intersects $\bar{D}$ transversally at 1 point, so $\pi_1(\bar{S}^0 \setminus \bar{D}) = 1$. In the remaining cases, one needs to write down the action of the $\pi_1(S^0)$ and see how it lifts to the universal cover of $\bar{S}^0 \setminus \bar{D}$.

Computing any entry of the table is an elementary task. Some computations are quick, but a few are quite tedious. It is unfortunately easy to miss or misdraw a -1-curve after performing many blow-ups, so anyone wishing to rely on a particular entry is advised to recheck it.

In the simply connected case, there are many isomorphisms between blow-ups, but this does not happen for the general case.

Lemma 25. *Let S be a Del Pezzo surface with cyclic quotient singularities such that* $|\pi_1(S^0)| > 1$. *There is a unique line in Table 2 such that S is a weighted blow-up of a surface on that line. (We do not claim that the blow-up itself is unique.)*

Proof. The blow-ups do not change the fundamental group [7, 7.3], and we create only A_1 and A_2 singularities since we can only have blow-ups $B_{m_1,\dots,m_r}T$ where $\sum m_i < \deg T - 1 \leq 3$. It turns out that the fundamental group and the collection of $A_i : i \geq 3$ singularities uniquely determine in which line of Table 2 the surface is. The Picard number and the $A_i : i \leq 2$ singularities now determine the number and type of blow-ups performed. $\square$

If $\deg T = 1$ (resp. $2, 3, 4$), then we get 1 (resp. $2, 4, 7$) blown-up surfaces, including T itself. Thus we get 39 deformation types of Del Pezzo surfaces with cyclic quotient singularities such that $|\pi_1(S^0)| > 1$. The 93 cases where $\pi_1(S^0) = 1$ were enumerated in [7, 7.6], giving a total of 132 deformation types.

Remark 26 (Existence of smooth Seifert bundles). Let S be one of the surfaces in Table 2 and $D \in |-K_S|$ a smooth elliptic curve. As in (20), we are considering Seifert bundles $Y(S, B, \frac{b}{m}D)$ where B is a Weil divisor class on S.

By (20.1), $Y(S, B, \frac{b}{m}D)$ is smooth iff B generates the local class group at every point.

Finding all such B requires a detailed computation of the divisor class group $\mathrm{Weil}(S)$ and the restriction map

$$\mathrm{Weil}(S)/\mathrm{Pic}(S) \to \sum_{p \in \mathrm{Sing}\, S} \mathrm{Weil}(p, S).$$

On a Del Pezzo surface of degree ≤ 7, the curves C with $(C \cdot K_S) = -1$ generate $\mathrm{Weil}(S)$. On the minimal desingularization $S' \to S$ these are the -1 curves. Thus if we have a description of S' as a blow-up of $\mathbb{P}^2$, we see all such curves by looking at lines through 2 blow-up points, conics through 5 blow-up points, etc. (See [11, Sec.26] for the complete list in degrees 2 and 1.) The description given in [5] gives exactly these blow-ups. See also [12, Fig. 1].

In some cases, $\mathrm{Weil}(S)/\mathrm{Pic}(S)$ is too small to get a surjection onto some $\mathrm{Weil}(p, S)$, and then there are no smooth Seifert bundles at all. This happens in 3 cases:

$$A_8, A_7 + A_1, A_7.$$

More surprising is the mildly singular $A_3 + 2A_1$ case, which again has no smooth Seifert bundle over it. On the minimal desingularization, the configuration of -1 and -2 curves is

$$\overset{-2}{\circ} - \overset{-1}{\circ} - \overset{-2}{\circ} - \overset{-2}{\circ} - \overset{-2}{\circ} - \overset{-1}{\circ} - \overset{-2}{\circ}$$

In order to generate both $\mathbb{Z}/2$ on the ends, we need to take the two -1-curves with odd coefficients, but then we get only twice the generator of the $\mathbb{Z}/4$ of the middle singularity.

The more complicated $A_5 + A_1$ case leads to the configuration

$$\overset{-2}{\circ} - \overset{-1}{\circ} - \overset{-2}{\circ} - \overset{-2}{\circ} - \overset{-2}{\circ} - \overset{-2}{\circ} - \overset{-2}{\circ}$$

plus an extra -1-curve. The -1-curve shown generates both local class groups, so there are smooth Seifert bundles, even with SE structure.

A glance at the diagrams for $A_5 + A_2 + A_1$ and for $2A_4$ in [12, Fig. 1] shows -1-curves that generate all local class groups.

As another concrete example, the group G_{27} operates freely on $\mathbb{C}^3$ outside the origin, thus $S^5/G_{27} \to \mathbb{P}^2/(\mathbb{Z}/3)^2$ is a smooth Seifert bundle.

Just for illustration, let us compute one simple case completely.

Example 27 (3A_2 case). We can write this as $S = \mathbb{P}^2/(\mathbb{Z}/3)$ by the action

$$(x:y:z) \mapsto (x:\varepsilon y:\varepsilon^2 z) \quad \text{where} \quad \varepsilon^3 = 1.$$

We can take $D = (x^3 + y^3 + z^3 = 0)$. The universal cover of $\mathbb{P}^2 \setminus D$ is the cubic

$$(x^3 + y^3 + z^3 = t^3) \subset \mathbb{P}^3.$$

We get a $(\mathbb{Z}/3)^2$-action generated by

$$(x:y:z:t) \mapsto (x:\varepsilon y:\varepsilon^2 z:t) \quad \text{and} \quad (x:y:z:t) \mapsto (x:y:z:\varepsilon t).$$

The 3 coordinate lines give the curves $A, B, C \subset \mathbb{P}^2/(\mathbb{Z}/3)$. These generate Weil$(S)$ subject to the relations $3A = 3B = 3C = A + B + C$. We can thus rewrite

$$\text{Weil}(S) = \mathbb{Z}[A] + \mathbb{Z}/3[A - B].$$

By explicit computations, the class $B_{uv} := uA + v(A - B)$ corresponds to a smooth Seifert bundle iff none of the numbers $u, v, u + v$ is divisible by 3. Whenever this holds, there is a smooth pre-SE Seifert bundle $L_{uv} \to (S, B_{uv}, (1 - \frac{1}{m})D)$ for any $(m, 3) = 1$. It has a Sasakian–Einstein metric for every $m \geq 4$.

Remark 28. As a consequence of the classification, [13, Thm, p. 184] concludes that the fundamental group of a Del Pezzo surface with Du Val singularities is Abelian.

This is easy to see directly as follows. Let $g : S' \to S$ be the universal cover. Pick a smooth elliptic curve $C \in |-K_S|$. Then $C' := g^{-1}(C) \in |-K_{S'}|$, thus it is also a smooth elliptic curve. Hence the fundamental group is the same as the kernel of the group homomorphism $C' \to C$, hence an Abelian group with at most 2 generators.

4 Klt[1] Conditions

Although we did not use it in the proof, it is instructive to see the relationship between finding S' and the klt condition for $(S^m, (1 - \frac{1}{r})D^m + \Delta^m + H^m)$.

Since Δ' is a nonnegative linear combination, all exceptional curves of $S' \to S^m$ have nonpositive discrepancy with respect to $(S^m, (1 - \frac{1}{r})D^m + \Delta^m + H^m)$. If the latter pair is klt, then, by an observation of Shokurov, there are only finitely many such curves on any birational model of S^m and they can be found explicitly (cf. [9, 2.36]).

The problem with using this result is that we do not know Δ^m and H^m, only the numerical class

$$\Delta^m + H^m \equiv - \left(K_{S^m} + (1 - \frac{1}{r})D^m \right).$$

The usual proofs of the above finiteness result start by taking a log resolution of $(S^m, (1 - \frac{1}{r})D^m + \Delta^m + H^m)$, and in essence we would need to understand all smooth surfaces dominating S^m.

The key part of the proofs in (16) and (20) is to describe all exceptional curves over S^m that have nonpositive discrepancy with respect to some $(S^m, (1 - \frac{1}{r})D^m + \Delta^m + H^m)$.

To this end, one can use the following result, which sharpens the klt conditions used in [7, Sec. 8].

Proposition 29. *Let S be a smooth surface, C a smooth curve on S, and D an effective $\mathbb{Q}$-divisor on S such that $C \not\subset \mathrm{Supp}\, D$. Let $p \in S$ be point and n a natural number. Then $(S, (1 - \frac{1}{n})C + cD)$ is klt at p for*

$$c < \min \left\{ \frac{1}{(C \cdot D)_p} + \frac{1}{n \cdot \mathrm{mult}_p D}, \frac{1}{\mathrm{mult}_p D} \right\}.$$

Proof. Choose local coordinates (x, y) such that $C = (y = 0)$ and let $f(x, y) = 0$ be an equation of mD for some m such that mD is an integral divisor.

Consider the local degree n cover $\pi : T \to S$ given by $y = z^n$. By [8, 20.3.2], $(S, (1 - \frac{1}{n})C + cD)$ is klt at p iff $(T, \frac{c}{m}(f(x, z^n) = 0))$ is klt at p.

We aim to apply a theorem of Varčenko [15] that gives a condition for $(T, \gamma(g(x, z) = 0))$ to be klt in terms of the Newton polygon of g in a suitable coordinate system, which is achieved after a series of coordinate changes of the form $(x, z) \mapsto (x - \alpha z^i, z)$ or $(x, z) \mapsto (x, z - \alpha x^i)$. The problem is that we can handle only those coordinate changes that are compatible with π. That is, those of the form $(x, z) \mapsto (x - \alpha z^{ni}, z)$. Thus we have to look carefully at the proof, not just the final result.

We state the result in the rather artificial form that we need. The reader should consult the proof given in [10, 6.40], especially pp. 172–3.

Lemma 30. *Write $g(x, z) = \sum b(i, j) x^i z^j$. Assume that one of the following holds.*

---

[1] Klt is short for Kawamata log terminal, cf. [8, Defn. 2.34].

1. *(Main case.) There are (i,j) and (i',j') such that $b(i,j) \neq 0$, $b(i',j') \neq 0$ and the line segment $[(i,j),(i',j')]$ contains a point (γ,γ) with $\gamma < c^{-1}$.*
2. *(Degenerate case.) There is (i,j) such that $b(i,j) \neq 0$ and $i,j < c^{-1}$.*

Then one of the following also holds:

3. *(Klt case.) $(T, c(g(x,z) = 0))$ is klt.*
4. *(Coordinate change.) There are natural numbers u,v,w and e such that $(u,v) = 1$, $e > w/(u+v)$ and*

$$\sum_{ui+vj=w} b(i,j)x^i y^j \quad \text{is divisible by} \quad (\alpha x^v + \beta y^u)^e.$$

In this case necessarily $u = 1$ or $v = 1$, there is a unique such factor, and the new coordinates are $(x, \alpha x^v + \beta y)$ if $u = 1$ and $(\alpha x + \beta y^u, y)$ if $v = 1$.

Set $a = (L \cdot D)_p$ and $d = \text{mult}_p D$. As before, write $f = \sum b(i,j)x^i y^j$. Then $b(am,0) \neq 0$ and $b(i,j) \neq 0$ for some $i + j = md$. Thus

$$f(x,z^n) = \sum b(i,j)x^i z^{nj}.$$

If $b(i,j) \neq 0$ for some $i + j = md$ and $j \geq md/(n+1)$, then the main case (30.1) applies using the line segment $[(am,0),(i,nj)]$. The maximum value of γ is achieved when $i = 0$, giving the condition $c < a^{-1} + (nd)^{-1}$.

Otherwise $b(i,j) \neq 0$ for some $i + j = md$ and $j < md/(n+1)$. Thus $i < md$, $nj < md$, and (30.2) applies, giving the condition $c < d^{-1}$.

Thus either $(T, \frac{c}{m}(f(x,z^n) = 0))$ is klt as required or we are in the coordinate change case (30.4). Thus we consider $\sum_{ui+vnj=w} b(i,j)x^i z^{nj}$. Set $U = u/(u,n)$, $V = vn/(u,n)$, $W = w/(u,n)$ and look at the irreducible factorization

$$\sum_{Ui+Vj=W} b(i,j)x^i y^j = \prod_\ell (\alpha_\ell x^V + \beta_\ell y^U)^{e_\ell}.$$

Correspondingly,

$$\sum_{ui+vnj=w} b(i,j)x^i z^{nj} = \prod_\ell (\alpha_\ell x^V + \beta_\ell z^{Un})^{e_\ell}.$$

Since (30.4) applies, either $\min\{V, Un\} = 1$, that is $V = 1$, or some $(\alpha_\ell x^V + \beta_\ell z^{Un})$ factors further. However, in the latter case we have $(V, Un) > 1$ factors of the same multiplicity $e_\ell > w/(v + un)$, which is impossible. Thus $V = 1$ and the coordinate change is of the form $(x,z) \mapsto (x - \gamma z^{Un}, z)$, which can be realized on S as $(x,y) \mapsto (x - \gamma y^U, y)$. □

The following consequences show that the klt condition is satisfied for the cases leading to $H_2(L,\mathbb{Z}) = (\mathbb{Z}/m)^2$ for $m \geq 7$ and for $H_2(L,\mathbb{Z}) = (\mathbb{Z}/3)^6$. This is why I expect to have only finitely many families for them.

Corollary 31. *Let $C \subset S$ be a smooth elliptic curve where S is $\mathbb{P}^2, \mathbb{P}^1 \times \mathbb{P}^1$ or $\mathbb{F}_2$. Let D be an effective $\mathbb{Q}$-divisor such that $D \equiv \frac{1}{n}C$. Then $(S, (1 - \frac{1}{n})C + D)$ is klt for $n \geq 7$.*

Proof. Start with the $S = \mathbb{P}^2$ case. If $H \subset \mathbb{P}^2$ is a line, then $C \equiv 3H$ and so $D \equiv \frac{3}{n}H$. Thus $(C \cdot D)_p \leq \frac{9}{n}$ and $\mathrm{mult}_p D \leq (H \cdot D) = \frac{3}{n}$. Since

$$\frac{n}{9} + \frac{1}{3} > 1 \quad \text{for } n \geq 7,$$

the rest follows from (29). The other two cases are similarly easy. □

Corollary 32. *Let $C \subset \mathbb{F}_4$ be a smooth curve of genus 3 such that $C \equiv 2E + 8F$. Let D be an effective $\mathbb{Q}$-divisor such that $D \equiv -(K + \frac{2}{3}C)$. Then $(\mathbb{F}_3, \frac{2}{3}C + D)$ is klt.*

Proof. Here $D \equiv \frac{2}{3}(E + F)$, thus

$$(C \cdot D)_p \leq \frac{2}{3} \cdot (2E + 8F) \cdot (E + F) = \frac{4}{3},$$

and $\mathrm{mult}_p D \leq \frac{2}{3}$. Since $\frac{3}{4} + \frac{1}{2} > 1$, the rest follows from (29). □

One can also use (29) to check the existence condition [4] for orbifold Kähler–Einstein metrics.

Corollary 33. *Let S be a Del Pezzo surface with quotient singularities and $C \in |-K_S|$ a smooth elliptic curve. Then $(S, (1 - \frac{1}{n})C)$ has an orbifold Kähler–Einstein metric whenever $n > \frac{2}{3}K_S^2$.*

Proof. Set $d = K_S^2$. Let D be an effective $\mathbb{Q}$-divisor such that $D \equiv \frac{1}{n}C$. We need to check that $(S, (1 - \frac{1}{n})C + \frac{2}{3}D)$ is klt. As in (31), this holds if

$$\frac{2}{3} < \min\left\{\frac{n}{d} + \frac{n}{nd}, \frac{n}{d}\right\} = \frac{n}{d}.$$

□

Acknowledgments I thank Ch. Boyer and K. Galicki for many comments and corrections and the University of Utah where some of this work was done. Partial financial support was provided by the NSF under grant number DMS-0500198.

References

1. D. Barden, *Simply connected five-manifolds*, Ann. Math. (2) 82 (1965), pp. 365–385. MR 0184241 (32 #1714).
2. C.P. Boyer & K. Galicki, *On Sasakian-Einstein geometry*, Int. J. Math. 11 (2000), 7, pp. 873–909. MR 2001k:53081.
3. C.P. Boyer, K. Galicki & J. Kollár, *Einstein metrics on spheres*, Ann. Math. 162 (2005), pp. 1–24.

4. J-P. Demailly & J. Kollár, *Semi-continuity of complex singularity exponents and Kähler-Einstein metrics on Fano orbifolds*, Ann. Sci. École Norm. Sup. (4) 34 (2001), 4, pp. 525–556. MR 2002e:32032.
5. M. Furushima, *Singular del Pezzo surfaces and analytic compactifications of 3-dimensional complex affine space* $\mathbb{C}^3$, Nagoya Math. J. 104 (1986), pp. 1–28. MR 88m:32051.
6. S. Kobayashi, *Topology of positively pinched Kähler manifolds*, Tôhoku Math. J. (2) 15 (1963), pp. 121–139. MR 27 #4185.
7. J. Kollár, *Einstein metrics on five-dimensional Seifert bundles*, J. Geom. Anal. 15 (2005), 3, pp. 445–476. MR 2190241.
8. ———, *Flips and abundance for algebraic threefolds*, Papers from the Second Summer Seminar on Algebraic Geometry held at the University of Utah, Salt Lake City, Utah, August 1991, Astérisque 211 (1992), Société Mathématique de France, Paris. MR 94f:14013.
9. J. Kollár & S. Mori, *Birational geometry of algebraic varieties*, Cambridge Tracts in Mathematics, vol. 134, Cambridge University Press, Cambridge, 1998. With the collaboration of C. H. Clemens and A. Corti. Translated from the 1998 Japanese original. MR 2000b:14018.
10. J. Kollár, K.E. Smith & A. Corti, *Rational and nearly rational varieties*, Cambridge Studies in Advanced Mathematics, vol. 92, Cambridge University Press, Cambridge, 2004. MR 2062787 (2005i:14063).
11. Y.I. Manin, *Cubic forms*, North-Holland Mathematical Library, vol. 4, North-Holland Publishing Co., Amsterdam, 1986. MR 833513 (87d:11037).
12. M. Miyanishi & D.-Q Zhang, *Gorenstein log del Pezzo surfaces of rank one*, J. Algebra, 118 (1988), 1, pp. 63–84. MR 89i:14033.
13. ———, *Gorenstein log del Pezzo surfaces. II*, J. Algebra, 156 (1993), 1, pp. 183–193. MR 94m:14045.
14. S. Smale, *On the structure of 5-manifolds*, Ann. Math. (2) 75 (1962), pp. 38–46. MR 25 #4544.
15. A.N. Varčenko, *Newton polyhedra and estimates of oscillatory integrals*, Funkcional. Anal. i Priložen. 10 (1976), 3, pp. 13–38. MR 0422257 (54 #10248).
16. C. Xu, *Notes on* π_1 *of Smooth Loci of Log Del Pezzo Surfaces*, arXiv:math.AG/0706.3957, 2007.
17. Q. Ye, *On Gorenstein log del Pezzo surfaces*, Japan. J. Math. (N.S.) 28 (2002), 1, pp. 87–136. MR 1933881 (2003h:14063).

Four-Manifolds, Curvature Bounds, and Convex Geometry

Claude LeBrun

Abstract Seiberg–Witten theory leads to a remarkable family of curvature estimates governing the Riemannian geometry of compact 4-manifolds. This chapter describes a more transparent and user-friendly repackaging of these estimates, formulated in terms of the convex hull of the set of monopole classes. New results are then obtained concerning the boundary cases of the reformulated curvature estimates.

1 Introduction

Seiberg–Witten theory leads to a remarkable family of curvature estimates governing the Riemannian geometry of compact 4-manifolds, and these, for example, imply interesting results concerning the existence and/or uniqueness of Einstein metrics on such spaces. The primary purpose of this chapter is to introduce a simplified, user-friendly repackaging of the information conveyed by the Seiberg–Witten equations into a single, easily understood numerical invariant that appears to play the starring role in the relevant curvature estimates. In addition, this article contains some new results concerning boundary cases of the curvature estimates that strengthen what was previously known.

The gist of the matter can be summarized as follows. Suppose that M is a smooth compact oriented 4-manifold with $b_+(M) \geq 2$. By considering a geometrically rich system of PDE called the Seiberg–Witten equations, one may then define a certain finite subset $\mathfrak{C} \subset H^2(M, \mathbb{R})$ that depends only on the orientation and smooth structure of M. The elements of $\mathfrak{C}$ are called *monopole classes*, and are, by definition, the first Chern classes of those spin^c structures on M for which the the Seiberg–Witten equations have solutions for all metrics. Now, while the elements of $\mathfrak{C}$ are all *integer* classes, we wish to focus here on the fact that $\mathfrak{C}$ sits in a real vector space, as

C. LeBrun
SUNY Stony Brook, Stony Brook, New York, USA

K. Galicki and S.R. Simanca (eds.), *Riemannian Topology and Geometric Structures on Manifolds,* Progress in Mathematics 271, DOI 10.1007/978-0-8176-4743-8,
© Springer Science+Business Media LLC 2009

this allows us to consider its convex hull **Hull**($\mathfrak{C}$). Because $\mathfrak{C}$ is finite, **Hull**($\mathfrak{C}$) is necessarily compact. We can therefore define a real-valued invariant of M by setting

$$\beta^2(M) = \max \left\{ \mathbf{v}^2 \mid \mathbf{v} \in \mathbf{Hull}(\mathfrak{C}) \right\}$$

when $\mathfrak{C} \neq \varnothing$, while setting $\beta^2(M) = 0$ if $\mathfrak{C} = \varnothing$. Here $\mathbf{v}^2 = \langle \mathbf{v} \smile \mathbf{v}, [M] \rangle$ denotes the *intersection pairing* of a class $\mathbf{v} \in H^2(M, \mathbb{R})$ with itself. Because $0 \in \mathbf{Hull}(\mathfrak{C})$ whenever $\mathfrak{C} \neq \varnothing$, one automatically has $\beta^2(M) \geq 0$; but, more importantly, there are actually many 4-manifolds M for which $\beta^2(M) > 0$. It is this last fact that gives the following result most of its interest:

Theorem A. *Let M be a compact oriented 4-manifold with $b_+ \geq 2$. Then any metric g on M satisfies the curvature estimates*

$$\int_M s^2 d\mu \geq 32\pi^2 \beta^2(M) \tag{1}$$

$$\int_M (s - \sqrt{6}|W_+|)^2 d\mu \geq 72\pi^2 \beta^2(M) \tag{2}$$

where s and W_+ respectively denote the scalar and Weyl curvatures of g. Moreover, if M carries a non-zero monopole class, equality occurs in either (1) or (2) if and only if g is Kähler–Einstein, with negative Einstein constant.

Now, in an important respect, Theorem A is ostensibly weaker than a result that the author has published elsewhere [28]. Indeed, as we shall see below, there is a "softer" invariant, called $\alpha^2(M)$, that can be defined in terms of $\mathfrak{C}(M)$ via a complicated minimax process; and naïve comparison of the definitions of α^2 and β^2 would lead one merely to expect that

$$\beta^2(M) \leq \alpha^2(M) \leq b_+(M)\beta^2(M).$$

Yet [28] inequalities (1) and (2) can still be shown to hold even when $\beta^2(M)$ is replaced by $\alpha^2(M)$, yielding an apparently stronger statement. Oddly enough, however, it seems that *in practice*, one consistently has

$$\alpha^2(M) = \beta^2(M),$$

so that modifying (1) or (2) in this manner effectively seems to yield no added punch. The fact that α^2 and β^2 typically coincide will only partially be explained here, via some simple results of distinctly limited scope. But the upshot is that the finite configuration $\mathfrak{C} \subset H^2(M, \mathbb{R})$ consistently displays some unanticipated geometrical properties that ought to be understood more precisely.

In yet a different direction, Theorem A contains some essentially new geometric information, because the stated characterization of the equality case of (2) was not previously known. The issue boils down to a problem in almost-Kähler geometry, and is eventually resolved by Theorem 4.10.

It should be pointed out that the convex hull of the set of monopole classes first explicitly arose[1] in the context of 3-manifold theory, where Kronheimer and Mrowka [22] proved that it coincides with the unit ball of the dual Thurston norm. Although this earlier theorem might superficially seem to bear little resemblance to the results proved here, the author has concluded, in retrospect, that it subconsciously represented an important paradigm during the conceptualization of the current work. He would therefore like to express his indebtedness to Kronheimer and Mrowka by drawing the reader's attention to their deep and beautiful paper.

2 Rudiments of 4-Dimensional Geometry

This article will make frequent reference to a constellation of basic facts regarding 4-dimensional geometry, which, though largely familiar to the *cognoscenti*, would completely confuse the neophyte if left unexplained. For clarity's sake, we will therefore begin with a quick review of the key points.

Many peculiar features of 4-dimensional geometry are directly attributable to the fact that the bundle of 2-forms over an oriented Riemannian 4-manifold (M, g) invariantly decomposes as the direct sum

$$\Lambda^2 = \Lambda^+ \oplus \Lambda^- \tag{3}$$

of the eigenspaces of the Hodge star operator

$$*: \Lambda^2 \to \Lambda^2.$$

The sections of Λ^+ are called *self-dual 2-forms*, whereas the sections of Λ^- are called *anti–self-dual 2-forms*. The decomposition (3) is, moreover, *conformally invariant*, in the sense that it is left unchanged if g is multiplied by an arbitrary smooth positive function. An arbitrary 2-form can thus be uniquely expressed as

$$\varphi = \varphi^+ + \varphi^-,$$

where $\varphi^\pm \in \Lambda^\pm$, and we then have

$$\varphi \wedge \varphi = \left(|\varphi^+|^2 - |\varphi^-|^2 \right) d\mu_g,$$

where $d\mu_g$ denotes the metric volume form associated with the fixed orientation.

The decomposition (3) in turn leads to a decomposition of the Riemann curvature tensor. Indeed, identifying the curvature tensor of g with the self-adjoint linear map

$$\mathcal{R} : \Lambda^2 \longrightarrow \Lambda^2$$
$$\varphi_{jk} \longmapsto \tfrac{1}{2} \varphi_{\ell m} R^{\ell m}{}_{jk}$$

---

[1] A dually related 4-manifold antecedent may be glimpsed in the seminorm J of [21].

we obtain a decomposition

$$
\mathcal{R} = \left(\begin{array}{c|c} W_+ + \frac{s}{12} & \mathring{r} \\ \hline \mathring{r} & W_- + \frac{s}{12} \end{array} \right). \tag{4}
$$

where $W_+ + \frac{s}{12} : \Lambda^+ \to \Lambda^+$, etc. Here W_+ is the trace-free piece of its block, and is called the *self-dual Weyl curvature* of (M, g); the anti–self-dual Weyl curvature W_- is defined analogously. Both of the objects are conformally invariant, in the sense that the tensors $(W_\pm)^j{}_{k\ell m}$ both remain unaltered if g is multiplied by any smooth positive function. Note that the scalar curvature s is understood to act in (4) by scalar multiplication, whereas the trace-free Ricci curvature $\mathring{r}_{jk} = R^\ell{}_{j\ell k} - \frac{s}{4} g_{jk}$ acts on 2-forms by

$$
\varphi_{jk} \mapsto \mathring{r}_{\ell[j} \varphi^\ell{}_{k]}.
$$

Next, let us suppose that (M, g) is a *compact* oriented Riemannian 4-manifold. The Hodge theorem then tells us that every de Rham class on M has a unique harmonic representative. In particular, we therefore have a canonical identification

$$
H^2(M, \mathbb{R}) = \{\varphi \in \Gamma(\Lambda^2) \mid d\varphi = 0, \, d*\varphi = 0\}.
$$

However, the Hodge star operator $*$ defines an involution of the right-hand side, giving rise to an eigenspace decomposition

$$
H^2(M, \mathbb{R}) = \mathcal{H}_g^+ \oplus \mathcal{H}_g^-, \tag{5}
$$

where

$$
\mathcal{H}_g^\pm = \{\varphi \in \Gamma(\Lambda^\pm) \mid d\varphi = 0\}
$$

are the spaces of self-dual and anti–self-dual harmonic forms. The intersection form

$$
\begin{array}{ccc}
H^2(M, \mathbb{R}) \times H^2(M, \mathbb{R}) & \longrightarrow & \mathbb{R} \\
(\mathbf{b}, \mathbf{c}) & \longmapsto & \mathbf{b} \cdot \mathbf{c} := \langle \mathbf{b} \smile \mathbf{c}, [M] \rangle
\end{array}
$$

becomes positive-definite when restricted to $\mathcal{H}_g^+$ and negative-definite when restricted to $\mathcal{H}_g^-$. Moreover, these two subspaces are mutually orthogonal with respect to the intersection form. The numbers $b_\pm(M) = \dim \mathcal{H}_g^\pm$ are therefore oriented homotopy invariants of M. Their difference

$$
\tau(M) = b_+(M) - b_-(M)
$$

is called the *signature* of M. By the Hirzebruch signature theorem, $\tau(M)$ coincides with $\langle \frac{1}{3} p_1(TM), [M] \rangle$, and so can be expressed as a curvature integral

$$\tau(M) = \frac{1}{12\pi^2} \int_M \left(|W^+|^2 - |W^-|^2 \right) d\mu \qquad (6)$$

for any Riemannian metric g on M. This, of course, is analogous to the generalized Gauss–Bonnet formula

$$\chi(M) = \frac{1}{8\pi^2} \int_M \left(\frac{s^2}{24} + |W^+|^2 + |W^-|^2 - \frac{|\mathring{r}|^2}{2} \right) d\mu \qquad (7)$$

for the Euler characteristic.

Lemma 2.1. *Let ψ be a closed 2-form on a compact oriented Riemannian 4-manifold (M, g). Let $\mathbf{v} = [\psi]$ denote the de Rham class of ψ, and use (5) to write*

$$\mathbf{v} = \mathbf{v}^+ + \mathbf{v}^-$$

where $\mathbf{v}^\pm \in \mathcal{H}_g^\pm$. Then

$$\int_M |\psi^+|^2 d\mu_g \geq (\mathbf{v}^+)^2,$$

with equality iff ψ^+ is a harmonic form.

Proof. Let ϕ be the unique harmonic form cohomologous to ψ. Since ϕ is then the de Rham representative of $\mathbf{v}$ of minimal L^2-norm, we therefore have

$$\int_M (|\psi^+|^2 + |\psi^-|^2) d\mu \geq \int_M (|\phi^+|^2 + |\phi^-|^2) d\mu \ ,$$

with equality iff $\psi = \phi$. However,

$$\int_M (|\psi^+|^2 - |\psi^-|^2) d\mu = \int_M (|\phi^+|^2 - |\phi^-|^2) d\mu \ ,$$

as $\int \psi \wedge \psi = \int \phi \wedge \phi$ by Stokes' theorem. Averaging these expressions, we therefore have

$$\int_M |\psi^+|^2 d\mu \geq \int_M |\phi^+|^2 d\mu = \int_M \phi^+ \wedge \phi^+ = (\mathbf{v}^+)^2 \ ,$$

with equality iff $\psi^+ = (\psi + *\psi)/2$ is closed. □

When using this result, it is important to remember that the decomposition (5) depends on the metric g, as consequently does the number $(\mathbf{v}^+)^2$. This makes it vital to better understand the natural map

$$\{\text{metrics on } M\} \longrightarrow Gr_{b_+}^+ [H^2(M, \mathbb{R})]$$
$$g \longmapsto \mathcal{H}_g^+$$

from the infinite-dimensional space of all metrics to the finite-dimensional Grassmannian of $b_+(M)$-dimensional subspaces of $H^2(M, \mathbb{R})$ on which the intersection form is positive-definite. This map is called the *period map* of M. It is easily seen to be invariant under both conformal rescaling and the identity component $Diff_0(M)$ of the diffeomorphism group. A beautiful result of Donaldson [9, p. 336] asserts that the period map is not only smooth but also is actually transverse to the set of planes containing any given element of positive self-intersection. This has the following useful consequence:

Proposition 2.2 (Donaldson). *Let (M,g) be any smooth compact oriented 4-manifold with $b_+(M) \geq 1$, and let $\mathbf{b} \in \mathcal{H}_g^+ \subset H^2(M,\mathbb{R})$ be the de Rham class of any non-zero harmonic self-dual 2-form on (M,g). Then there is a smooth family of Riemannian metrics $g_{\mathbf{t}}$, $\mathbf{t} \in B_\varepsilon(0) \subset \mathcal{H}_g^-$, with $g_0 = g$, such that $(\mathbf{b}+\mathbf{t}) \in \mathcal{H}_{g_{\mathbf{t}}}^+$ for each and every $\mathbf{t}$.*

As the above discussion makes clear, the Hodge Laplacian

$$\Delta_d = dd^* + d^*d = -*d*d - d*d*$$

is an operator of fundamental geometric importance. It is thus worth pointing out that if ψ is a self-dual 2-form, then $\Delta_d \psi$ is also self-dual, and can, moreover, be re-expressed by means of the Weitzenböck formula [8]

$$\Delta_d \psi = \nabla^* \nabla \psi - 2W_+(\psi, \cdot) + \frac{s}{3}\psi . \tag{8}$$

Taking the L^2 inner product with ψ, we therefore have

$$\int_M \left(|\nabla \psi|^2 - 2W_+(\psi, \psi) + \frac{s}{3}|\psi|^2 \right) d\mu \geq 0,$$

as Δ_d is a nonnegative operator. On the other hand, since $W_+ : \Lambda^+ \to \Lambda^+$ is self-adjoint and trace-free,

$$|W_+(\psi, \psi)| \leq \sqrt{\frac{2}{3}}|W_+||\psi|^2,$$

so it follows that any self-dual 2-form ψ satisfies

$$\int_M |\nabla \psi|^2 d\mu \geq \int_M \left(-2\sqrt{\frac{2}{3}}|W_+| - \frac{s}{3} \right) |\psi|^2 d\mu. \tag{9}$$

Moreover, assuming that $\psi \not\equiv 0$, equality holds iff ψ is closed, belongs to the lowest eigenspace of W_+ at each point, and the two largest eigenvalues of W_+ are everywhere equal. Of course, this last assertion crucially depends on the fact [1, 3] that if $\Delta_d \psi = 0$ and $\psi \not\equiv 0$, then $\psi \neq 0$ on a dense subset of M.

A rather special set of techniques can be applied when (M,g) happens to admit a closed self-dual 2-form $\omega \in \mathcal{H}_g^+$ with constant point-wise norm $|\omega|_g \equiv \sqrt{2}$. In this case, there is an almost-complex structure $J : TM \to TM$, $J^2 = 1$, defined by

$$g(J\cdot, \cdot) = \omega(\cdot, \cdot),$$

and this almost-complex structure then acts on TM in a g-preserving fashion. The triple (M, g, ω) is then said to be an *almost-Kähler 4-manifold*. Because J allows one to to think of TM as a complex vector bundle, it is only natural to look for a connection on its anticanonical line bundle $L = \wedge^2 T_J^{1,0} \cong \Lambda_J^{0,2}$ in order to use the Chern–Weil theorem to express $c_1^{\mathbb{R}}(M, J)$ as

$$c_1^{\mathbb{R}}(L) = [\frac{i}{2\pi} F] \in H_{DR}^2(M, \mathbb{R}) ,$$

where F is the curvature of the relevant connection on L. A particular choice of Hermitian connection on L was first introduced by Blair [7], and is so geometrically natural that it was later rediscovered by Taubes [32] for entirely different reasons. The curvature $F_B = F_B^+ + F_B^-$ of this *Blair connection* is given [10, 29] by

$$iF_B^+ = \frac{s + s^*}{8} \omega + W^+(\omega)^\perp \tag{10}$$

$$iF_B^- = \frac{s - s^*}{8} \hat{\omega} + \mathring{\rho} \tag{11}$$

where the so-called *star-scalar curvature* is given by

$$s^* = s + |\nabla \omega|^2 = 2W_+(\omega, \omega) + \frac{s}{3} ,$$

while $W^+(\omega)^\perp$ is the component of $W^+(\omega)$ orthogonal to ω,

$$\mathring{\rho}(\cdot, J\cdot) = \frac{\mathring{r} + J^* \mathring{r}}{2} ,$$

and where the anti–self-dual 2-form $\hat{\omega} \in \Lambda^-$ is defined only on the open set where $s^* - s \neq 0$, and satisfies $|\hat{\omega}| \equiv \sqrt{2}$.

An important special case occurs when $\nabla \omega = 0$. This happens precisely when J is integrable, and g is a *Kähler metric* on the complex surface (M, J). In this case, $s = s^*$, ω is an eigenvector of the W_+, and r is J-invariant, so that iF_B just becomes the Ricci form ρ defined by $\rho(\cdot, \cdot) = r(J\cdot, \cdot)$. In fact, ω is an eigenvector of W_+ with eigenvalue $s/6$, whereas the elements of $\omega^\perp = \mathfrak{Re}\Lambda_J^{2,0}$ are eigenvectors of eigenvalue $-s/12$.

Kähler manifolds with scalar curvature $s = \mathrm{const} < 0$ will play an important role in this paper. By the above discussion, they belong to the following broader class of almost-Kähler manifolds:

Definition 2.3. An *almost-Kähler 4-manifold* (M^4, g, ω) will be said to be *saturated if*

- $s + s^*$ *is a negative constant;*
- ω *belongs to the lowest eigenspace of* $W_+ : \Lambda^+ \to \Lambda^+$ *everywhere; and*
- *the two largest eigenvalues of* $W_+ : \Lambda^+ \to \Lambda^+$ *are everywhere equal.*

3 The Seiberg–Witten Equations

This section is intended both to fix our terminological conventions and to provide streamlined proofs of the key preliminary curvature estimates. We note that, whereas the results in this section can generally be found elsewhere in the literature [17, 26–28], some of the proofs given here are both simpler and more detailed than those published heretofore.

We begin with a discussion of *spin^c structures*. If M is any smooth oriented 4-manifold, its second Stieffel–Whitney class $w_2(TM) \in H^2(M, \mathbb{Z}_2)$ is always [15, 18] in the image of the natural homomorphism

$$H^2(M, \mathbb{Z}) \rightarrow H^2(M, \mathbb{Z}_2),$$

and we can therefore always find Hermitian line bundles $L \rightarrow M$ such that

$$c_1(L) \equiv w_2(TM) \bmod 2.$$

For any such L, and for any Riemannian metric g on M, one can then find rank-2 Hermitian vector bundles $\mathbb{V}_\pm$ that formally satisfy

$$\mathbb{V}_\pm = \mathbb{S}_\pm \otimes L^{1/2},$$

where $\mathbb{S}_\pm$ are the locally defined left- and right-handed spinor bundles of (M, g). Such a choice of $\mathbb{V}_\pm$, up to isomorphism, is called a spin^c structure $\mathfrak{c}$ on M. Moreover, if $H_1(M, \mathbb{Z})$ does not contain elements of order 2, then $\mathfrak{c}$ is completely determined by

$$c_1(L) = c_1(\mathbb{V}_\pm) \in H^2(M, \mathbb{Z}),$$

which is called the first Chern class of the spin^c structure $\mathfrak{c}$.

Every unitary connection A on L induces a connection

$$\nabla_A : \Gamma(\mathbb{V}_+) \rightarrow \Gamma(\Lambda^1 \otimes \mathbb{V}_+),$$

and composition of this with the natural *Clifford multiplication* homomorphism

$$\Lambda^1 \otimes \mathbb{V}_+ \rightarrow \mathbb{V}_-$$

gives one a spin^c version

$$D_A : \Gamma(\mathbb{V}_+) \rightarrow \Gamma(\mathbb{V}_-)$$

of the Dirac operator [16, 23]. This is an elliptic first-order differential operator, and in many respects it closely resembles the usual Dirac operator of spin geometry. In particular, one has the so-called Weitzenböck formula

$$\langle \Phi, D_A^* D_A \Phi \rangle = \frac{1}{2} \Delta |\Phi|^2 + |\nabla_A \Phi|^2 + \frac{s}{4} |\Phi|^2 + 2 \langle -i F_A^+, \sigma(\Phi) \rangle \qquad (12)$$

for any $\Phi \in \Gamma(\mathbb{V}_+)$, where F_A^+ is the self-dual part of the curvature of A, and where $\sigma : \mathbb{V}_+ \to \Lambda^+$ is a natural real-quadratic map satisfying

$$|\sigma(\Phi)| = \frac{1}{2\sqrt{2}}|\Phi|^2.$$

Equation (12) is a natural generalization of the Weitzenböck formula used by Lichnerowicz [30] to prove that metrics with $s > 0$ cannot exist when M is spin and $\tau(M) \neq 0$. Unfortunately, however, one cannot hope to derive interesting geometric information about the Riemannian metric g by just using (12) for an arbitrary connection A, since one would have no control at all over the F_A^+ term. Witten [33], however, had the brilliant insight that one could remedy this by considering both Φ and A as unknowns, subject to the *Seiberg–Witten equations*

$$D_A\Phi = 0 \tag{13}$$
$$-iF_A^+ = \sigma(\Phi). \tag{14}$$

These equations are nonlinear, but they become an elliptic first-order system once one imposes the "gauge-fixing" condition

$$d^*(A - A_0) = 0, \tag{15}$$

where A_0 is an arbitrary "background" connection on L, and $i(A - A_0)$ is simply treated as a real-valued 1-form on M. This eliminates the natural action of the "gauge group" of automorphisms of the Hermitian line bundle $L \to M$.

Because the Seiberg–Witten equations are nonlinear, one cannot use something like an index formula to predict that they must have solutions. Nonetheless, there exist spin^c structures on many 4-manifolds for which there is at least one solution for every metric g. This situation is conveniently described by the following terminology [20]:

Definition 3.1. *Let M be a smooth compact oriented 4-manifold with $b_+ \geq 2$. An element $\mathbf{a} \in H^2(M, \mathbb{R})$ is called a* **monopole class** *of M iff there exists a spin^c structure $\mathfrak{c}$ on M with*

$$c_1^{\mathbb{R}}(L) = \mathbf{a}$$

for which the Seiberg–Witten equations (13–14) have a solution for every Riemannian metric g on M.

When the gauge-fixing condition (15) is imposed, the Seiberg–Witten equations amount to saying that (Φ, A) belongs to the preimage of zero for a Fredholm map of Banach spaces. This so-called *monopole map* turns out to behave roughly like a proper map of finite-dimensional spaces [5]. When the "expected dimension" of the moduli space of solutions modulo gauge equivalence, as determined by the Fredholm index of the monopole map, is zero, Witten [33] discovered that one can define an invariant analogous to the degree of a map between finite-dimensional manifolds of the same dimension. More recently, Bauer and Furuta [4,5] discovered that the monopole map determines a well-defined class in an equivariant cohomo-topy group. Either of these invariants can be used [17] to detect the presence of a monople class. Moreover, these invariants are often nontrivial; for example, a celebrated result of Taubes [32] shows that if (M, ω) is a symplectic 4-manifold with $b_+ \geq 2$, then Witten's invariant is non-zero for the spin^c structure canonically detemined by ω, so that $\pm c_1(M, \omega)$ are both monopole classes. On the other hand, Kronheimer's results [20] may indicate that some 4-manifolds admit monopole classes that are not detected by any known invariant. For a discussion of the $b_+ = 1$ case, see Section 5.

Equations (13–14) are precisely chosen so as to imply the Weitzenböck formula

$$0 = 2\Delta|\Phi|^2 + 4|\nabla_A\Phi|^2 + s|\Phi|^2 + |\Phi|^4. \tag{16}$$

In particular, these Seiberg–Witten equations can never admit a solution (Φ, A) with $\Phi \not\equiv 0$ relative to a metric g with $s > 0$. This leads, in particular, to a cornucopia of simply connected nonspin 4-manifolds that do not admit positive-scalar-curvature metrics — in complete contrast with the situation in higher dimensions [14]. Even more strikingly, the Seiberg–Witten equations actually imply integral estimates for the scalar curvature [25, 33]:

Proposition 3.2. *Let (M, g) be a smooth compact oriented Riemannian manifold, and let $\mathbf{a} \in H^2(M, \mathbb{R})$ be a monopole class of M. Then the scalar curvature s of g satisfies*

$$\int_M s^2 d\mu_g \geq 32\pi^2(\mathbf{a}^+)^2.$$

If $\mathbf{a}^+ \neq 0$, moreover, equality occurs iff there is an integrable complex structure J with $c_1^{\mathbb{R}}(M, J) = \mathbf{a}$ such that (M, g, J) is a Kähler manifold of constant negative scalar curvature.

Proof. By Definition 3.1, there must be a spin^c structure with $c_1^{\mathbb{R}}(L) = \mathbf{a}$ for which the Seiberg–Witten equations (13–14) have a solution (Φ, A) on (M, g). However, given such a solution, Φ satisfies the Weitzenböck formula (16) with respect to g and A, and integrating this then reveals that

$$0 = \int [4|\nabla_A\Phi|^2 + s|\Phi|^2 + |\Phi|^4] d\mu.$$

Hence

$$\int (-s)|\Phi|^2 d\mu \geq \int |\Phi|^4 d\mu,$$

and applying the Cauchy–Schwarz inequality to the left-hand side yields

$$\left(\int s^2 d\mu\right)^{1/2} \left(\int |\Phi|^4 d\mu\right)^{1/2} \geq \int |\Phi|^4 d\mu.$$

Equation (14) therefore tells us that

$$\int s^2 d\mu \geq \int |\Phi|^4 d\mu = 8 \int |F_A^+|^2 d\mu.$$

However, because $iF_A/2\pi$ represents $\mathbf{a}$ in de Rham cohomology, Lemma 2.1 tells us that

$$\int |F_A^+|^2 d\mu \geq 4\pi^2(\mathbf{a}^+)^2 .$$

It follows that

$$\int s^2 d\mu \geq 32\pi^2(\mathbf{a}^+)^2 ,$$

exactly as claimed.

If equality holds, the inequalities in the above argument must all be equalities. Hence $\nabla_A \Phi = 0$, and $iF_A^+ = -\sigma(\Phi)$ is therefore a parallel self-dual 2-form with de Rham class $2\pi\mathbf{a}^+$. When this cohomology class is non-zero, this form cannot vanish, and we therefore conclude that (M, g) is Kähler. Inspection of (16) then reveals that s must be a negative constant. Moreover, $\Phi \otimes \Phi$ is then a non-zero section of $\Lambda^{2,0} \otimes L$ with respect to the relevant complex structure, so $c_1^{\mathbb{R}}(M, J) = c_1^{\mathbb{R}}(L) = \mathbf{a}$. Conversely, any such Kähler metric would saturate the inequality because the self-dual part of the Ricci form of any Kähler metric on a Kähler surface is $s\omega/4$, where $\omega = g(J\cdot, \cdot)$ is the Kähler form. $\qquad\square$

Proposition 3.3. *Let M be a compact oriented 4-manifold with $b_+(M) \geq 2$. If there is a non-zero monopole class $\mathbf{a} \in H^2(M, \mathbb{R}) - \{\mathbf{0}\}$, then M does not admit metrics of scalar curvature $s \geq 0$.*

Proof. Let M be a smooth compact 4-manifold with $b_+(M) \geq 2$, and suppose that $\mathbf{a} \in H^2(M, \mathbb{R}) - \mathbf{0}$ is a non-zero monopole class. By definition, this means that there is a spin^c structure on M with $c_1^{\mathbb{R}}(L) = \mathbf{a}$ for which the Seiberg–Witten equations have some solution (Φ, A) with respect to any metric g on M. But if the metric in question has $s \geq 0$, the Weitzenböck formula (16) says that

$$0 = 2\Delta|\Phi|^2 + 4|\nabla_A \Phi|^2 + s|\Phi|^2 + |\Phi|^4$$

so that $s \geq 0$ implies that

$$0 \geq \int |\Phi|^4 d\mu_g$$

and we therefore have $\Phi \equiv 0$. But this implies that $F_A^+ = i\sigma(\Phi) \equiv 0$, too, so that $\mathbf{a} = [\frac{i}{2\pi} F_A] \in \mathcal{H}_g^-$. In particular, if g has scalar curvature $s \geq 0$ and if $\mathbf{b} \in \mathcal{H}_g^+$, then $\mathbf{a} \cdot \mathbf{b} = 0$.

Next, suppose that we had some metric g on M with strictly positive scalar curvature $s > 0$. Choose some $\mathbf{b} \in \mathcal{H}_g^+$ with $\mathbf{b}^2 = 1$. The argument in the previous paragraph tells us that $\mathbf{a} \in \mathcal{H}_g^-$, so that the integer class $\mathbf{a} \neq \mathbf{0}$ must satisfy $\mathbf{a}^2 \leq -1$. However, Proposition 2.2 now tells us we can now find a smooth 1-parameter family of metrics $g_t, t \in (-\varepsilon, \varepsilon)$, such that $g_0 = g$, and such that $\mathbf{b} + t\mathbf{a} \in \mathcal{H}_{g_t}^+$ for all t. Since we have assumed that g has $s > 0$, the same is necessarily true of all the metrics g_t for sufficiently small t, and we thus certainly have a contradiction, because the argument of the previous paragraph would now tell us that $\mathbf{a} \cdot (\mathbf{b} + t\mathbf{a}) = t\mathbf{a}^2$ would have to vanish for all small values of t. It follows that M cannot admit any metrics of positive scalar curvature.

Finally, let us suppose instead that g is a metric on M with $s \geq 0$. Since M is now known not to admit metrics of positive scalar curvature, g must then have $s \equiv 0$, since otherwise [2, 6] we would be able to produce a metric of strictly positive scalar curvature by conformally rescaling it. We may now proceed much as in the previous case. Once again, $s \equiv 0$ implies that $\mathbf{a} \in \mathcal{H}_g^-$. Again, choose some $\mathbf{b} \in \mathcal{H}_g^+$ with $\mathbf{b}^2 = 1$, and observe that, once again, there exists a family of metrics g_t, $t \in (-\varepsilon, \varepsilon)$ with $g_0 = g$ for which $\mathbf{b} + t\mathbf{a} \in \mathcal{H}_{g_t}^+$. But this time, we invoke a theorem of Koiso [6, 19] on the Yamabe problem with parameters, and thereby construct a smooth family of constant-scalar-curvature, unit-volume metrics $\tilde{g}_t$ by conformally rescaling each g_t. The conformal invariance of (3) then tells us that we still have $\mathbf{b} + t\mathbf{a} \in \mathcal{H}_{\tilde{g}_t}^+$. Since the family $\tilde{g}_t$ is smooth, the value $s_{\tilde{g}_t}$ of its scalar curvature is therefore a smooth function of t. But since M does admit metrics of positive scalar curvature, and since $s_{\tilde{g}_0} = 0$, this smooth function must have a maximum at $t = 0$. Hence there is a positive constant C such that

$$0 \geq s_{\tilde{g}_t} \geq -Ct^2$$

for all sufficiently small t, and we therefore have

$$C^2 t^4 \geq s_{\tilde{g}_t}^2 = \int_M s_{\tilde{g}_t}^2 \, d\mu_{\tilde{g}_t}$$

for t in the same range. However, Proposition 3.2 and the Cauchy–Schwarz inequality tell us that

$$\int_M s_{\tilde{g}_t}^2 \, d\mu_{\tilde{g}_t} \geq 32\pi^2 (\mathbf{a}_{\tilde{g}_t}^+)^2 \geq 32\pi^2 \frac{[\mathbf{a} \cdot (\mathbf{b} + t\mathbf{a})]^2}{(\mathbf{b} + t\mathbf{a})^2} = 32\pi^2 \frac{t^2 |\mathbf{a}^2|^2}{1 - t^2 |\mathbf{a}^2|} \geq 32\pi^2 t^2$$

so we conclude that $(\text{const})t^4 \geq t^2$ for all small t, which is certainly a contradiction. Thus no metric with $s \geq 0$ can exist, and we are done. $\qquad\square$

Definition 3.4. *For any smooth compact oriented 4-manifold M with $b_+ \geq 2$, we set*

$$\mathfrak{C}(M) = \left\{ \text{monopole classes } \mathbf{a} \in H^2(M, \mathbb{R}) \right\}.$$

We will often abbreviate $\mathfrak{C}(M)$ as $\mathfrak{C}$ when no confusion seems likely to result.

Lemma 3.5. *For any smooth compact oriented 4-manifold M with $b_+ \geq 2$,*

$$\mathfrak{C}(M) = -\mathfrak{C}(M).$$

That is, $\mathbf{a} \in H^2(M, \mathbb{R})$ is a monopole class iff $-\mathbf{a} \in H^2(M, \mathbb{R})$ is a monopole class, too.

Proof. Let g be any metric on M, and let $\mathbb{V}_\pm$ be the twisted spin bundles of some spin^c structure $\mathfrak{c}$. Then the conjugate vector bundles $\overline{\mathbb{V}}_\pm$ are the twisted spin bundles of a second spin^c structure $\overline{\mathfrak{c}}$, since we have natural isomorphisms

$$\overline{\mathbb{V}}_\pm \cong \mathbb{V}_\pm \otimes L^{-1}, \quad L = \det(\mathbb{V}_\pm)$$

induced by the wedge and inner products. Since we locally have

$$\mathbb{V}_{\pm} = \mathbb{S}_{\pm} \otimes L^{1/2}$$
$$\overline{\mathbb{V}}_{\pm} = \mathbb{S}_{\pm} \otimes L^{-1/2}$$

as bundles with connection, it follows that

$$\overline{D_A \Phi} = D_{\bar{A}} \overline{\Phi}$$

for any $\Phi \in \Gamma(\mathbb{V}_+)$ and any Hermitian connection A on L, where $\bar{A}$ denotes the dual connection on L^{-1} induced by A. Moreover, because, the associated antilinear map

$$\mathbb{S}_+ \to \mathbb{S}_+$$

acts by multiplying by the quaternion j, we also have

$$\sigma(\overline{\Phi}) = -\sigma(\Phi).$$

Since $F_{A^*} = -F_A$, it follows that if (Φ, A) is a solution of (13–14) with respect to $(g, \mathfrak{c})$, then $(\overline{\Phi}, \bar{A})$ is a solution of (13–14) with respect to $(g, \bar{\mathfrak{c}})$. If, for every metric g on M, there is a solution of the Seiberg–Witten equations for the spin^c structure $\mathfrak{c}$, the same is therefore also true of $\bar{\mathfrak{c}}$. Since $c_1(\bar{\mathfrak{c}}) = c_1(\overline{\mathbb{V}}_+) = -c_1(\mathbb{V}_+) = -c_1(\mathfrak{c})$, it follows that the set of monopole classes is invariant under multiplication by -1. $\square$

A particularly important consequence of Proposition 3.2 is the following fundamental fact [17]:

Proposition 3.6. *Let M be any smooth compact oriented 4-manifold with $b_+(M) \geq 2$. Then $\mathfrak{C}(M)$ is a finite set.*

Proof. First, observe that one can always find a basis $\{e_j \mid j = 1, \ldots, b_2\}$ for $H^2(M, \mathbb{R})$, together with a collection of metrics g_j such that $e_j \in \mathcal{H}^+_{g_j}$. Indeed, let $g = g_1 = \cdots = g_{b_+}$ be any metric, let $e_1, \ldots, e_{b_+}$ be any basis for $\mathcal{H}^+_g$, and let $e_{b_+ + 1}, \ldots, e_{b_2}$ then be small perturbations of e_1 by linearly independent elements of $\mathcal{H}^-_g$, while using Proposition 2.2 to find compatible metrics $g_{b_+ + 1}, \ldots, g_{b_2}$. Alternatively, one can simply take the e_j to be any collection of rational classes with $e_j^2 > 0$ that span $H^2(M, \mathbb{R})$, and then cite a remarkable recent construction of Gay and Kirby [12, Theorem 1], which shows that any rational cohomology class with positive self-intersection can be be represented by a closed 2-form that is self-dual with respect to some metric. Given this data, we now introduce a constant for each j by setting

$$\kappa_j = \left(\frac{e_j^2}{32\pi^2} \int_M s_{g_j}^2 \, d\mu_{g_j} \right)^{1/2}.$$

Let $L_j : H^2(M, \mathbb{R}) \to \mathbb{R}$ be the linear functionals $L_j(x) = e_j \cdot x$. Since the intersection form is positive-definite on each $\mathcal{H}^+_{g_j}$, the Cauchy–Schwarz inequality and Proposition 3.2 together imply that any monopole class $\mathbf{a} \in H^2(M, \mathbb{R})$ must satisfy

$$\left|L_j(\mathbf{a})\right| = \left|e_j \cdot \mathbf{a}\right| = \left|e_j \cdot \mathbf{a}_{g_j}^+\right| \leq \sqrt{e_j^2}\sqrt{(\mathbf{a}^+)^2} \leq \kappa_j$$

for each j. Hence $\mathfrak{C} \subset H^2(M, \mathbb{R})$ is contained in the parallelepiped

$$\left\{x \in H^2(M, \mathbb{R}) \mid |L_j(x)| \leq \kappa_j \; \forall j = 1, \ldots, b_2(M)\right\},$$

which is a compact set. But $\mathfrak{C} \subset H^2(M, \mathbb{Z})/\text{torsion}$, and is therefore also discrete. Hence $\mathfrak{C}$ is finite, as claimed. $\qquad\square$

We now introduce a generalization of the Seiberg–Witten equations. Let (M, g) be a smooth oriented Riemannian 4-manifold, let $\mathfrak{c}$ be a spin^c-structure on M, and let $f : M \to \mathbb{R}^+$ be a smooth positive function. Then we will say that (Φ, A) solves the *rescaled Seiberg–Witten equations* if

$$D_A\Phi = 0 \tag{17}$$
$$-iF_A^+ = f\sigma(\Phi) \tag{18}$$

Lemma 3.7. *Let M be a smooth compact 4-manifold with $b_+ \geq 2$, and let $\mathbf{a} \in H^2(M, \mathbb{R})$ be a monopole class. Then, for any smooth metric g and any smooth positive function f, there is a solution of the rescaled Seiberg–Witten equations (17–18) for some spin^c structure on M with $c_1^{\mathbb{R}}(L) = \mathbf{a}$.*

Proof. Consider the conformally related metric $\hat{g} = f^{-2}g$. Because $\mathbf{a}$ is a monopole class, there must then be a solution $(\hat{\Phi}, A)$ of the Seiberg–Witten equations wtih respect to $\hat{g}$ and some spin^c structure with $c_1^{\mathbb{R}}(L) = \mathbf{a}$. However, the Dirac equation (13) is conformally invariant. More precisely, $\hat{\Phi}$ uniquely determines [23, 31] a solution Φ of (13) with respect to g, such that $|\Phi|_g = f^{-3/2}|\hat{\Phi}|_{\hat{g}}$, and such that $\sigma_g(\Phi) = f^{-1}\sigma_{\hat{g}}(\hat{\Phi})$. Hence (Φ, A) satisfies (17–18) with respect to g. $\qquad\square$

Given a solution (Φ, A) of (17–18), substitution into (12) yields the Weitzenböck formula

$$0 = 2\Delta|\Phi|^2 + 4|\nabla_A\Phi|^2 + s|\Phi|^2 + f|\Phi|^4.$$

Multiplying by $|\Phi|^2$ and integrating, we then obtain an inequality

$$0 \geq \int_M \left[4|\Phi|^2|\nabla_A\Phi|^2 + s|\Phi|^4 + f|\Phi|^6\right]d\mu \tag{19}$$

and we will now use this to prove our next result.

Proposition 3.8. *Let (M, g) be a smooth compact oriented Riemannian manifold, and let $\mathbf{a} \in H^2(M, \mathbb{R})$ be a monopole class of M. Then the scalar curvature s and self-dual Weyl curvature W_+ of g satisfy*

$$\int_M (s - \sqrt{6}|W_+|)^2 d\mu_g \geq 72\pi^2(\mathbf{a}^+)^2.$$

If $\mathbf{a}^+ \neq 0$, moreover, equality holds iff there is a symplectic form ω, where $[\omega]$ is a negative multiple of $\mathbf{a}^+$ and $c_1^{\mathbb{R}}(M, \omega) = \mathbf{a}$, such that (M, g, ω) is a saturated almost-Kähler manifold in the sense of Definition 2.3.

Proof. For any smooth function $f > 0$ on M, Lemma 3.7 guarantees that the corresponding rescaled Seiberg–Witten equations (17–18) must have some solution (Φ, A). Set $\psi = 2\sqrt{2}\sigma(\Phi)$, and observe that the definition of σ then implies that

$$|\Phi|^4 = |\psi|^2, \qquad 4|\Phi|^2|\nabla_A\Phi|^2 \geq |\nabla\psi|^2.$$

Thus inequality (19) tells us that

$$0 \geq \int_M \left[|\nabla\psi|^2 + s|\psi|^2 + f|\psi|^3 \right] d\mu.$$

However, inequality (9) also tells us that

$$\int_M |\nabla\psi|^2 d\mu \geq \int_M \left(-2\sqrt{\frac{2}{3}}|W_+| - \frac{s}{3} \right) |\psi|^2 d\mu,$$

and combining these facts yields

$$0 \geq \int_M \left[\left(\frac{2}{3}s - 2\sqrt{\frac{2}{3}}|W_+| \right) |\psi|^2 + f|\psi|^3 \right] d\mu.$$

Set $\varphi = \frac{3}{2}\psi = 3\sqrt{2}\sigma(\Phi)$. We then have

$$0 \geq \int_M \left[\left(s - \sqrt{6}|W_+| \right) |\varphi|^2 + f|\varphi|^3 \right] d\mu.$$

Rewriting this as

$$\int_M \left[-\left(s - \sqrt{6}|W_+| \right) f^{-2/3} \right] \left(f^{2/3}|\varphi|^2 \right) d\mu \geq \int_M f|\varphi|^3 d\mu$$

and applying the Hölder inequality to the left-hand side then yields

$$\left[\int_M \left| s - \sqrt{6}|W_+| \right|^3 f^{-2} d\mu \right]^{1/3} \left[\int_M f|\varphi|^3 d\mu \right]^{2/3} d\mu \geq \int_M f|\varphi|^3 d\mu,$$

which is to say that

$$\int_M \left| s - \sqrt{6}|W_+| \right|^3 f^{-2} d\mu \geq \int_M f|\varphi|^3 d\mu.$$

But the Hölder inequality also tells us that

$$\left(\int_M f^4 d\mu \right)^{1/3} \left(\int_M f|\varphi|^3 d\mu \right)^{2/3} \geq \int_M f^{4/3} \left[f^{2/3}|\varphi|^2 \right] d\mu ,$$

where equality holds only if $|\varphi|$ is a constant multiple of f. Hence

$$\left(\int_M f^4 d\mu\right)^{1/3} \left(\int_M \left|s - \sqrt{6}|W_+|\right|^3 f^{-2} d\mu\right)^{2/3} \geq \int_M f^2 |\varphi|^2 d\mu \, .$$

But as $f\varphi = 3\sqrt{2}f\sigma(\Phi) = 3\sqrt{2}(-iF_A^+)$, we also have

$$\int_M f^2 |\varphi|^2 d\mu = 18 \int_M |F_A^+|^2 d\mu \geq 18(2\pi a^+)^2 = 72\pi^2 (a^+)^2$$

by Lemma 2.1, $iF_A \in 2\pi c_1^{\mathbb{R}}(L) = 2\pi a$. Thus

$$\left(\int_M f^4 d\mu_g\right)^{1/3} \left(\int_M \left|s - \sqrt{6}|W_+|\right|^3 f^{-2} d\mu_g\right)^{2/3} \geq 72\pi^2 (a^+)^2 \qquad (20)$$

for any smooth positive function f on M.

Now choose a sequence of smooth positive functions f_j on M with

$$f_j \searrow \sqrt{\left|s - \sqrt{6}|W_+|\right|}$$

uniformly on M. Because the inequality $f_j^2 \geq \left|s - \sqrt{6}|W_+|\right|$ implies

$$\int_M f_j^4 d\mu \geq \left(\int_M f_j^4 d\mu_g\right)^{1/3} \left(\int_M \left|s - \sqrt{6}|W_+|\right|^3 f_j^{-2} d\mu_g\right)^{2/3},$$

we then have

$$\int_M f_j^4 d\mu \geq 72\pi^2 (a^+)^2$$

by applying (20). But since

$$\int_M \left(s - \sqrt{6}|W_+|\right)^2 d\mu = \lim_{j\to\infty} \int_M f_j^4 d\mu,$$

this shows that

$$\int_M \left(s - \sqrt{6}|W_+|\right)^2 d\mu \geq 72\pi^2 (a^+)^2, \qquad (21)$$

as desired.

Finally, we analyze the equality case. Suppose that g is a metric such that equality holds in (21). Then g must in particular minimize

$$\mathcal{A}(g) = \int (s_g - \sqrt{6}|W_+|_g)^2 d\mu_g$$

in its conformal class. However, if u is any smooth positive function, and if $\hat{g} = u^2 g$, then

$$A(u^2g) = \int (s_g + 6u^{-1}\Delta_g u - \sqrt{6}|W_+|_g)^2 d\mu_g$$

so that, for the 1-parameter family of metrics given by

$$g_t = (1+tF)^2 g$$

one has

$$\frac{d}{dt}A(g_t)\Big|_{t=0} = 12\int [\Delta_g F](s_{\hat{g}} - \sqrt{6}|W_+|_{\hat{g}})d\mu_g.$$

If g minimizes A in its conformal class, we must therefore have

$$\Delta_g\left(s - \sqrt{6}|W_+|\right) = 0$$

in the weak (or distributional) sense. Elliptic regularity [13] therefore tells us that $s - \sqrt{6}|W_+|$ is smooth, and integrating by parts

$$\int \left|\nabla\left(s - \sqrt{6}|W_+|\right)\right|^2 d\mu = \int \left(s - \sqrt{6}|W_+|\right)\left[\Delta\left(s - \sqrt{6}|W_+|\right)\right] d\mu = 0$$

therefore shows that

$$s - \sqrt{6}|W_+| = \text{constant}.$$

Assuming $\mathbf{a}^+ \neq 0$, moreover, Proposition 3.3 tells us this constant must be negative. With this proviso, we can then set

$$f = \sqrt{\left|s - \sqrt{6}|W_+|\right|},$$

and equality in (21) then implies that equality occurs in (20) for this choice of $f > 0$. But then, for this choice of f, we must therefore have equality at every step of the proof of (20). Since this f is constant, it thus follows that $\varphi = 3\sqrt{2}\sigma(\Phi)$ is a closed self-dual 2-form of non-zero constant length. Setting $\omega = \sqrt{2}\varphi/|\varphi|$, it follows that (M, g, ω) is an almost-Kähler manifold. Moreover, since $\psi = \frac{2}{3}\varphi$ belongs to the lowest eigenspace of W_+ at each point, while the two largest eigenvalues of W_+ must be equal at every point, we have

$$|W_+| = \sqrt{\frac{3}{2}}\frac{[-W_+(\omega, \omega)]}{|\omega|^2} = -\frac{1}{2}\sqrt{\frac{3}{2}}W_+(\omega, \omega)$$

so that

$$s + s^* = s + \left[\frac{s}{3} + 2W_+(\omega, \omega)\right] = \frac{4}{3}\left(s - \sqrt{6}|W_+|\right),$$

which we already know to be a negative constant. The almost-Kähler manifold (M, g, ω) is therefore saturated in the sense of Definition 2.3. Moreover, since $\Phi \otimes \Phi$ is a non-zero section of $\Lambda_J^{2,0} \otimes L$, we have $c_1^{\mathbb{R}}(M, \omega) = c_1^{\mathbb{R}}(L) = \mathbf{a}$. Moreover, by construction, ω is a negative multiple of $iF_A^+/2\pi$, which is therefore the harmonic representative of $\mathbf{a}^+$.

Conversely, if (M, g, ω) is an almost-Kähler manifold with $b_+ \geq 2$, then $\mathbf{a} = c_1^{\mathbb{R}}(M, \omega)$ is a monopole class by Taubes' theorem [32], and in the saturated case our formula (10) then shows not only that the harmonic representative of $\mathbf{a}^+$ is given by $iF_B^+/2\pi$, where F_B is the curvature of the Blair connection, but also moreover that equality occurs in (21) for this choice of monopole class. The proposition therefore follows.   □

4 Monopoles and Convex Hulls

In the previous section, we saw that monopole classes lead to nontrivial lower bounds for the L^2-norms of certain curvature expressions. Unfortunately, however, these lower bounds still depend on the image of g under the period map and so are not yet uniform in the metric. We will now remedy this, using some simple tricks from convex geometry.

We begin by establishing a notational convention:

Definition 4.1. *Let $\mathbf{V}$ be a vector space over $\mathbb{R}$, and let $S \subset \mathbf{V}$. Then $\mathbf{Hull}(S) \subset \mathbf{V}$ will denote the* convex hull *of S, meaning the smallest convex subset of $\mathbf{V}$ which contains S.*

Lemma 4.2. *Let M be a smooth compact oriented 4-manifold with $b_+ \geq 2$, and let $\mathfrak{C} = \mathfrak{C}(M) \subset H^2(M, \mathbb{R})$ be its set of non-zero monopole classes. Then $\mathbf{Hull}(\mathfrak{C}) \subset H^2(M, \mathbb{R})$ is compact. Moreover, $\mathbf{Hull}(\mathfrak{C})$ is symmetric, in the sense that $\mathbf{Hull}(\mathfrak{C}) = -\mathbf{Hull}(\mathfrak{C})$.*

Proof. By definition, $\mathbf{Hull}(\mathfrak{C})$ is the smallest convex subset of $H^2(M, \mathbb{R})$ that contains $\mathfrak{C}(M)$. However, since $\mathfrak{C}(M)$ is a finite subset, say $\{\mathbf{a}_1, \ldots, \mathbf{a}_n\}$, we can explicitly express this convex hull as

$$\mathbf{Hull}(\mathfrak{C}) = \left\{ \sum_{j=1}^{n} t_j \mathbf{a}_j \,\middle|\, t_j \in [0, 1], \ \sum_{j=1}^{n} t_j = 1 \right\},$$

since the set on the right is certainly a convex subset containing the $\mathbf{a}_j$, and conversely is necessarily contained in any convex subset containing these points. In particular, this means that $\mathbf{Hull}(\mathfrak{C})$ is the image of the standard $(n-1)$-simplex

$$\triangle^{n-1} = \left\{ (t_1, \cdots, t_n) \in [0, 1]^n \,\middle|\, \sum_j t_j = 1 \right\}$$

under the continuous map

$$(t_1, \cdots, t_n) \longmapsto \sum_{j=1}^{n} t_j \mathbf{a}_j,$$

and, as $\triangle^{n-1}$ is compact, it follows that $\mathbf{Hull}(\mathfrak{C})$ is, too.

On the other hand, Lemma 3.5 tells us that $\mathfrak{C}(M)$ is symmetric. Hence

$$\mathbf{Hull}(\mathfrak{C}) = \mathbf{Hull}(-\mathfrak{C}) = -\mathbf{Hull}(\mathfrak{C})$$

and $\mathbf{Hull}(\mathfrak{C})$ is therefore symmetric, too. $\square$

Let us now consider the self-intersection function

$$Q : H^2(M, \mathbb{R}) \longrightarrow \mathbb{R}$$
$$\mathbf{v} \longmapsto \mathbf{v}^2 ,$$

where $\mathbf{v}^2$ is of course just short-hand for $\mathbf{v} \cdot \mathbf{v} = \langle \mathbf{v} \smile \mathbf{v}, [M] \rangle$. Notice that Q is a polynomial function, and therefore continuous. Since $\mathbf{Hull}(\mathfrak{C})$ is compact by Lemma 4.2, it thus follows that $Q|_{\mathbf{Hull}(\mathfrak{C})}$ necessarily achieves its maximum. We are thus entitled to make the following definition:

Definition 4.3. *Let M be a smooth compact oriented 4-manifold with $b_+ \geq 2$, and let $\mathbf{Hull}(\mathfrak{C}) \subset H^2(M, \mathbb{R})$ denote the convex hull of the set $\mathfrak{C} = \mathfrak{C}(M)$ of monopole classes of M. If $\mathfrak{C} \neq \varnothing$, we define*

$$\beta^2(M) = \max \left\{ \mathbf{v}^2 \mid \mathbf{v} \in \mathbf{Hull}(\mathfrak{C}) \right\}.$$

If, on the other hand, $\mathfrak{C} = \varnothing$, we instead set $\beta^2(M) = 0$.

Proposition 4.4. *For any smooth M^4 with $b_+ \geq 2$, $\beta^2(M) \geq 0$.*

Proof. If $\mathfrak{C} = \varnothing$, we have $\beta^2(M) = 0$ by Definition 4.3. Otherwise, let $\mathbf{a} \in \mathfrak{C}$, and observe that $-\mathbf{a} \in \mathfrak{C}$, too, by Lemma 3.5. Thus $\mathbf{0} = \frac{1}{2}\mathbf{a} + \frac{1}{2}(-\mathbf{a}) \in \mathbf{Hull}(\mathfrak{C})$. Hence

$$\beta^2(M) = \max \left\{ \mathbf{v}^2 \mid \mathbf{v} \in \mathbf{Hull}(\mathfrak{C}) \right\} \geq \mathbf{0}^2 = 0,$$

exactly as claimed. $\square$

Proposition 4.5. *Let M be a smooth compact oriented 4-manifold with $\mathfrak{C}(M) \neq \varnothing$. Then, for any Riemannian metric g on M, there is a monopole class $\mathbf{a} \in \mathfrak{C}(M)$ such that*

$$(\mathbf{a}^+)^2 \geq \beta^2(M).$$

Proof. Let $\mathbf{v} \in \mathbf{Hull}(\mathfrak{C})$ be a maximum point of Q, so that $\mathbf{v}^2 = \beta^2(M)$ by Definition 4.3. Let $\Pi : H^2(M, \mathbb{R}) \to \mathcal{H}_g^+$ denote the orthogonal projection map. Since Π is a linear map, we automatically have $\mathbf{Hull}(\Pi(\mathfrak{C})) = \Pi(\mathbf{Hull}(\mathfrak{C}))$. However, since the intersection form is positive definite on $\mathcal{H}_g^+$, $Q|_{\mathcal{H}_g^+}$ has positive-definite Hessian, and the maximum of Q on a line segment $\overline{\mathbf{pq}} \subset \mathcal{H}_g^+$ can therefore never occur at an interior point. The maximum points of $Q|_{\Pi(\mathbf{Hull}(\mathfrak{C}))}$ must therefore all belong to $\Pi(\mathfrak{C})$. In particular, there must be a monopole class $\mathbf{a} \in \mathfrak{C}$ such that

$$(\mathbf{a}^+)^2 = Q(\Pi(\mathbf{a})) \geq Q(\Pi(\mathbf{v})) = (\mathbf{v}^+)^2.$$

On the other hand,

$$\mathbf{v}^2 = (\mathbf{v}^+)^2 - |(\mathbf{v}^-)^2|,$$

so we therefore have

$$(\mathbf{a}^+)^2 \geq (\mathbf{v}^+)^2 \geq \mathbf{v}^2 = \beta^2(M),$$

and the monopole class $\mathbf{a}$ therefore fulfills our desideratum.

The first part of Theorem A now follows immediately:

Theorem 4.6. *Let M be a smooth compact oriented 4-manifold with $b_+ \geq 2$. Then any metric g on M satisfies curvature estimates* (1) *and* (2):

$$\int_M s^2 d\mu \geq 32\pi^2 \beta^2(M)$$

$$\int_M (s - \sqrt{6}|W_+|)^2 d\mu \geq 72\pi^2 \beta^2(M)$$

Proof. For any metric g on M, Proposition 4.5 tells us that there is a monopole class $\mathbf{a}$ such that $(\mathbf{a}^+)^2 \geq \beta^2(M)$. Proposition 3.2 therefore tells us that

$$\int_M s^2 d\mu \geq 32\pi^2 (\mathbf{a}^+)^2 \geq 32\pi^2 \beta^2(M),$$

while Proposition 3.8 tells us that

$$\int_M (s - \sqrt{6}|W_+|)^2 d\mu \geq 72\pi^2 (\mathbf{a}^+)^2 \geq 72\pi^2 \beta^2(M),$$

and the theorem therefore follows. $\square$

To prove Theorem A, we therefore merely need to understand the equality cases of the curvature estimates (1) and (2). To do this, we will first need the following simple observation:

Lemma 4.7. *Suppose that (M, g) is a Riemannian manifold with $b_+ \geq 2$, and that M carries a non-zero monopole class. If equality occurs in either* (1) *or* (2)*, then $\beta^2(M) > 0$.*

Proof. If equality were to hold in (1) or (2), and if we also had $\beta^2(M) = 0$, the metric in question would necessarily have $s \geq 0$. But Proposition 3.3 says that no such metric can exist in the presence of a non-zero monopole class. The claim thus follows by contradiction. $\square$

We will also need the following basic fact:

Lemma 4.8. *If M is a a smooth compact oriented 4-manifold with $b_+ > 1$, and if g is a Kähler–Einstein metric on M with negative scalar curvature, then equality is achieved in* (1) *by g.*

Proof. For any compact Kähler surface (M, J) with $b_+ > 1$, the classical Seiberg–Witten invariant is well-defined and non-zero [33] for the spin^c structure determined by J, and $\mathbf{a} = c_1^{\mathbb{R}}(M, J)$ is therefore a monopole class. Hence $c_1^{\mathbb{R}}(M, J) \in \mathfrak{C} \subset \mathbf{Hull}(\mathfrak{C})$, and

$$\beta^2(M) = \max\{\mathbf{v}^2 \mid \mathbf{v} \in \mathbf{Hull}(\mathfrak{C})\} \geq c_1^2(M).$$

On the other hand, the Ricci form $\rho = r(J\cdot, \cdot)$ represents $2\pi c_1^{\mathbb{R}}(M, J)$ in de Rham cohomology, and just equals $s\omega/4$ in the Kähler–Einstein case. Thus, since the volume form of a Kähler surface is given by $\omega^2/2$, we have

$$\int_M s^2 d\mu = \int \frac{(s\omega)^2}{2} = 8 \int \rho \wedge \rho = 32\pi^2 c_1^2(M) .$$

Proposition 3.2 therefore tells us that

$$32\pi^2 c_1^2(M) = \int_M s^2 d\mu \geq 32\pi^2 \beta^2(M) \geq 32\pi^2 c_1^2(M),$$

and equality must thus hold at every step. Hence $\beta^2(M) = c_1^2(M)$, and equality is achieved in (1) by g, as claimed. □

Lemma 4.9. *Let M be a compact oriented 4-manifold with $b_+ \geq 2$ that carries a non-zero monopole class. Then whenever equality holds in (1) for a metric g on M, equality holds in (2), too.*

Proof. If equality holds in (1), we have

$$32\pi^2 \beta^2(M) = \int_M s^2 d\mu ,$$

so by Propositions 3.2 and 4.5, there is a monopole class $\mathbf{a}$ such that

$$32\pi^2 \beta^2(M) = \int_M s^2 d\mu \geq 32\pi^2 (\mathbf{a}^+)^2 \geq 32\pi^2 \beta^2(M) > 0,$$

and equality must therefore hold throughout. But Proposition 3.2 then asserts that there exists a complex structure J such that (M, g, J) is a Kähler manifold of constant negative scalar curvature.

Now any Kähler metric on a complex surface automatically satisfies $|W_+|^2 = s^2/24$, so that $s - \sqrt{6}|W_+| = \frac{3}{2}s$ wherever $s \leq 0$. Our negative-scalar-curvature Kähler metric g thus satisfies

$$\int_M (s - \sqrt{6}|W_+|)^2 d\mu = \left(\frac{3}{2}\right)^2 \int_M s^2 d\mu = 72\pi^2 \beta^2(M),$$

and therefore also achieves equality in (2), as claimed. □

We now analyze the boundary case of (2).

Theorem 4.10. *Let M be a compact oriented 4-manifold with $b_+ \geq 2$ which carries a non-zero monopole class, and suppose that g is a metric on M such that equality holds in (2):*

$$\int_M (s - \sqrt{6}|W_+|)^2 d\mu = 72\pi^2 \beta^2(M).$$

Then g is Kähler–Einstein, with negative Einstein constant.

Proof. Let $\mathbf{v} \in \mathbf{Hull}(\mathfrak{C})$ be a point where $\mathbf{v}^2 = \mathbf{v} \cdot \mathbf{v}$ achieves its maximum value, namely $\beta^2(M)$. Let $\mathbf{a}_1, \ldots, \mathbf{a}_n \in \mathfrak{C}$ be a list of all the monopole classes, and express $\mathbf{v} \in \mathbf{Hull}(\mathfrak{C})$ as

$$\mathbf{v} = \sum_{j=1}^{n} t_j \mathbf{a}_j$$

where the coefficients $t_j \in [0,1]$ satisfy $\sum_j t_j = 1$; and after permuting the $\mathbf{a}_j$ as necessary, we may henceforth assume that $t_j > 0$ iff $j \leq m$, where m is some integer, $1 \leq m \leq n$. By Propositions 3.8 and 4.5,

$$\frac{1}{72\pi^2} \int_M (s - \sqrt{6}|W_+|)^2 d\mu \geq \max \{(\mathbf{a}_j^+)^2 \mid j = 1, \ldots, n\}$$
$$\geq (\mathbf{v}^+)^2 \geq \mathbf{v}^2 = \beta^2(M)$$

and our hypotheses therefore imply that equality holds at every step. In particular, it follows that $\mathbf{v} = \mathbf{v}^+$ and that $\max_j (\mathbf{a}_j^+)^2 = \beta^2(M)$. Since the intersection form is positive definite on $\mathcal{H}_g^+$, the Cauchy–Schwarz inequality therefore tells us that

$$\mathbf{v} \cdot \mathbf{a}_j^+ \leq \sqrt{(\mathbf{a}_j^+)^2} \sqrt{(\mathbf{v})^2} \leq \beta^2(M),$$

for all j, with equality iff $\mathbf{a}_j^+ = \mathbf{v}$. Since

$$\beta^2(M) = \mathbf{v} \cdot \mathbf{v} = \mathbf{v} \cdot \mathbf{v}^+$$
$$= \mathbf{v} \cdot \left(\sum_{j=1}^{m} t_j \mathbf{a}_j^+ \right)$$
$$= \sum_{j=1}^{m} t_j \left(\mathbf{v} \cdot \mathbf{a}_j^+ \right)$$
$$\leq \sum_{j=1}^{m} t_j \, \beta^2(M)$$
$$= \beta^2(M) \left(\sum_{j=1}^{m} t_j \right)$$
$$= \beta^2(M),$$

we must therefore have $\mathbf{a}_j^+ = \mathbf{v}$ for every $j = 1, \ldots, m$.

For each $j = 1, \ldots, m$, we therefore have $(\mathbf{a}_j^+)^2 = \beta^2(M)$. Moreover, $\beta^2(M) > 0$ by Lemma 4.7. Our hypotheses thus imply that

$$\int (s - \sqrt{6}|W_+|)^2 d\mu = 72\pi^2 (\mathbf{a}_j^+)^2 > 0,$$

and Proposition 3.8 therefore tells us that there is a g-compatible symplectic form ω_j such that $[\omega_j]$ is a negative multiple of $\mathbf{a}_j^+ = \mathbf{v}$, and such that $c_1^{\mathbb{R}}(M, \omega_j) = \mathbf{a}_j$ for each $j = 1, \ldots, m$. Since $[\omega_j]^2/2 = \mathrm{Vol}(M, g)$ for each j, it follows that $[\omega_1] = \cdots = [\omega_m] \in H^2(M, \mathbb{R})$. But each ω_j is harmonic with respect to g, and the harmonic representative of any de Rham class is unique. Hence $\omega_1 = \cdots = \omega_m$. But since $c_1^{\mathbb{R}}(M, \omega_j) = \mathbf{a}_j$, this implies that $\mathbf{a}_1 = \cdots = \mathbf{a}_m$. Hence $m = 1$, and

$$\mathbf{v} = \sum_{j=1}^{m} t_j \mathbf{a}_j = \mathbf{a}_1 = c_1^{\mathbb{R}}(M, \omega_1).$$

Let us now simplify our notation by setting $\omega = \omega_1$. Since $-[\omega] \propto \mathbf{v} = c_1(M, \omega)$, the curvature of any connection on the anti-canonical line bundle L of (M, ω) must be cohomologous to a constant negative multiple of ω. However, we saw in (10–11) that the curvature $F_B = F_B^+ + F_B^-$ of the Blair connection on L is given by

$$iF_B^+ = \frac{s + s^*}{8} \omega + W^+(\omega)^\perp$$

$$iF_B^- = \frac{s - s^*}{8} \hat{\omega} + \mathring{\rho}$$

where $W^+(\omega)^\perp$ is the component of $W^+(\omega)$ orthogonal to ω,

$$\mathring{\rho}(\cdot, J \cdot) = \frac{\mathring{r} + J^* \mathring{r}}{2},$$

and where the bounded anti–self-dual 2-form $\hat{\omega} \in \Lambda^-$ is defined only on the open set where $s^* - s \neq 0$, and satisfies $|\hat{\omega}| \equiv \sqrt{2}$. Here, the star-scalar curvature s^* once again means the important quantity

$$s^* = s + |\nabla \omega|^2 = 2W_+(\omega, \omega) + \frac{s}{3}.$$

Since Proposition 3.8 tells us that (M, g, ω) is saturated, $s + s^*$ is constant and $W^+(\omega)^\perp = 0$. Hence F_B^+ is closed, and therefore $*F_B = 2F_B^+ - F_B$ is closed, too. Thus F_B is harmonic. But we also know that F_B is cohomologous to a constant multiple of ω, which is itself a self-dual harmonic form. Hence $F_B^- \equiv 0$, and

$$\mathring{\rho} \equiv \frac{s^* - s}{8} \hat{\omega}.$$

This shows that

$$|\mathring{r}|^2 \geq \frac{(s^* - s)^2}{16}$$

at every point of M, with equality precisely at those points at which the Ricci tensor r is J-invariant.

On the other hand, W_+ has eigenvalues $(-\lambda/2, -\lambda/2, \lambda)$, where

$$\lambda = \frac{1}{2}W_+(\omega, \omega) = \frac{3s^* - s}{12},$$

so

$$|W_+|^2 = \frac{(3s^* - s)^2}{96}.$$

Hence

$$
\begin{aligned}
4\pi^2(2\chi + 3\tau)(M) &= \int_M \left(\frac{s^2}{24} + 2|W_+|^2 - \frac{|\mathring{r}|^2}{2} \right) d\mu \\
&= \int_M \left(\frac{s^2}{24} + \frac{2(3s^* - s)^2}{96} - \frac{|\mathring{r}|^2}{2} \right) d\mu \\
&\leq \int_M \left(\frac{s^2}{24} + \frac{2(3s^* - s)^2}{96} - \frac{(s^* - s)^2}{32} \right) d\mu \\
&= \frac{1}{32} \int_M \left(s^2 - 2ss^* + 5(s^*)^2 \right) d\mu
\end{aligned}
$$

with equality iff $|\mathring{r}|^2 \equiv (s^* - s)^2/16$. On the other hand, since $F_B = F_B^+$,

$$
\begin{aligned}
4\pi^2(2\chi + 3\tau)(M) &= 4\pi^2 c_1^2(M) \\
&= \int_M \left(\frac{s+s^*}{8}\omega \right) \wedge \left(\frac{s+s^*}{8}\omega \right) \\
&= \frac{1}{32} \int_M \left(s^2 + 2ss^* + (s^*)^2 \right) d\mu .
\end{aligned}
$$

We therefore have

$$\int_M \left(s^2 - 2ss^* + 5(s^*)^2 \right) d\mu \geq \int_M \left(s^2 + 2ss^* + (s^*)^2 \right) d\mu,$$

which we can rewrite as

$$\int_M 4s^*(s^* - s)d\mu \geq 0; \qquad (22)$$

moreover, equality can hold only if $|\mathring{r}|^2 \equiv (s^* - s)^2/16$. However, since (M, g, ω) is saturated, $s^* + s$ is a negative constant, and $W_+(\omega, \omega) \leq 0$; hence $s^* \leq s/3$, and $s^* \leq (s+s^*)/4$ is therefore negative everywhere. On the other hand, $s^* - s = |\nabla\omega|^2 \geq 0$ on any almost-Kähler manifold. Hence

$$s^*(s^* - s) \leq 0$$

everywhere on M, with equality only at points where $s = s^*$. The inequality (22) therefore implies that

$$|\nabla \omega|^2 = s - s^* \equiv 0.$$

Hence (M, g, ω) is Kähler. But equality in (22) only holds if $|\mathring{r}|^2 \equiv (s^* - s)^2/16$, so we moreover must have $\mathring{r} \equiv 0$, and we therefore conclude that (M, g) is Kähler–Einstein, as promised. $\square$

Our main result now follows easily:

Proof of Theorem A. Theorem 4.6 shows that (1) and (2) hold for any metric on any 4-manifold with $b_+ \geq 2$. On the other hand, assuming there is at least one non-zero monopole class, Theorem 4.10 shows that any metric for which equality holds in (2) must be Kähler–Einstein. Lemma 4.9 thus implies that any metric for which equality holds in (1) must be Kähler–Einstein, too. Finally, Lemmas 4.8 and 4.9 show that equality actually does hold in (1) and (2) when the metric is Kähler–Einstein. $\square$

Of course, the method used here to treat the boundary case of (1) has a Rube Goldberg feel to it, as it proceeds by reducing an easy problem to a harder one. However, it is not difficult to winnow a simple, direct treatment of this case out of the above discussion. Details are left to the interested reader.

5 Theory and Practice

One apparent weakness of our definition of $\beta^2(M)$ is that there is no obvious way of exactly determining the entire set $\mathfrak{C}(M)$ of all monopole classes of a given 4-manifold M. However, we *do* have various criteria that serve to show that certain classes actually belong to $\mathfrak{C}(M)$. Thus, if $\mathfrak{S} \subset \mathfrak{C}(M)$ is some collection of known monopole classes, we then have

$$\beta^2(M) \geq \max\{\mathbf{v}^2 \mid \mathbf{v} \in \mathbf{Hull}(\mathfrak{S})\}.$$

It is thus relatively easy to find lower bounds for β^2, even without knowing $\mathfrak{C}(M)$ exactly.

At the same time, our curvature estimates (1) and (2) provide upper bounds for $\beta^2(M)$ for each metric g on M. By taking an infimum of such upper bounds for a carefully chosen sequence of metrics g_j on M, one can, in practice, often determine $\beta^2(M)$ by showing that it is simultaneously no less than and no greater than some target value.

Example. Let X be a minimal complex surface of general type with $b_+ > 1$, and let $M = X \# k\overline{\mathbb{CP}}_2$ be its blow-up at k points. Then

$$\beta^2(M) = c_1^2(X).$$

Indeed, if $E_1,\ldots,E_k$ are generators for the various copies of $H^2(\overline{\mathbb{CP}}_2,\mathbb{Z})$, then $\pm c_1(X)\pm E\pm\cdots\pm E_k$ are [11,26] the first Chern classes of various complex structures of Kähler type on M, and so are monopole classes [33]. Hence $c_1(X)\in$ **Hull**$(\mathfrak{C}(M))$, and hence $\beta^2(M)\geq c_1^2(X)$. However, by approximating the Kähler–Einstein orbifold metric on the pluricanonical model for X, one can construct [26] sequences of metrics g_j on M with $\int s^2 d\mu \searrow 32\pi^2 c_1^2(X)$. Thus (1) implies that we also have $c_1^2(X)\geq\beta^2(M)$, and the claim follows. $\diamond$

Example. Let X, Y, and Z be simply connected, minimal complex surfaces of general type with $h^{2,0}$ odd. Let $M = X\#Y\#Z\#k\overline{\mathbb{CP}}_2$. Then

$$\beta^2(M) = c_1^2(X) + c_1^2(Y) + c_1^2(Z).$$

Indeed, using the Bauer–Furuta invariant [4,5], one can show that

$$\pm c_1(X)\pm c_1(Y)\pm c_1(Z)\pm E_1\pm\cdots\pm E_k\in\mathfrak{C}(M).$$

Hence $\mathbf{v} = c_1(X)+c_1(Y)+c_1(Z)\in$ **Hull**$(\mathfrak{C}(M))$, and

$$\beta^2(M)\geq[c_1(X)+c_1(Y)+c_1(Z)]^2 = c_1^2(X)+c_1^2(Y)+c_1^2(Z).$$

On the other hand, there exist [17] sequences of metrics g_j on M with $\int s^2 d\mu \searrow 32\pi^2[c_1^2(X)+c_1^2(Y)+c_1^2(Z)]$, so (1) therefore shows that we also have $c_1^2(X)+c_1^2(Y)+c_1^2(Z)\geq\beta^2(M)$. The claim therefore follows.

Similar techniques can also be used for connected sums involving two or four surfaces of general type. $\diamond$

Example. Let N be any oriented 3-manifold, and let $M = N\times S^1$. Then $\beta^2(M) = 0$, because one has $\int s^2 d\mu \searrow 0$ for product metrics on M with shorter and shorter S^1 factors. However, such manifolds often carry [20] many monopole classes, although these all belong to the isotropic subspace $H^2(N)\hookrightarrow H^2(N\times S^1)$. $\diamond$

By the arguments detailed in [24,27], the estimates (1) and (2) have the following interesting consequences:

Theorem 5.1. *Let M be a smooth compact oriented 4-manifold with $b_+(M)\geq 2$. If M admits an Einstein metric g, then*

$$(2\chi - 3\tau)(M)\geq\frac{1}{3}\beta^2(M).$$

Moreover, if M carries a non-zero monopole class, equality occurs if and only if (M,g) is a compact quotient $\mathbb{CH}_2/\Gamma$ of the complex hyperbolic plane, equipped with a constant multiple of its standard Kähler–Einstein metric.

Theorem 5.2. *Let M be a smooth compact oriented 4-manifold with $b_+(M) \geq 2$. If M admits an Einstein metric g, then*

$$(2\chi + 3\tau)(M) \geq \frac{2}{3}\beta^2(M),$$

with equality only if both sides vanish, in which case g must be a hyper-Kähler metric, and M must be diffeomorphic to K3 or T^4.

Theorem 5.3. *Let M be a smooth compact oriented 4-manifold with $b_+(M) \geq 2$. Then any metric g on M satisfies*

$$\int_M |r|^2 d\mu \geq 8\pi^2 \left[2\beta^2 - (2\chi + 3\tau)\right](M),$$

with equality iff g is Kähler–Einstein.

Now Proposition 4.4 entitles us to introduce the following definition:

Definition 5.4. *If M is any smooth compact oriented 4-manifold with $b_+(M) \geq 2$, we set $\beta(M) := \sqrt{\beta^2(M)}$.*

This invariant provides a natural yardstick with which to measure the Yamabe invariants of 4-manifolds:

Theorem 5.5. *Let M be a smooth compact oriented 4-manifold with $b_+(M) \geq 2$. If M carries at least one non-zero monopole class, then the Yamabe invariant of M satisfies*

$$\mathcal{Y}(M) \leq -4\sqrt{2}\pi\beta(M).$$

We remark in passing that if M does not admit a metric of positive scalar curvature, its Yamabe invariant $\mathcal{Y}(M)$ is just the supremum of the scalar curvatures of unit-volume constant-scalar-curvature metrics on M. This result is therefore an immediate consequence of (1). More intriguingly, though, Theorem 5.5 is actually sharp; equality actually holds [17, 26] for large classes of 4-manifolds, including those discussed previously.

Now, although we have seen that considering the convex hull of the set of monopole classes leads to an elegant invariant $\beta^2(M)$ that seems remarkably well adapted to the study of the curvature of 4-manifolds, it is still unclear whether this approach is optimal in all circumstances. Indeed, the basic forms of our estimates, seen in Propositions 3.2 and 3.8, involve the numbers $(\mathbf{a}^+)^2$ for the various monopole classes, and one can therefore [28] define an invariant that simply tries to make optimal use of this information. Indeed, consider the open Grassmannian $\mathbf{Gr} = Gr_{b_+}^+[H^2(M, \mathbb{R})]$ of all maximal linear subspaces $\mathbf{H}$ of the second cohomology on which the restriction of the intersection pairing is positive definite. Each element $\mathbf{H} \in \mathbf{Gr}$ then determines an orthogonal decomposition

$$H^2(M, \mathbb{R}) = \mathbf{H} \oplus \mathbf{H}^\perp$$

with respect to the intersection form. Given a monopole class $\mathbf{a} \in \mathfrak{C}$ and a positive subspace $\mathbf{H} \in \mathbf{Gr}$, we may then define $\mathbf{a}^+$ to be the orthogonal projection of $\mathbf{a}$ into $\mathbf{H}$. Using this, we can now define [28] yet another oriented-diffeomorphism invariant.

Definition 5.6. *Let M be a smooth compact oriented manifold with $b_+ \geq 2$. If $\mathfrak{C} = \varnothing$, set $\alpha^2(M) = 0$. Otherwise, we set*

$$\alpha^2(M) = \inf_{\mathbf{H} \in \mathbf{Gr}} \left[\max_{a \in \mathfrak{C}} \; (\mathbf{a}^+)^2 \right].$$

Propositions (3.2) and (3.8) then easily imply that (1) and (2) still hold when $\beta^2(M)$ is replaced by $\alpha^2(M)$. Moreover, the proof of Proposition 4.5 shows that one always has

$$\alpha^2(M) \geq \beta^2(M).$$

On the other hand, we have also seen that (1) and (2) are sharp for large classes of manifolds, such as those discussed previously. Thus $\alpha^2 = \beta^2$ in all these cases. It is therefore only natural for us to ask whether this is a general phenomenon. In this direction, however, we can only give some partial results. We begin with the following:

Lemma 5.7. *Let M be a smooth oriented 4-manifold with $b_+ \geq 2$. Then*

$$\alpha^2(M) = 0 \Longleftrightarrow \beta^2(M) = 0.$$

Proof. The $\Longrightarrow$ direction is obvious, as $\alpha^2 \geq \beta^2 \geq 0$. Conversely, if $\beta^2 = 0$, the intersection form must be negative-semidefinite on **span** $(\mathfrak{C})$. Write this subspace as $\mathbf{N} \oplus \mathbf{I}$, where the intersection form is negative-definite on $\mathbf{N}$ and vanishes on $\mathbf{I}$. We can then choose a sequence $\mathbf{H}_j \in \mathbf{Gr}$ that are all orthogonal to $\mathbf{N}$ and that decompose orthogonally as $\mathbf{P} \oplus \mathbf{J}_j$, where $\mathbf{P}$ is orthogonal to $\mathbf{I}$ and $\mathbf{J}_j \to \mathbf{I}$. Then each monopole class satisfies $(\mathbf{a}^+)^2 \to 0$ for this sequence. It thus follows that $\alpha^2 = 0$, as claimed. $\square$

Next, we point out the following:

Proposition 5.8. *Let M be a smooth oriented 4-manifold with $b_+ \geq 2$. Suppose, moreover, that there is a linear subspace $\mathbf{L} \subset H^2(M, \mathbb{R})$ on which the intersection form is of Lorentzian $(+- \cdots -)$ type, with*

$$\mathfrak{C}(M) \subset \mathbf{L} \subset H^2(M, \mathbb{R}).$$

Then $\alpha^2(M) = \beta^2(M)$.

Proof. By Lemma 5.7, we may assume that $\beta^2(M) > 0$. Let $\mathbf{v} \in \mathbf{Hull}(\mathfrak{C}) \subset \mathbf{L}$ be an element with $\mathbf{v}^2 = \beta^2(M) > 0$. Now, because $(1 - t)\mathbf{v} + t\mathbf{a} \in \mathbf{Hull}(\mathfrak{C})$ for any $\mathbf{a} \in \mathfrak{C}$ and any $t \in [0, 1]$, we therefore have

$$\mathbf{v}^2 \geq [(1 - t)\mathbf{v} + t\mathbf{a}]^2 = \mathbf{v}^2 + 2t(\mathbf{v} \cdot \mathbf{a} - \mathbf{v}^2) + O(t^2)$$

for all small positive t, and it therefore follows that

$$\mathbf{v} \cdot \mathbf{a} \leq \mathbf{v}^2$$

for all monopole classes $\mathbf{a}$. Since $\mathfrak{C}(M)$ is invariant under multiplication by -1, it moreover follows that

$$|\mathbf{v} \cdot \mathbf{a}| \leq \mathbf{v}^2 \ \forall \mathbf{a} \in \mathfrak{C}(M).$$

Now let $\mathbf{P} \subset \mathbf{L}^{\perp}$ be a maximal positive subspace, and set $\mathbf{H} = \mathbf{P} \oplus \mathrm{span} \ (\mathbf{v})$. Then for this choice of $\mathbf{H} \in \mathbf{Gr}$ we, have

$$\mathbf{a}^+ = \frac{\mathbf{v} \cdot \mathbf{a}}{\mathbf{v}^2} \mathbf{v}$$

and hence

$$(\mathbf{a}^+)^2 = \frac{(\mathbf{v} \cdot \mathbf{a})^2}{\mathbf{v}^2} \leq \mathbf{v}^2 = \beta^2(M)$$

for all $\mathbf{a} \in \mathfrak{C}$. Hence

$$\alpha^2(M) = \inf_{\mathbf{H} \in \mathbf{Gr}} \left[\max_{a \in \mathfrak{C}} \ (\mathbf{a}^+)^2 \right] \leq \beta^2(M).$$

But we also know that $\beta^2 \leq \alpha^2$, so it follows that $\alpha^2 = \beta^2$, as claimed. $\square$

Example. If (M, J) is a compact complex surface of Kähler type with $b_+ > 1$, we may take $\mathbf{L} = H^{1,1}(M, \mathbb{R}) \subset H^2(M, \mathbb{R})$. Since an argument due to Witten [33] shows that solutions of the Seiberg–Witten equations can exist with respect to a Kähler metric only when $c_1(L)$ is a $(1,1)$-class, it follows that any monopole class must belong to $\mathbf{L}$. This provides one explanation of why $\alpha^2 = \beta^2$ for complex algebraic surfaces. $\Diamond$

In light of Proposition 5.8, the reader may wonder why we have systematically excluded the $b_+ = 1$ case throughout our discussion, especially as the entire theory does nominally work in this setting and even has some nontrivial consequences. Such an approach, however, would simply ignore most of what has been learned from the chamber-dependent theory [11, 26] of Seiberg–Witten invariants on manifolds with $b_+ = 1$. To avoid this pitfall, it seems wiser to instead introduce the following set of different definitions to treat this case. Let M be a smooth compact oriented 4-manifold with $b_+(M) = 1$. The set of cohomology classes $\mathbf{b} \in H^2(M, \mathbb{R})$ with $\mathbf{b}^2 > 0$ thus consists of two connected components. We arbitrarily label members of one component *future-pointing* and those in the other *past-pointing*. Let $\mathbb{K} \subset H^2(M, \mathbb{R})$ be the closure of the set of past-pointing vectors. We now declare $\mathbf{a} \in H^2(M, \mathbb{R})$ to be a *retroactive class* if there is at least one metric g such that

$$(\mathbf{a}^+)_g \in \mathbb{K} \tag{23}$$

and if, for some spin^c structure with $c_1 \equiv \mathbf{a}$ mod torsion, there exists a solution of the Seiberg–Witten equations *for every metric g satisfying* (23). Let $\mathfrak{D} \subset H^2(M, \mathbb{R})$ denote the set of retroactive classes, and set

$$\mathbb{A} = \mathbb{K} \cap \overline{\mathbf{Hull}(\mathfrak{D})}.$$

For any metric g, a variant of Proposition 3.2 shows that $\mathfrak{D}$ is contained in a half-space of the form $\{\mathbf{v} \mid \mathbf{b} \cdot \mathbf{v} \geq \kappa\}$, where $\mathbf{b} \in \mathcal{H}_g^+$ is future-pointing and κ is a nonpositive constant. In particular, the closed convex set $\mathbb{A}$ is bounded, and hence compact.

Definition 5.9. *Let M be a smooth compact oriented 4-manifold with* $b_+(M) = 1$, *and let* $\mathbb{A} = \mathbb{A}(M) \subset H^2(M, \mathbb{R})$ *be the compact convex set defined above. If* $\mathbb{A} \neq \varnothing$, *we set*

$$\beta^2(M) = \max\left\{\mathbf{v}^2 \mid \mathbf{v} \in \mathbb{A}\right\}.$$

If, on the other hand, $\mathbb{A} = \varnothing$, *we set* $\beta^2(M) = 0$.

Given any metric g, a minor modification of the proof of Proposition 3.6 shows that the discrete set $\{\mathbf{a} \in \mathfrak{D} \mid (\mathbf{a}^+)_g \in \mathbb{K}\}$ is covered by a finite union of parallelepipeds and is therefore finite. Consequently, by an argument parallel to the proof of Proposition 4.5, there is, for any metric g and any $\mathbf{v} \in \mathbb{A}$, a retroactive class $\mathbf{a}$ satisfying (23) for which $(\mathbf{a}^+)_g^2 \geq \mathbf{v}^2$. But to say that $\mathbf{a}$ is a retroactive class means that (23) implies the existence of a solution of the Seiberg–Witten equations (13–14) with $c_1 \equiv \mathbf{a}$; and since (23) is conformally invariant, it also implies the existence of a solution of the rescaled Seiberg–Witten equations (17–18) for any f. The proofs of Propositions 3.2 and 3.8 thus apply. Choosing $\mathbf{v} \in \mathbb{A}$ so that $\mathbf{v}^2 = \beta^2(M)$, we thus conclude that every metric satisfies

$$\int_M s^2 d\mu_g \geq 32\pi^2 (\mathbf{a}^+)_g^2 \geq 32\pi^2 \mathbf{v}^2 = 32\pi^2 \beta^2(M)$$

and

$$\int_M (s - \sqrt{6}|W_+|)^2 d\mu_g \geq 72\pi^2 (\mathbf{a}^+)_g^2 \geq 72\pi^2 \mathbf{v}^2 = 72\pi^2 \beta^2(M).$$

In other words, we have successfully extended our definition of $\beta^2(M)$ in a manner that guarantees that the curvature estimates (1) and (2) now also hold for smooth compact 4-manifolds with $b_+ = 1$. If $\beta^2(M) > 0$, moreover, one can still show that equality in (1) or (2) occurs iff g is Kähler–Einstein, with negative scalar curvature.

Example. Let X be a minimal complex surface of general type with $b_+ = 1$, and let $M = X \# k\overline{\mathbb{CP}}_2$ be obtained from X by blowing up k points. Let $E_1, \ldots, E_k$ denote the Poincaré duals of the exceptional divisors introduced by the blow-up procedure. If we conventionally label $-c_1(X)$ as a future-pointing vector, then the set of retroactive classes satisfies

$$\mathfrak{D} \supset \{c_1(X) \pm E_1 \pm \cdots \pm E_k\}$$

because each of these classes has non-zero Seiberg–Witten invariant in the appropriate chamber [11, 26]. Hence

$$c_1(X) \in \mathbf{Hull}(\mathfrak{D}) \cap \mathbb{K} \subset \mathbb{A}$$

and we therefore have

$$c_1^2(X) \leq \max_{\mathbf{v} \in \mathbb{A}} \mathbf{v}^2 = \beta^2(M).$$

On the other hand, there is [26] a sequence of metrics on M, obtained by approximating the Kähler–Einstein orbifold metric on the pluricanonical model, with $\int s^2 d\mu \searrow 32\pi^2 c_1^2(X)$. Thus, inequality (1) shows that $c_1^2(X) \geq \beta^2(M)$, and it follows that

$$\beta^2(M) = c_1^2(X).$$

It follows [27] that the estimates (1) and (2) are now sharp for all complex surfaces of general type, whatever the value of b_+. $\diamond$

We leave it as an exercise for the interested reader to define a $b_+ = 1$ analog of α^2 in terms of retroactive classes. If this is done in the most natural manner, the proof of Proposition 5.8 will then show, *mutatis mutandis*, that this invariant necessarily coincides with our $b_+ = 1$ version of β^2.

We now return to the setting of $b_+ \geq 2$, and conclude our comparison of α^2 and β^2 with the following simple result:

Proposition 5.10. *Let M be a smooth oriented 4-manifold with $b_+ \geq 2$, and suppose that there is a collection of mutually orthogonal linear subspaces $\mathbf{L}_j \subset H^2(M,\mathbb{R}), j = 1,\dots,\ell$, on each of which the intersection form is of Lorentzian $(+-\cdots-)$ type. Moreover, suppose that*

$$\mathfrak{C}(M) = \mathfrak{C}_1 \times \cdots \times \mathfrak{C}_\ell \subset \mathbf{L}_1 \oplus \cdots \oplus \mathbf{L}_\ell,$$

for some subsets

$$\mathfrak{C}_j \subset \mathbf{L}_j, \quad j = 1,\dots,\ell.$$

Then $\alpha^2(M) = \beta^2(M)$.

Proof. Fix a maximal positive subspace $\mathbf{P} \subset (\mathbf{L}_1 \oplus \cdots \oplus \mathbf{L}_\ell)^\perp$, and consider choices of $\mathbf{H} \in \mathbf{Gr}$ of the form $\mathbf{H} = \mathbf{P} \oplus \mathbf{span}\{e_1,\dots,e_\ell\}$, where $e_j \in \mathbf{L}_j$ is a non-zero time-like vector. If the intersection form on $\mathbf{span}(\mathfrak{C}_j)$ is negative-definite, moreover choose $e_j \in \mathbf{L}_j$ to be orthogonal to this subspace. If, on the other hand, $\mathbf{span}(\mathfrak{C}_j)$ is Lorentzian, set $e_j = \mathbf{v}_j$, where $\mathbf{v}_j$ maximizes $\mathbf{v}^2$ on $\mathbf{Hull}(\mathfrak{C}_j)$. Finally, if the intersection form is degenerate on $\mathbf{span}(\mathfrak{C}_j)$, choose $\mathbf{v}_j \in \mathbf{Hull}(\mathfrak{C}_j)$ to be a non-zero null vector, and consider a sequence of different possible e_j converging to $\mathbf{v}_j$. In this way, one obtains a sequence of choices of $\mathbf{H}$ for which $\max(\mathbf{a}^+)^2 \to \sum(\mathbf{v}_j)^2 = \beta^2(M)$. Hence $\alpha^2 \leq \beta^2 \leq \alpha^2$, and $\alpha^2(M) = \beta^2(M)$, as claimed. $\square$

This result gives a partial explanation of why $\alpha^2 = \beta^2$ for the connected sums of complex surfaces we have considered, since the set of known monopole classes in this case constitutes a configuration of the described type, where the Lorentzian subspaces in question are given by $H^{1,1}$ of the various summands. Of course, this explanation still remains less than entirely satisfactory, since we cannot be absolutely certain that we currently have a full catalog of the monopole classes of these spaces.

Finally, let us point out that one *cannot* hope to prove that $\alpha^2 = \beta^2$ if $\mathfrak{C}$ is simply replaced with an arbitrary finite, centrally symmetric set in an arbitrary finite-dimensional vector space with indefinite inner product. For example, let us just consider $\mathbb{R}^3$ equipped with the $(++-)$ inner product $dx^2 + dy^2 - dz^2$, and consider the candidate for "$\mathfrak{C}$" given by

$$\left\{ \pm(1,0,1),\ \pm\left(\frac{\sqrt{3}}{2}, -\frac{1}{2}, 1\right),\ \pm\left(-\frac{\sqrt{3}}{2}, -\frac{1}{2}, 1\right) \right\},$$

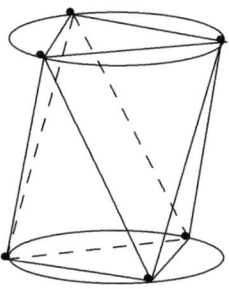

Because the elements of this configuration "$\mathfrak{C}$" are all null vectors, one can use Proposition 5.8 "upside-down" to show that the analog of α^2 must then equal 1. On the other hand, a simple symmetry argument shows that the analog of β^2 equals $\frac{3}{4}$ for this configuration. Of course, this choice of "$\mathfrak{C}$" does not consist of integer points, but one can easily remedy this by rational approximation and rescaling.

The upshot is that whereas, one definitely has $\alpha^2(M) = \beta^2(M)$ for a remarkably large and interesting array of examples, this statement can generally hold only to the degree that the set $\mathfrak{C}$ of monopole classes satisfies some manifestly nontrivial geometric constraints. The precise extent to which these constraints hold or fail remains to be determined. It is hoped that some interested reader will find the challenge of fully fathoming this mystery both stimulating and fruitful.

Acknowledgments It is a pleasure to thank Peter Kronheimer and the anonymous referee for their helpful comments, which have led to a number of improvements in the manuscript. This work was supported in part by NSF grant DMS-0604735.

References

1. N. Aronszajn, A. Krzywicki, and J. Szarski, *A unique continuation theorem for exterior differential forms on Riemannian manifolds*, Ark. Mat., 4 (1962), pp. 417–453.

2. T. Aubin, *Some Nonlinear Problems in Riemannian Geometry*, Springer Monographs in Mathematics, Springer-Verlag, Berlin, 1998.

3. C. Bär, *On nodal sets for Dirac and Laplace operators*, Comm. Math. Phys., 188 (1997), pp. 709–721.

4. S. Bauer, *A stable cohomotopy refinement of Seiberg-Witten invariants. II*, Invent. Math., 155 (2004), pp. 21–40.

5. S. Bauer and M. Furuta, *A stable cohomotopy refinement of Seiberg-Witten invariants. I*, Invent. Math., 155 (2004), pp. 1–19.

6. A. Besse, *Einstein Manifolds*, Springer-Verlag, Berlin, 1987.

7. D. E. Blair, *The "total scalar curvature" as a symplectic invariant and related results*, in Proceedings of the 3rd Congress of Geometry (Thessaloniki, 1991), Thessaloniki, 1992, Aristotle Univ. Thessaloniki, pp. 79–83.

8. J.-P. Bourguignon, *Les variétés de dimension 4 à signature non nulle dont la courbure est harmonique sont d'Einstein*, Invent. Math., 63 (1981), pp. 263–286.

9. S. K. Donaldson, *Connections, cohomology and the intersection forms of 4-manifolds*, J. Differential Geom., 24 (1986), pp. 275–341.

10. T. C. Drăghici, *On some 4-dimensional almost Kähler manifolds*, Kodai Math. J., 18 (1995), pp. 156–168.

11. R. Friedman and J. Morgan, *Algebraic surfaces and Seiberg-Witten invariants*, J. Alg. Geom., 6 (1997), pp. 445–479.

12. D. T. Gay and R. Kirby, *Constructing symplectic forms on 4-manifolds which vanish on circles*, Geom. Topol., 8 (2004), pp. 743–777 (electronic).

13. D. Gilbarg and N. S. Trudinger, *Elliptic Partial Differential Equations of Second Order*, Springer-Verlag, Berlin, second ed., 1983.

14. M. Gromov and H. B. Lawson, *The classification of simply connected manifolds of positive scalar curvature*, Ann. Math., 111 (1980), pp. 423–434.

15. F. Hirzebruch and H. Hopf, *Felder von Flächenelementen in 4-dimensionalen Mannigfaltigkeiten*, Math. Ann., 136 (1958), pp. 156–172.

16. N. Hitchin, *Harmonic spinors*, Adv. Math., 14 (1974), pp. 1–55.

17. M. Ishida and C. LeBrun, *Curvature, connected sums, and Seiberg-Witten theory*, Comm. Anal. Geom., 11 (2003), pp. 809–836.

18. T. P. Killingback and E. G. Rees, $Spin^c$ *structures on manifolds*, Classical Quantum Gravity, 2 (1985), pp. 433–438.

19. N. Koiso, *A decomposition of the space $\mathcal{M}$ of Riemannian metrics on a manifold*, Osaka J. Math., 16 (1979), pp. 423–429.

20. P. B. Kronheimer, *Minimal genus in $S^1 \times M^3$*, Invent. Math., 135 (1999), pp. 45–61.

21. P. B. Kronheimer and T. S. Mrowka, *Recurrence relations and asymptotics for four-manifold invariants*, Bull. Am. Math. Soc. (N.S.), 30 (1994), pp. 215–221.

22. ———, *Scalar curvature and the Thurston norm*, Math. Res. Lett., 4 (1997), pp. 931–937.

23. H. B. Lawson and M. Michelsohn, *Spin Geometry*, Princeton University Press, Princeton, NJ, 1989.

24. C. LeBrun, *Einstein metrics and Mostow rigidity*, Math. Res. Lett., 2 (1995), pp. 1–8.

25. ———, *Polarized 4-manifolds, extremal Kähler metrics, and Seiberg-Witten theory*, Math. Res. Lett., 2 (1995), pp. 653–662.

26. ———, *Four-manifolds without Einstein metrics*, Math. Res. Lett., 3 (1996), pp. 133–147.

27. ———, *Ricci curvature, minimal volumes, and Seiberg-Witten theory*, Inv. Math., 145 (2001), pp. 279–316.

28. ———, *Einstein metrics, four-manifolds, and differential topology*, in Surveys in differential geometry, Vol. VIII (Boston, MA, 2002), Surv. Differ. Geom., VIII, International Press of Boston, Somerville, MA, 2003, pp. 235–255.

29. ———, *Einstein metrics, symplectic minimality, and pseudo-holomorphic curves*, Ann. Global Anal. Geom., 28 (2005), pp. 157–177.

30. A. Lichnerowicz, *Spineurs harmoniques*, C.R. Acad. Sci. Paris, 257 (1963), pp. 7–9.

31. R. Penrose and W. Rindler, *Spinors and space-time. Vol. 1*, Cambridge University Press, Cambridge, 1984. Two-spinor calculus and relativistic fields.

32. C. H. Taubes, *The Seiberg-Witten invariants and symplectic forms*, Math. Res. Lett., 1 (1994), pp. 809–822.

33. E. Witten, *Monopoles and four-manifolds*, Math. Res. Lett., 1 (1994), pp. 809–822.

The 1-Nullity of Sasakian Manifolds

Philippe Rukimbira

Abstract On a closed, $(2n+1)$-dimensional Sasakian manifold, we show that either the dimension of the 1-nullity distribution $N(1)$ is less than or equal to n, or $N(1)$ is the entire tangent bundle TM. In the latter case, the Sasakian manifold M is isometric to a quotient of the Euclidean sphere under a finite group of isometries.

1 Introduction

Contact, non-Sasakian manifolds whose characteristic vector field lies in the (k,μ)-nullity distribution have been fully classified by Boecks [6]. One of the main goals of this chapter is to describe the leaves of the 1-nullity distribution and the topology of Sasakian manifolds using variational calculus.

We collect some preliminaries on contact metric geometry in Section 2 and define the k-nullity distribution of a Riemannian manifold in Section 3. Section 4 deals with Sasakian manifolds in general. Using variational calculus techniques, we prove the main theorem of this paper there.

2 Preliminaries

A *contact form* on a $(2n+1)$-dimensional manifold M is a 1-form α such that $\alpha \wedge (d\alpha)^n$ is a volume form on M. There is always a unique vector field Z, the characteristic vector field of α, which is determined by the equations $\alpha(Z) = 1$ and $d\alpha(Z,X) = 0$ for arbitrary X. The distribution $D_p = \{V \in T_pM : \alpha(V) = 0\}$ is called the contact distribution of α. Clearly, D is a symplectic vector bundle with symplectic form $d\alpha$.

P. Rukimbira
Department of Mathematics, Florida International University, Miami, Florida, USA

K. Galicki and S.R. Simanca (eds.), *Riemannian Topology and Geometric Structures on Manifolds,* Progress in Mathematics 271, DOI 10.1007/978-0-8176-4743-8,
© Springer Science+Business Media LLC 2009

On a contact manifold (M,α,Z), there is also a nonunique Riemannian metric g and a partial complex operator J adapted to α in the sense that the identities

$$2g(X,JY) = d\alpha(X,Y), \quad J^2X = -X + \alpha(X)Z$$

hold for any vector fields X, Y on M. We have adopted the convention for exterior derivative so that

$$d\alpha(X,Y) = X\alpha(Y) - Y\alpha(X) - \alpha([X,Y]).$$

The tensors α, Z, J, and g are called contact metric structure tensors, and the manifold M with such a structure will be called a *contact metric manifold* [4]. We will use the notation (M,α,Z,J,g) to denote a contact metric manifold M with specified structure tensors. Assuming that (M,g) is a complete Riemannian manifold, let ψ_t, $t \in \mathbb{R}$ denote the 1-parameter group of diffeomorphism generated by Z. The group ψ_t preserves the contact form α, that is, $\psi_t^*\alpha = \alpha$. If ψ_t is also a 1-parameter group of isometries of g, then the contact metric manifold is called a *K-contact manifold*. By ∇, we shall denote the Levi–Civita covariant derivative operator of g. On a K-contact manifold, one has the identity

$$\nabla_X Z = -JX,$$

valid for any tangent vector X. On a general contact metric manifold, the identity

$$\nabla_X Z = -JX - JhX$$

is satisfied, where $hX = \frac{1}{2}(L_Z J)X$. If the identity

$$(\nabla_X J)Y = g(X,Y)Z - \alpha(Y)X$$

is satisfied for any vector fields X and Y on M, then the contact metric structure (M,α,Z,J,g) is called a *Sasakian* structure. A submanifold N in a contact manifold (M,α,Z,J,g) is said to be invariant if Z is tangent to N and JX is tangent to N whenever X is. An invariant submanifold is a contact submanifold.

3 The *k*-Nullity Distribution

For a real number k, the *k-nullity distribution* of a Riemannian manifold (M,g) is the subbundle $N(k)$ defined at each point $p \in M$ by

$$N_p(k) = \{H \in T_pM \,|\, R(X,Y)H = k(g(Y,H)X - g(X,H)Y) \,\forall X,Y \in T_pM\},$$

where R denotes the Riemann curvature tensor given by the formula

$$R(X,Y)H = \nabla_X\nabla_Y H - \nabla_Y\nabla_X H - \nabla_{[X,Y]}H$$

for arbitrary vector fields X, Y and H on M. If H lies in $N(k)$, then the sectional curvatures of all plane sections containing H are equal to k. By R_k, we denote the tensor field defined for arbitrary vector fields X, Y, H by

$$R_k(X,Y)H = R(X,Y)H - k\{g(Y,H)X - g(X,H)Y\}.$$

R_k satisfies similar identities as the curvature tensor R, mainly:

(i) $g(R_k(X,Y)H,V) = -g(R_k(X,Y)V,H),$
(ii) $g(R_k(X,Y)H,V) = g(R_k(X,H)Y,V),$
(iii) $\nabla_X R_k(Y,H)V + \nabla_Y R_k(H,X)V + \nabla_H R_k(X,Y)V = 0.$

Now, let X, Y, V be tangent vectors at $p \in M$. Extend X, Y and V into local vector fields such that at p, one has $\nabla X = 0 = \nabla Y = \nabla V$. Let H, W be two vector fields in the nullity distribution of R_k, that is,

$$R_k(X,Y)H = 0 = R_k(X,Y)W$$

for any X, Y on M. Using identity (iii), one obtains:

$$\begin{aligned}
0 &= g(\nabla_H R_k(X,Y)V + \nabla_X R_k(Y,H)V + \nabla_Y R_k(H,X)V, W) \\
&= g(\nabla_H(R_k(X,Y)V) + \nabla_X(R_k(Y,H)V) + \nabla_Y(R_k(H,X)V) - R_k(Y, \nabla_X H)V \\
&\quad - R_k(\nabla_Y H, X)V + Others, W) \\
&= Zg(R_k(X,Y)V,W) - g(R_k(X,Y)V, \nabla_H W) + Xg(R_k(Y,H)V,W) \\
&\quad - g(R_k(Y,H)V, \nabla_X W) + Yg(R_k(H,X)V,W) - g(R_k(H,X)V, \nabla_Y W) \\
&\quad - g(R_k(Y, \nabla_X H)V, W) - g(R_k(\nabla_Y H, X)V, W) + g(Others, W).
\end{aligned}$$

"Others" stands for terms vanishing at p. Applying identities (i) and (ii), and evaluating at p, we obtain

$$0 = g(R_k(X,Y)\nabla_H W, V),$$

for arbitrary X, Y and V. This means that $\nabla_H W$ also belongs to the k nullity distribution whenever H and W do. The above argument proves that $N(k)$ is an integrable subbundle with totally geodesic leaves of constant curvature k [12]. Hence, if $k > 0$ and $\dim N(k) > 1$, then each leaf of $N(k)$ is a compact manifold ([9], Corollary 19.5).

In [3], the authors introduced a class of contact metric manifolds (M, α, Z, J, g) for which the characteristic vector field Z belongs to the (k, μ)-nullity distribution for some real numbers k and μ. This means that the Riemannian curvature tensor R satisfies

$$R(X,Y)Z = k(\alpha(Y)X - \alpha(X)Y) + \mu(\alpha(Y)hX - \alpha(X)hY),$$

for all vector fields X and Y on M, where $h = \frac{1}{2}L_Z J$. From now on, when $\mu = 0$, the (k, μ)-nullity distribution will be referred to as the k-nullity distribution.

On a contact metric $(2n + 1)$-dimensional manifold M, $n > 1$, Blair and Koufogiorgos showed that if the characteristic vector field Z lies in the (k, μ)-nullity distribution, then $k \leq 1$. If $k < 1$ and $k \neq 0$, then the dimension of the (k, μ)-nullity distribution is equal to 1 [1]. The corresponding result for $n = 1$ is due to R. Sharma [11]. If $k = 0$, then M is locally $E^{n+1} \times S^n(4)$ and Z is tangent to the Euclidean factor giving that the dimension of the $(0, \mu)$-nullity distribution is equal to $n + 1$ [5]. If $k = 1$, the contact metric structure is Sasakian, and we wish to investigate the dimension of $N(1)$ on a Sasakian manifold. Contact, non-Sasakian, manifolds whose characteristic vector field lies in the (k, μ)-nullity distribution have been fully classified by Boeckx in [6]. First, we will describe the leaves of the 1-nullity distribution on a Sasakian manifold.

Proposition 1. *Let (M, α, Z, J, g) be a closed Sasakian manifold. If the dimension of $N(1)$ is bigger than 1, then each leaf of $N(1)$ is a closed Sasakian submanifold that is isometric to a quotient of a Euclidean sphere under a finite group of isometries. In particular, if the dimension of $N(1)$ is bigger than 1, then the Sasakian manifold (M, α, g) is quasi-regular, i.e., all integral curves of Z are circles.*

Proof. Let N be a leaf of $N(1)$. Since the leaf is a totally geodesic submanifold and Z is tangent to it, one has that $JX = -\nabla_X Z$ is tangent to the leaf for any X tangent to it. So, N is an invariant contact submanifold of the Sasakian manifold M and therefore it is also Sasakian. Since N is complete of constant curvature 1, it is isometric to a quotient of a Euclidean sphere under a finite group of Euclidean isometries [13]. $\square$

To simplify notations, we will denote the dimension of $N(k)$ by $\dim N(k)$.

Remark 2. The second part in the statement of Proposition 1 makes it clear that Theorem 4.3 of [10] is incorrect as stated. If $\dim N(1) > 1$, then the Sasakian manifold has necessarily rank 1 and cannot therefore have isolated characteristics.

4 The 1-Nullity of Sasakian Manifolds

On compact Sasakian manifolds, one has the following lemma the first part of which is due to Binh and Tamássy [2].

Lemma 3. *Let (M, α, Z, J, g) be a closed $(2n + 1)$-dimensional Sasakian manifold and $N \subset M$ a $(2r + 1)$-dimensional invariant submanifold. Let $\gamma(t)$ be a normal geodesic issuing from $\gamma(0) = x \in N$ in a direction perpendicular to N. Then, there exist orthonormal vectors $E_i \in T_x N$, $i = 1, 2, \ldots, r$ such that their parallel translated $E_i(t)$ along $\gamma(t)$, completed with $JE_i(t)$, form a vector system that is orthonormal and parallel along $\gamma(t)$. Moreover, denoting by V the unit tangent vector field along γ, each of $E_i(t)$, $JE_i(t)$ is simultaneously orthogonal to V and JV.*

Proof. We will prove only the second part of the lemma, referring to the papers [2] or [10] for the rest of the proof. Since $JE_i(t)$ is orthogonal to V and V is orthogonal to Z along γ, one has also

$$g(E_i(t), JV) = -g(JE_i(t), V) = 0,$$
$$g(JE_i(t), JV) = g(E_i(t), V) = 0.$$

This completes the proof of the second part of Lemma 1. □

Given a closed Sasakian manifold (M, α, g), we shall denote by G_2TM the Grassmannian bundle of tangent 2-planes on M and by H, the function "sectional curvature" defined on G_2TM. Let $-\kappa^2$ denote the minimum value for the directional derivative of H over G_2TM, that is $-\kappa^2$ is the minimum value of $dH(v)$, where v runs through all unit tangent vectors on G_2TM with a metric naturally induced form the Sasakian metric g. We now state the main theorem:

Theorem 4. *Let (M, α, Z, J, g) be a closed $(2n+1)$-dimensional Sasakian manifold. Then either $\dim N(1) \leq n$ or $\dim N(1) = 2n+1$.*

Proof. Suppose N_1 and N_2 are two distinct $2r+1$-dimensional leaves of $N(1)$, where $2r+1 > 1$. Denoting by T the distance between N_1 and N_2, there exist a minimal geodesic $\gamma(t)$, $0 \leq t \leq T$ from N_1 to N_2 such that $\gamma(0) \in N_1$, $\gamma(T) \in N_2$, and $\gamma(t)$ realizes the distance between the two compact leaves. Let $V(t)$ be the unit tangent vector to the geodesic $\gamma(t)$. Then $V(0)$ is orthogonal to N_1 and $V(T)$ is orthogonal to N_2. Let E_i, JE_i, $i = 1, 2, \ldots, r$ be an orthonormal basis for the contact distribution at $\gamma(0) \in N_1$ (recall N_1 is a contact submanifold). Let $E_i(t)$ denote the parallel translation of E_i from $\gamma(0)$ to $\gamma(t)$. Then $E_i(t)$, $JE_i(t)$, $i = 1, 2, \ldots, r$ is a parallel orthonormal frame field along $\gamma(t)$ as stated in Lemma 1 and proved in [2] or [10]. Suppose now that $2r+1 > n$. Then $2r - (2n - 2r - 2) = 2(2r+1) - 2n > 0$. This means that the vector space spanned by E_i, JE_i, $i = 1, 2, \ldots, r$ has dimension $2r$, which is bigger than $2n - 2r - 2$, where $2n - 2r - 2$ is the dimension of the orthogonal complement of V and JV in the normal space of N_2. Since the vector space spanned by E_i and JE_i is orthogonal to V and JV by Lemma 1, one can find at least one unit vector $F \in T_{\gamma(0)}N_1$ that is a linear combination of the $E_i(0)$, $JE_i(0)$, $i = 1, 2, \ldots, r$ such that its parallel translated $F(T) \in T_{\gamma(T)}N_2$.

We interrupt the proof for a needed lemma.

Lemma 5. *Suppose N_1 and N_2 are at distance T from each other, $0 \leq T \leq \frac{1}{2\kappa^2}$. Then the sectional curvature $K(V, F)(t)$ along γ satisfies:*

$$K(V, F)(t) > 0.$$

Proof. Since $K(V, F)(0) = 1$, one has for $0 \leq t \leq T \leq \frac{1}{2\kappa^2}$ (and H denoting the function "sectional curvature" defined on G_2TM):

$$K(V, F)(t) = 1 + \int_0^t \frac{dH}{ds} ds > 1 + \int_0^t -\kappa^2 \, ds \geq 1 - \kappa^2 t \geq \frac{1}{2} > 0,$$

completing the proof of Lemma 2. □

With Lemma 2 at hand, we continue the proof of the main theorem.

The vector field $F(t)$ along $\gamma(t)$ provides a variation $\gamma_s(t)$ of the geodesic $\gamma(t)$ with endpoints in N_1 and N_2. Let $V_s(t)$ denote the tangent vector to the curves in such a variation. Then the arclength functional $L(s)$ is given by:

$$L(s) = \int_0^T \|V_s(t)\| dt .$$

One has $L'(0) = 0$ because $\gamma(t)$ is a minimal geodesic. Furthermore, by Synge's formula for the second variation [8], one has

$$L_F''(0) = \sigma_{N_2}(F,F)(T) - \sigma_{N_1}(F,F)(0) - \int_0^T g(R(F,V)V,F)(t)dt ,$$

where σ_{N_i} is the second fundamental form of the submanifold N_i, and $g(R(F,V)V,F)$ $= K(V,F)(t)$ is the sectional curvature of the plane spanned by F and V. Recalling that $N_i, i = 1,2$ is totally geodesic and the curvature $K(V,F)(t) > 0$ for $0 \leq t \leq T \leq \frac{1}{2\kappa^2}$ by Lemma 2, one concludes that $\sigma_{N_1}(F,F)(0) = 0 = \sigma_{N_2}(F,F)(T)$ and

$$L_F''(0) = -\int_0^T K(V,F)(t)dt < 0 .$$

Therefore, the second variation of L in the direction F along $\gamma(t)$ is strictly negative in contradiction with the minimality of the geodesic $\gamma(t)$. Thus, we have established that if $\dim N(1) > n$, then $N(1)$ can have only one leaf, which has to be the manifold M itself.   □

An immediate consequence of the main theorem is that for a 5-dimensional, closed Sasakian manifold M, either the dimension of its 1-nullity distribution is 1, or M is of constant sectional curvature 1.

Next we give an example of a 7-dimensional Riemannian manifold whose 1-nullity distribution has dimension 3.

If a Riemannian manifold (M^{2n+1}, g) admits three contact metric structures (α_i, Z_i, J_i, g), $i = 1,2,3$ satisfying the following for a circular permutation (i,j,k) of $(1,2,3)$,

$$J_k = J_iJ_j - \alpha_j \otimes Z_i = -J_jJ_i + \alpha_i \otimes Z_j ,$$
$$Z_k = J_iZ_j = -J_jZ_i, \quad \alpha_k = \alpha_i \circ J_j = -\alpha_j \circ J_i ,$$

then the manifold is said to have a 3-Sasakian structure [4]. The homogeneous 7-dimensional Riemannian manifold $M^7 = SU(3)/U(1)$ carries a 3-Sasakian structure whose contact metric is not of constant sectional curvature ([7], page 253). Let Z_1, Z_2, Z_3 be 3 orthonormal characteristic vector fields in the 3-Sasakian structure. Each one of these belongs to the 1-nullity distribution of M^7, so the dimension of $N(1)$ is at least equal to 3 in this case. But since the manifold M^7 is not of constant sectional curvature, it follows from Theorem 4 that the dimension of $N(1)$ must be exactly 3.

To our knowledge, there are at present no known examples of 1-nullity distributions of dimension $n \geq 5$ on closed, Sasakian manifolds of dimension $2n + 1$. In fact, it is widely believed that on such a manifold, the dimension of $N(1)$ can take only one of the values 1, 3, or $2n + 1$.

References

1. C. Baikoussis, D.E. Blair & T. Koufogiorgos, *A decomposition of the curvature tensor of a contact manifold with* ξ *in* $N(k)$, Mathematics Technical Report, University of Ioannina (Greece) (1992).
2. T.Q. Binh & L. Tamássy, *Galloway's compactness theorem on Sasakian manifolds,* Aequationes Math. 58 (1999), 118–124.
3. D.E. Blair, T. Koufogiorgos & B.T. Papantoniou, *Contact metric manifolds satisfying a nullity condition*, Israel J. Math. 91 (1995), 189–214.
4. D. Blair, *Riemannian geometry of contact and symplectic manifolds*, Progress in Mathematics 203, Birkhäuser, Boston, 2002.
5. D.E. Blair, *Two remarks on contact metric structures*, Tôhoku Math. J. 29 (1977), 319–324.
6. E. Boeckx, *A full classification of contact metric* (k, μ)-*spaces*, Illinois J. Math. 44 (2000), 212–219.
7. C.P. Boyer, K. Galicki & B.M. Mann, *Quaternionic reduction and Einstein manifolds*, Commun. Anal. Geom. 1 (1993), 229–279.
8. T. Frankel, *On the fundamental group of a compact minimal submanifold*, Ann. Math. 83 (1966), 68–73.
9. J. Milnor, *Morse Theory*, Annals of Mathematics Studies **51**, Princeton University Press, Princeton, NJ, 1968.
10. P. Rukimbira, *Rank and k-nullity of contact manifolds*, Int. J. Math. Math. Sci. 20 (2004), 1025–1034.
11. R. Sharma, *On the curvature of contact metric manifolds*, J. Geom. 53 (1995), 179–190.
12. S. Tanno, *Some differential equations on Riemannian manifolds*, J. Math. Soc. Japan 30 (1978), 509–531.
13. J.A. Wolf, *Spaces of constant curvature*, McGraw-Hill, New York, 1967.

New Results in Sasaki–Einstein Geometry

James Sparks

Abstract This article is a summary of some of the author's work on Sasaki–Einstein geometry. A rather general conjecture in string theory known as the AdS/CFT correspondence relates Sasaki–Einstein geometry, in low dimensions, to superconformal field theory; properties of the latter are therefore reflected in the former, and vice versa. Despite this physical motivation, many recent results are of independent geometrical interest and are described here in purely mathematical terms: explicit constructions of infinite families of both quasi-regular and irregular Sasaki–Einstein metrics; toric Sasakian geometry; an extremal problem that determines the Reeb vector field for, and hence also the volume of, a Sasaki–Einstein manifold; and finally, obstructions to the existence of Sasaki–Einstein metrics. Some of these results also provide new insights into Kähler geometry, and in particular new obstructions to the existence of Kähler–Einstein metrics on Fano orbifolds.

1 Introduction

Sasaki–Einstein geometry is the odd-dimensional cousin of Kähler–Einstein geometry. In fact the latter, for positive Ricci curvature, is strictly contained in the former; Sasaki–Einstein geometry is thus a generalization of Kähler–Einstein geometry. The author's initial interest in this subject stemmed from a rather general conjecture in string theory known as the AdS/CFT correspondence [31]. This is probably the most important conceptual development in theoretical physics in recent years. AdS/CFT conjecturally relates quantum gravity, in certain backgrounds, to ordinary quantum field theory without gravity. Moreover, the relation between the two theories is *holographic*: the quantum field theory resides on the *boundary* of the region in which gravity propagates.

J. Sparks
Department of Mathematics, Harvard University, Cambridge, Massachusetts, USA
and
Jefferson Physical Laboratory, Harvard University, Cambridge, Massachusetts, USA

K. Galicki and S.R. Simanca (eds.), *Riemannian Topology and Geometric Structures on Manifolds,* Progress in Mathematics 271, DOI 10.1007/978-0-8176-4743-8,
© Springer Science+Business Media LLC 2009

In a particular setting, the AdS/CFT correspondence relates Sasaki–Einstein geometry, in dimensions five and seven, to superconformal field theory, in dimensions four and three, respectively. Superconformal field theories are very special types of quantum field theories: they possess superconformal symmetry, and hence in particular conformal symmetry. The five-dimensional case of this correspondence is currently understood best. One considers a ten-dimensional Riemannian product $(B \times L, g_B + g_L)$, where (L, g_L) is a Sasaki–Einstein five-manifold and (B, g_B) is five-dimensional *hyperbolic space*. We may present B as the open unit ball $B = \{x \in \mathbb{R}^5 \mid \|x\| < 1\}$ with metric

$$g_B = \frac{4 \sum_{i=1}^5 \mathrm{d}x_i \otimes \mathrm{d}x_i}{(1 - \|x\|^2)^2} . \tag{1.1}$$

Here $\|x\|$ denotes the Euclidean norm of $x \in \mathbb{R}^5$. We may naturally compactify (B, g_B), in the sense of Penrose, by adding a conformal boundary at infinity. One thus considers the closed unit ball $\bar{B} = \{x \in \mathbb{R}^5 \mid \|x\| \leq 1\}$ equipped with the metric

$$g_{\bar{B}} = f^2 g_B \tag{1.2}$$

where f is a smooth function on $\bar{B}$ that is positive on B and has a simple zero on $\partial \bar{B} = S^4$. For example, $f = 1 - \|x\|^2$ induces the standard metric on S^4, but there is no natural choice for f. Thus S^4 inherits only a *conformal structure*. The isometric action of $SO(1,5)$ on (B, g_B) extends to the action of the conformal group $SO(1,5)$ on the four-sphere. The AdS/CFT correspondence[1] conjectures that type IIB string theory, which is supposed to be a theory of quantum gravity, propagating on $(B \times L, g_B + g_L)$ is equivalent to a four-dimensional superconformal field theory that resides on the boundary four-sphere. Indeed, the ten-dimensional manifold is a *supersymmetric* solution to type IIB supergravity; in differentio-geometric terms, this means that there exists a solution to a certain Killing spinor equation. The AdS/CFT correspondence thus in particular implies a correspondence between Sasaki–Einstein manifolds in dimension five and superconformal field theories in four dimensions: for each Sasaki–Einstein five-manifold (L, g_L), we obtain a different superconformal field theory. The AdS/CFT correspondence then naturally maps geometric properties of (L, g_L) to properties of the dual superconformal field theory.

I should immediately emphasize, however, that this article is aimed at geometers, rather than theoretical physicists. Unfortunately, explaining AdS/CFT to a mathematical audience is beyond the scope of this chapter. Instead I shall focus mainly on the new geometrical results obtained by the author. The paper is based on a talk given at the conference "Riemannian Topology: Geometric Structures on Manifolds" in Albuquerque, New Mexico.

The outline of the article is as follows. Section 2 contains a brief review of Sasakian geometry, in the language of Kähler cones. Section 3 summarizes the properties of several infinite families of Sasaki–Einstein manifolds that were constructed

---

[1] The Lorentzian version of hyperbolic space is known as anti–de Sitter spacetime (AdS). The acronym CFT stands for conformal field theory.

in [13, 14, 21–23, 33], focusing on the (most physically interesting) case of dimension five. Section 4 reviews toric Sasakian geometry, as developed in [34]. Section 5 is a brief account of an extremal problem that determines the Reeb vector field for a Sasaki–Einstein metric [34, 35]. This is understood completely for toric Sasaki–Einstein manifolds, whereas the general case currently contains some technical gaps. Finally, Section 6 reviews some obstructions to the existence of Sasaki–Einstein metrics [25]. A far more detailed account of Sasakian geometry, together with many other beautiful results in Sasaki–Einstein geometry not described here, may be found in the book [7].

2 Sasakian Geometry

Whereas Kähler geometry [27] has been studied intensively for more than seventy years, Sasakian geometry [39] has, in contrast, received relatively little attention. Sasakian geometry was originally defined in terms of metric-contact geometry, but this does not really emphasize its relation to Kähler geometry. The following is a good definition.

Definition. A compact Riemannian manifold (L, g_L) is *Sasakian* if and only if its metric cone $(X_0 = \mathbb{R}_+ \times L, g = \mathrm{d}r^2 + r^2 g_L)$ is a *Kähler cone*.

Here $r \in (0, \infty)$ may be regarded as a coordinate on the positive real line $\mathbb{R}_+$. The reason for the subscript on X_0 will become apparent later. Note that (L, g_L) is isometrically embedded $\iota : L \to X_0$ into the cone with image $\{r = 1\}$. By definition, (X_0, g) is a noncompact Kähler manifold, with Kähler form

$$\omega = \frac{1}{4} \mathrm{dd}^c r^2 \ . \tag{2.1}$$

Here $\mathrm{d}^c = \mathcal{J} \circ \mathrm{d} = i(\bar{\partial} - \partial)$, as usual, with $\mathcal{J}$ the complex structure tensor on X_0. The square of the radial function r^2 thus serves as a global Kähler potential on the cone. It is not difficult to verify that the homothetic vector field $r\partial/\partial r$ is holomorphic, and that

$$\xi = \mathcal{J} \left(r \frac{\partial}{\partial r} \right) \tag{2.2}$$

is holomorphic and also a *Killing vector field*: $\mathcal{L}_\xi g = 0$. ξ is known as the *Reeb vector field*. It is tangent to the surfaces of constant r, and thus defines a vector field on L, which, in a standard abuse of notation, we also denote by ξ. Another important object is the one-form

$$\eta = \mathrm{d}^c \log r \ . \tag{2.3}$$

This is homogeneous degree zero[2] under $r\partial/\partial r$ and pulls back via ι^* to a one-form, which we also denote by η, on L. In fact, η is precisely a *contact one-form* on L; that is, $\eta \wedge (\mathrm{d}\eta)^{n-1}$ is a volume form on L. Here $n = \dim_{\mathbb{C}} X_0$, or equivalently $2n - 1 = \dim_{\mathbb{R}} L$. The pair $\{\eta, \xi\}$ then satisfy, either on the cone or on L, the relations

$$\eta(\xi) = 1, \qquad \mathrm{d}\eta(\xi, \cdot) = 0 . \tag{2.4}$$

This is the usual definition of the Reeb vector field in contact geometry. Indeed, the open cone (X_0, ω) is the *symplectization* of the contact manifold (L, η), where one regards $\omega = \frac{1}{2}\mathrm{d}(r^2\eta)$ as a symplectic form on X_0.

The square norm of ξ, in the cone metric g, is $\|\xi\|_g^2 = r^2$, and thus in particular ξ is nowhere zero on X_0. It follows that the orbits of ξ define a foliation of L. It turns out that the metric transverse to these orbits g_T is also a Kähler metric. Thus Sasakian structures are sandwiched between two Kähler structures: the Kähler cone of complex dimension n and the transverse Kähler structure of complex dimension $n-1$.

Consideration of the orbits of ξ leads to a global classification of Sasakian structures. Suppose that all the orbits of ξ close. This means that ξ generates an isometric $U(1)$ action on (L, g_L). Such Sasakian structures are called *quasi-regular*. Since ξ is nowhere zero, this action must be locally free: the isotropy subgroup at any point must be finite and therefore isomorphic to a cyclic group $\mathbb{Z}_m \subset U(1)$. If the isotropy subgroups for all points are trivial, then the $U(1)$ action is free, and the Sasakian structure is called *regular*. We use the term *strictly quasi-regular* for a quasi-regular Sasakian structure that is not regular. In either case there is a quotient $V = L/U(1)$, which is generally an orbifold. The isotropy subgroups descend to the local orbifold structure groups in the quotient space $V = L/U(1)$; thus V is a manifold when the Sasakian structure is regular. The transverse Kähler metric descends to a Kähler metric on the quotient, so that (V, g_V) is a Kähler manifold or orbifold. Indeed, (V, g_V) may be regarded as the *Kähler reduction* of the Kähler cone (X_0, ω) with respect to the $U(1)$ action, which is Hamiltonian with Hamiltonian function $\frac{1}{2}r^2$.

If the orbits of ξ do not all close, the Sasakian structure is said to be *irregular*. The generic orbit is $\mathbb{R}$, and in this case one cannot take a meaningful quotient. The closure of the orbits of ξ defines an Abelian subgroup of the isometry group of (L, g_L). Since L is compact, the isometry group of (L, g_L) is compact, and the closure of the orbits of ξ therefore defines a torus $\mathbb{T}^s$, $s > 1$, which acts isometrically on (X_0, g) or (L, g_L). Thus irregular Sasakian manifolds have at least a $\mathbb{T}^2$ isometry group.

The main focus of this article will be *Sasaki–Einstein* manifolds. A simple calculation shows that[3]

$$\mathrm{Ric}(g) = \mathrm{Ric}(g_L) - 2(n-1)g_L = \mathrm{Ric}(g_T) - 2ng_T \tag{2.5}$$

---

[2] That is, $\mathcal{L}_{r\partial/\partial r}\eta = 0$.

[3] Notice the slight abuse of notation here: we are regarding all tensors in this equation as tensors on X_0.

where $\mathrm{Ric}(\cdot)$ denotes the Ricci tensor of a given metric. Thus the Kähler cone (X_0, g) is Ricci-flat if and only if (L, g_L) is Einstein with positive scalar curvature $2(n-1)(2n-1)$, if and only if the transverse metric is Einstein with positive scalar curvature $4n(n-1)$. (L, g_L) is then said to be a Sasaki–Einstein manifold. Notice that (X_0, g) is then a Calabi–Yau cone, and in the quasi-regular case the circle quotient of (L, g_L) is a Kähler–Einstein manifold or orbifold of positive Ricci curvature. The converse is also true: give a positively curved Kähler–Einstein manifold or orbifold (V, g_V), there exists a Sasaki–Einstein metric[4] on the total space of a $U(1)$ principal (orbi-)bundle over V. This was proved in general by Boyer and Galicki in [4].

3 Explicit Constructions of Sasaki–Einstein Manifolds

Explicit examples of Sasaki–Einstein manifolds were, until recently, quite rare. In dimension five, the only simply-connected[5] examples that were known in explicit form were the round sphere and a certain homogeneous metric on $S^2 \times S^3$ [42], known as $T^{1,1}$ in the physics literature. These are both regular, being circle bundles over $\mathbb{CP}^2$ and $\mathbb{CP}^2 \times \mathbb{CP}^1$ with their standard Kähler–Einstein metrics, respectively. In fact, regular Sasaki–Einstein manifolds are classified [18]. This follows since smooth Kähler–Einstein surfaces with positive Ricci curvature have been classified by Tian and Yau [43,44]. The result is that the base may be taken to be a Del Pezzo surface obtained by blowing up $\mathbb{CP}^2$ at k generic points with $3 \leq k \leq 8$; although proven to exist, the Kähler–Einstein metrics on these Del Pezzo surfaces are not known explicitly. More recently, Boyer, Galicki, and collaborators have produced vast numbers of quasi-regular Sasaki–Einstein metrics using existence results of Kollár for Kähler–Einstein metrics on Fano orbifolds, together with the $U(1)$ lifting mentioned at the end of the previous section. For a review of their work, see [6,7].

Until 2004, no explicit examples of nontrivial strictly quasi-regular Sasaki–Einstein manifolds were known, and it was not known whether or not *irregular* Sasaki–Einstein manifolds even existed. In fact, Cheeger and Tian conjectured in [10] that they did not exist. The following theorem disproves this conjecture:

Theorem 3.1 ([22]). *There exist a countably infinite number of Sasaki–Einstein metrics $Y^{p,q}$ on $S^2 \times S^3$, labeled naturally by $p, q \in \mathbb{N}$ where $\gcd(p, q) = 1$, $q < p$. $Y^{p,q}$ is quasi-regular if and only if $4p^2 - 3q^2$ is the square of a natural number, otherwise it is irregular. In particular, there are infinitely many strictly quasi-regular and irregular Sasaki–Einstein metrics on $S^2 \times S^3$.*

---

[4] In the orbifold case, this lifting may or may not be an orbifold. If $\{\Gamma_\alpha\}$ denote the local orbifold structure groups of V, then the data that define an orbibundle over V with structure group G include elements of $\mathrm{Hom}(\Gamma_\alpha, G)$ for each α, subject to certain gluing conditions. A moment's thought shows that the total space of a G principal orbibundle over V is smooth if and only if all these maps are *injective*.

[5] Positively curved Einstein manifolds have finite fundamental group [37].

These metrics were constructed explicitly in [22], based on supergravity constructions by the same authors in [21]. The metrics are cohomogeneity one, meaning that the generic orbit under the action of the isometry group has real codimension one. The Lie algebra of this group is $\mathfrak{su}(2) \times \mathfrak{u}(1) \times \mathfrak{u}(1)$. The volumes of the metrics are given by the formula

$$\frac{\text{vol}[Y^{p,q}]}{\pi^3} = \frac{q^2(2p + \sqrt{4p^2 - 3q^2})}{3p^2(3q^2 - 2p^2 + \sqrt{4p^2 - 3q^2})} . \tag{3.1}$$

The result that $Y^{p,q}$ is diffeomorphic to $S^2 \times S^3$ follows from Smale's classification of 5-manifolds [41]. Interestingly, the cone X_0 corresponding to $Y^{2,1}$ is the open complex cone over the first Del Pezzo surface [32]. Note that the first Del Pezzo surface was missing from the list of Tian and Yau; it cannot admit a Kähler–Einstein metric since its Futaki invariant is non-zero. The Ricci-flat Kähler cone metric is in fact *irregular* for $Y^{2,1}$. We shall return to this in Section 6. Recently, Conti [12] has classified cohomogeneity one Sasaki–Einstein 5-manifolds: they are precisely the set $\{Y^{p,q}\}$. The construction of the above metrics also easily extends to higher dimensions [11, 23, 24]. This leads to the following

Corollary 3.2. *There exist countably infinitely many strictly quasi-regular and irregular Sasaki–Einstein structures in every (odd) dimension greater than 3.*

This corollary should be contrasted with several other results. It is known that for fixed dimension there are finitely many (deformation classes of) Fano manifolds [28]; thus there are only finitely many positively curved Kähler–Einstein structures in each dimension, and hence finitely many *regular* Sasaki–Einstein structures in each odd dimension. On the other hand, these may occur in continuous families. This is already true for Del Pezzo surfaces with $k \geq 5$ blow-ups, which have a complex structure moduli space of complex dimension $2(k-4)$. The quasi-regular existence results of Boyer and Galicki also produce examples with, sometimes quite large, moduli spaces [6]. It is currently unknown whether or not there exist continuous families of *irregular* Sasaki–Einstein structures.

Perhaps surprisingly, there also exist explicit cohomogeneity two Sasaki–Einstein 5-manifolds. The following subsumes Theorem 3.1:

Theorem 3.3 ([13, 14, 33]). *There exist a countably infinite number of Sasaki–Einstein metrics $L^{a,b,c}$ on $S^2 \times S^3$, labeled naturally by $a, b, c \in \mathbb{N}$ where $a \leq b$, $c \leq b$, $d = a + b - c$, $\gcd(a,b,c,d) = 1$, $\gcd(\{a,b\}, \{c,d\}) = 1$. Here the latter means that each of the pair $\{a,b\}$ must be coprime to each of $\{c,d\}$. Moreover, $L^{p-q,p+q,p} = Y^{p,q}$. The metrics are generically cohomogeneity two, generically irregular, and generically have volumes that are the product of quartic irrational numbers with π^3.*

The condition under which the metrics are (strictly) quasi-regular is not simple to determine in general. The quartic equation with integer coefficients that is satisfied by $\text{vol}[L^{a,b,c}]/\pi^3$ is written down explicitly in [13]. For integers (a,b,c) not satisfying some of the coprime conditions, one obtains Sasaki–Einstein orbifolds.

We conclude this section with a comment on how the metrics in Theorem 3.3 were constructed. In [13, 14], the local form of the metrics was found by writing down the Riemannian forms of known black hole metrics, and then taking a certain "BPS" limit. The initial family of metrics are local Einstein metrics, and in the limit one obtains a local family of Sasaki–Einstein metrics. One then determines when these local metrics extend to complete metrics on compact manifolds, and this is where the integers (a, b, c) enter. The same metrics were independently discovered in a slightly different manner in [33]. In the latter reference, the local Kähler–Einstein metrics in dimension four are constructed first. It turns out that these are precisely the *orthotoric* Kähler–Einstein metrics in [1]. The construction again easily extends to higher dimensions [13, 14]; the metrics are generically cohomogeneity $n - 1$.

4 Toric Sasakian Geometry

In this section, we summarize some of the results in [34] on toric Sasakian geometry. This probably warrants a:

Definition. A Sasakian manifold (L, g_L) is said to be *toric* if there exists an effective, holomorphic, and Hamiltonian action of the torus $\mathbb{T}^n$ on the corresponding Kähler cone (X_0, g). The Reeb vector field ξ is assumed to lie in the Lie algebra of the torus $\xi \in \mathfrak{t}_n$.

The Hamiltonian condition means that there exists a $\mathbb{T}^n$-invariant moment map

$$\mu : X_0 \to \mathfrak{t}_n^* . \tag{4.1}$$

The condition on the Reeb vector field implies that the image is a strictly convex rational polyhedral cone [17, 29]. Symplectic toric cones with Reeb vector fields not satisfying this condition form a short list and have been classified in [29]. The main result of this section is Proposition 4.1, which describes, in a rather explicit form, the space of toric Sasakian metrics on the link of a fixed affine toric variety X. Any toric Sasakian manifold is of this form, with the open Kähler cone $X_0 = X \setminus \{p\}$ being the smooth part of X, with p an isolated singular point.

We begin by fixing a *strictly convex rational polyhedral cone* $\mathcal{C}^*$ in $\mathbb{R}^n$, where the latter is regarded as the dual Lie algebra of a torus $\mathfrak{t}_n^* \cong \mathbb{R}^n$ with a particular choice of basis:

$$\mathcal{C}^* = \{y \in \mathfrak{t}_n^* \mid \langle y, v_a \rangle \geq 0, \forall a = 1, \ldots, d\} . \tag{4.2}$$

The strictly convex condition means that $\mathcal{C}^*$ is a cone over a convex polytope of dimension $n - 1$. It follows that necessarily $n \leq d \in \mathbb{N}$. The rational condition on $\mathcal{C}^*$ means that the vectors $v_a \in \mathfrak{t}_n \cong \mathbb{R}^n$ are rational. In particular, one can normalize the v_a so that they are primitive vectors in $\mathbb{Z}^n \cong \ker\{\exp : \mathfrak{t}_n \to \mathbb{T}^n\}$. The v_a are thus the inward-pointing primitive normal vectors to the bounding hyperplanes of the polyhedral cone C^*. We may alternatively define C^* in terms of its generating vectors $\{u_\alpha \in \mathbb{Z}^n\}$:

$$C^* = \left\{ \sum_\alpha \lambda_\alpha u_\alpha \mid \lambda_\alpha \geq 0 \right\} . \tag{4.3}$$

The primitive vectors u_α generate the one-dimensional faces, or *rays*, of the polyhedral cone C^*.

Define the linear map

$$A : \mathbb{R}^d \to \mathbb{R}^n$$
$$e_a \mapsto v_a \tag{4.4}$$

where $\{e_a\}$ denotes the standard orthonormal basis of $\mathbb{R}^d$. Let $\Lambda \subset \mathbb{Z}^n$ denote the lattice spanned by $\{v_a\}$ over $\mathbb{Z}$. This is of maximal rank, since C^* is strictly convex. There is an induced map of tori

$$\mathbb{T}^d \cong \mathbb{R}^d/2\pi\mathbb{Z}^d \to \mathbb{R}^n/2\pi\mathbb{Z}^n \cong \mathbb{T}^n \tag{4.5}$$

where the kernel is a compact Abelian group $\mathcal{A}$, with $\pi_0(\mathcal{A}) \cong \Gamma \cong \mathbb{Z}^n/\Lambda$.

Using this data, we may construct the following Kähler quotient:

$$X = \mathbb{C}^d//\mathcal{A} . \tag{4.6}$$

Here we equip $\mathbb{C}^d$ with its standard flat Kähler structure ω_{flat}. $\mathcal{A} \subset \mathbb{T}^d$ acts holomorphically and Hamiltonianly on $(\mathbb{C}^d, \omega_{\text{flat}})$. We then take the Kähler quotient (4.6) at level zero. The origin of $\mathbb{C}^d$ projects to a singular point in X, and the induced Kähler metric g_{can} on its complement X_0 is a cone. Moreover, the quotient torus $\mathbb{T}^d/\mathcal{A} \cong \mathbb{T}^n$ acts holomorphically and Hamiltonianly on $(X_0, \omega_{\text{can}})$, with moment map

$$\mu : X_0 \to \mathfrak{t}_n^*; \qquad \mu(X) = C^* . \tag{4.7}$$

The quotient (4.6) may be written explicitly as follows. One computes a primitive basis for the kernel of A over $\mathbb{Z}$ by finding all solutions to

$$\sum_a Q_I^a v_a = 0 \tag{4.8}$$

with $Q_I^a \in \mathbb{Z}$, and such that for each I the $\{Q_I^a \mid a = 1, \ldots, d\}$ have no common factor. The number of solutions, which are indexed by I, is $d - n$ since A is surjective; this latter fact again follows since C^* is strictly convex. One then has

$$X = \mathcal{K}/\mathcal{A} \equiv \mathbb{C}^d//\mathcal{A} \tag{4.9}$$

with

$$\mathcal{K} \equiv \left\{ (Z_1, \ldots, Z_d) \in \mathbb{C}^d \mid \sum_a Q_I^a |Z_a|^2 = 0 \right\} \subset \mathbb{C}^d \tag{4.10}$$

where Z_a denote standard complex coordinates on $\mathbb{C}^d$, and the charge matrix Q_I^a specifies the torus embedding $\mathbb{T}^{d-n} \subset \mathbb{T}^d$.

It is a standard fact that the space X is an affine toric variety; that is, X is an affine variety equipped with an effective holomorphic action of the *complex* torus $\mathbb{T}_{\mathbb{C}}^n \cong (\mathbb{C}^*)^n$, which has a dense open orbit. Let

$$\mathcal{C} = \{\xi \in \mathfrak{t}_n \mid \langle \xi, y \rangle \geq 0, \forall y \in \mathcal{C}^*\} . \tag{4.11}$$

This is the *dual cone* to $\mathcal{C}^*$, which is also a convex rational polyhedral cone by Farkas' theorem. In the algebro-geometric language, the cone $\mathcal{C}$ is precisely the *fan* for the affine toric variety X. We have $X_0 = \mathbb{R}_+ \times L$ with L compact. If one begins with a general strictly convex rational polyhedral cone $\mathcal{C}^*$, the link L will be an orbifold; in order that L be a smooth manifold, one requires the moment polyhedral cone $\mathcal{C}^*$ to be *good* [29]. This puts certain additional constraints on the vectors v_a; the reader is referred to [34] for the details. Note that L inherits a canonical Sasakian metric from the Kähler quotient metric g_{can} on X_0.

Let $\partial/\partial\phi_i$, $i = 1, \ldots, n$, be a basis[6] for $\mathfrak{t}_n$, where $\phi_i \in [0, 2\pi)$ are coordinates on the real torus $\mathbb{T}^n$. Then we have the following:

Proposition 4.1 [34]). *The space of toric Kähler cone metrics on the smooth part of an affine toric variety X_0 is a product*

$$\mathcal{C}_{\text{int}} \times \mathcal{H}^1(\mathcal{C}^*)$$

where $\xi \in \mathcal{C}_{\text{int}} \subset \mathfrak{t}_n$ labels the Reeb vector field, with $\mathcal{C}_{\text{int}}$ the open interior of $\mathcal{C}$, and $\mathcal{H}^1(\mathcal{C}^*)$ denotes the space of homogeneous degree one functions on $\mathcal{C}^*$ that are smooth up to the boundary (together with the convexity condition below).

Explicitly, on the dense open image of $\mathbb{T}_{\mathbb{C}}^n$ we have

$$g = G_{ij} dy^i dy^j + G^{ij} d\phi_i d\phi_j \tag{4.12}$$

where

$$G_{ij} = \frac{\partial^2 G}{\partial y^i \partial y^j} \tag{4.13}$$

with matrix inverse G^{ij}, and the function

$$G(y) = G_{\text{can}}(y) + G_\xi(y) + h(y) \tag{4.14}$$

is required to be strictly convex with $h(y) \in \mathcal{H}^1(\mathcal{C}^*)$ and

$$G_{\text{can}}(y) = \frac{1}{2} \sum_{a=1}^d \langle y, v_a \rangle \log\langle y, v_a \rangle$$

$$G_\xi(y) = \frac{1}{2} \langle \xi, y \rangle \log\langle \xi, y \rangle - \frac{1}{2} \left(\sum_{a=1}^d \langle v_a, y \rangle \right) \log \left(\sum_{a=1}^d \langle v_a, y \rangle \right) .$$

---

[6] By a standard abuse of notation, we identify vector fields on X_0 with corresponding elements of the Lie algebra.

In particular, the canonical metric on X_0, induced from Kähler reduction of the flat metric on $\mathbb{C}^d$, is given by setting $G(y) = G_{\text{can}}(y)$. This function has a certain singular behavior at the boundary ∂C^* of the polyhedral cone; this is required precisely so that the metric compactifies to a smooth metric on X_0.

The space of Reeb vector fields is the interior of the cone $\mathcal{C}$. One can show [35] that for $\xi \in \partial \mathcal{C}$, the vector field ξ must vanish somewhere on X_0. Specifically, the bounding facets of $\mathcal{C}$ correspond to the generating rays of $\mathcal{C}^*$ under the duality map between cones; ξ being in a bounding facet of $\mathcal{C}$ implies that the corresponding vector field then vanishes on the inverse image, under the moment map, of the dual generating ray of $\mathcal{C}^*$. However, since the Reeb vector field is nowhere vanishing, we see that the boundary of $\mathcal{C}$ is a singular limit of Sasakian metrics on X_0.

Fixing a particular choice of Kähler cone metric on X_0, the image of $L = \{r = 1\}$ under the moment map is

$$\mu(L) = \left\{ y \in C^* \mid \langle y, \xi \rangle = \tfrac{1}{2} \right\} \equiv H(\xi) . \tag{4.15}$$

The hyperplane $\langle y, \xi \rangle = \tfrac{1}{2}$ is called the *characteristic hyperplane* [5]. This intersects the moment cone C^* to form a compact n-dimensional polytope $\Delta(\xi) = \mu(\{r \le 1\})$, bounded by ∂C^* and the compact $(n-1)$-dimensional polytope $H(\xi)$. In particular, the image $H(\xi)$ of L under the moment map depends only on the Reeb vector field ξ, and not on the choice of homogeneous degree one function h in Proposition 4.1. Moreover, the *volume* of a toric Sasakian manifold (L, g_L) is [34]

$$\text{vol}[g_L] = 2n(2\pi)^n \, \text{vol}[\Delta(\xi)] \tag{4.16}$$

where $\text{vol}[\Delta(\xi)]$ is the *Euclidean* volume of $\Delta(\xi)$.

Finally in this section, we introduce the notion of a *Gorenstein singularity*:

Definition. An analytic space X with isolated singular point p and smooth part $X \setminus \{p\} = X_0$ is said to be *Gorenstein* if there exists a smooth nowhere zero holomorphic $(n,0)$-form Ω on X_0.

We shall refer to Ω as a *holomorphic volume form*. X being Gorenstein is a necessary condition for X_0 to admit a Ricci-flat Kähler metric and hence for the link L to admit a Sasaki–Einstein metric. Indeed, the Ricci-form $\rho = \text{Ric}(\mathcal{J}\cdot, \cdot)$ is a curvature two-form for the holomorphic line bundle $\Lambda^{n,0}$. The Ricci-flat Kähler condition implies

$$\frac{i^n}{2^n}(-1)^{n(n-1)/2}\Omega \wedge \bar{\Omega} = \frac{1}{n!}\omega^n . \tag{4.17}$$

For affine toric varieties, it is again well-known that X being Gorenstein is equivalent to the existence of a basis for the torus $\mathbb{T}^n$ for which $v_a = (1, w_a)$ for each $a = 1, \ldots, d$, and $w_a \in \mathbb{Z}^{n-1}$.

Example ([32,33]). From Theorems 3.1 and 3.3, one sees that $L^{a,b,c}$, which contain $Y^{p,q}$ as a subset, have a holomorphic Hamiltonian action of $\mathbb{T}^3$ on the corresponding Kähler cones and are thus toric Sasaki–Einstein manifolds. One finds that the image of the cone under the moment map is always a four–sided polyhedral cone ($d = 4$) in $\mathbb{R}^3$. The charge matrix Q is

$$Q = (a, b, -c, -a - b + c) \, . \tag{4.18}$$

The Gorenstein condition is reflected by the fact that the sum of the components of Q is zero. In particular, for $Y^{p,q}$ (which is $a = p - q$, $b = p + q$, $c = p$), we have

$$v_1 = [1, 0, 0], \quad v_2 = [1, 1, 0], \quad v_3 = [1, p, p], \quad v_4 = [1, p - q - 1, p - q] \, . \tag{4.19}$$

It is relatively straightforward to see that the affine toric Gorenstein singularities for $L^{a,b,c}$ are the most general such that are generated by four rays.

5 A Variational Problem for the Reeb Vector Field

In this section, we consider the following problem: given a Gorenstein singularity (X, Ω), with isolated singular point and smooth set $X_0 = \mathbb{R}_+ \times L$, what is the Reeb vector field for a Ricci-flat Kähler cone metric on X_0, assuming it exists? We shall go quite a long way in answering this question and give a complete solution for affine toric varieties; in general, more work still remains to be done.

The strategy is to set up a variational problem on a space of Sasakian metrics on L, or equivalently a space of Kähler cone metrics on X_0. To this end, we suppose that X_0 is equipped with an effective holomorphic action of the torus $\mathbb{T}^s$ for some s; this is clearly necessary. Indeed, for irregular Sasakian metrics, one requires $s > 1$, as commented in Section 2. We then assume we are given a space of Kähler cone metrics $\mathcal{S}(X_0)$ on X_0 such that:

- The torus $\mathbb{T}^s$ acts Hamiltonianly on each metric $g \in \mathcal{S}(X_0)$.
- The Reeb vector field for each metric lies in the Lie algebra $\mathfrak{t}_s$ of $\mathbb{T}^s$.

We shall continue to denote Kähler cone metrics by g, the corresponding Sasakian metric by g_L, and regard either as elements of $\mathcal{S}(X_0)$. The second condition above ensures that the torus action is of *Reeb type* [5, 17]. We then have the following:

Proposition 5.1 ([35]). *The* volume *of the link* (L, g_L), *as a functional on the space* $\mathcal{S}(X_0)$, *depends only on the Reeb vector field* ξ *for the Sasakian metric* $g_L \in \mathcal{S}(X_0)$.

It follows that vol may be regarded as a function on the space of Reeb vector fields:

$$\mathrm{vol} : \mathcal{R}(X_0) \to \mathbb{R}_+ \tag{5.1}$$

where

$$\mathcal{R}(X_0) = \{\xi \in \mathfrak{t}_s \mid \xi = \text{Reeb vector field for some } g_L \in \mathcal{S}(X_0)\} \, . \tag{5.2}$$

The first and second derivatives are given by

Proposition 5.2.

$$\mathrm{d}\,\mathrm{vol}(Y) = -n \int_L \eta(Y)\mathrm{d}\mu \tag{5.3}$$

$$\mathrm{d}^2\,\mathrm{vol}(Y,Z) = n(n+1) \int_L \eta(Y)\eta(Z)\mathrm{d}\mu \ . \tag{5.4}$$

Here Y,Z are holomorphic Killing vector fields in $\mathfrak{t}_s$, η is the contact one-form for the Sasakian metric, and $\mathrm{d}\mu$ is the Riemannian measure on (L, g_L).

Note that for toric Sasakian metrics (for which the torus $\mathbb{T}^s$ has maximal dimension: $s = n$), we already noted Proposition 5.1 in the previous section — see equation (4.16). Note also that (5.4) shows that vol is a strictly convex function of ξ.

Proposition 5.1 is proved roughly as follows. Suppose one has two Kähler cone metrics on X_0 with the same homothetic vector field $r\partial/\partial r$. It is then straightforward to show that the Kähler potentials differ by a multiplicative factor $\exp\varphi$, where φ is a basic homogeneous degree zero function: $\mathcal{L}_\xi\varphi = 0 = \mathcal{L}_{r\partial/\partial r}\varphi$, where recall that $\xi = \mathcal{J}(r\partial/\partial r)$. One then shows that the volume is independent of φ.

This is precisely analogous to the situation in Kähler geometry where one fixes a *Kähler class*. Suppose that (M, ω) is a *compact* Kähler manifold with Kähler class $[\omega] \in H^{1,1}(M)$. Then any other Kähler metric on M in the same Kähler class is given by $\omega + i\partial\bar\partial\varphi$ for some smooth real function φ. The volume of (M, ω) clearly depends only on $[\omega]$.

Indeed, one can push the analogy further. A choice of Reeb vector field on X_0 should be regarded as a choice of *polarization*.[7] The space of quasi-regular Reeb vector fields is dense in the space of all Reeb vector fields: quasi-regular Reeb vector fields correspond to rational vectors in the Lie algebra $\mathfrak{t}_s \cong \mathbb{R}^s$, and these are dense since the rationals are dense in the reals. For ξ quasi-regular, the $U(1)$ quotient is a Kähler orbifold (V, ω_V). Changing the polarization ξ thus changes the quotient V, in contrast to the Kähler setting in the last paragraph where M is fixed and the Kähler class changes. It is also straightforward to show that the space of Reeb vector fields ξ forms a *cone*: if ξ is a Reeb vector field, then $c\xi$ is also a Reeb vector field for another Kähler cone metric on X_0, for any constant $c > 0$. Thus the space $\mathcal{R}(X_0)$ of Reeb polarizations forms a cone, analogous to the Kähler cone in Kähler geometry: we saw this explicitly in the previous section on toric Sasakian manifolds, where the space of Reeb vector fields is the interior $\mathcal{C}_{\mathrm{int}}$ of the polyhedral cone $\mathcal{C}$.

We now suppose that X is also Gorenstein. This is necessary for the existence of a Ricci-flat Kähler metric on X_0. We then introduce the subspace $\mathcal{S}(X_0, \Omega)$ as the space of metrics in $\mathcal{S}(X_0)$ for which the Reeb vector field ξ satisfies

$$\mathcal{L}_\xi\Omega = in\Omega \ . \tag{5.5}$$

Equivalently, Ω should be homogeneous degree n under $r\partial/\partial r$. This is again clearly a necessary condition for a Ricci-flat Kähler cone metric, cf. (4.17).

---

[7] This terminology was introduced in [8].

Recall now that Einstein metrics on L are critical points of the *Einstein–Hilbert action*:

$$\mathcal{I} : \text{Metrics}(L) \to \mathbb{R}$$
$$g_L \mapsto \int_L [s(g_L) + 2(n-1)(3-2n)] \, d\mu \tag{5.6}$$

where $s(g_L)$ is the scalar curvature of g_L. We then have the following proposition:

Proposition 5.3 ([35]). *The Einstein–Hilbert action, as a functional on $\mathcal{S}(X_0)$, depends only on the Reeb vector field ξ. It may thus be regarded as a function of ξ. Moreover, for Sasakian metrics $g_L \in \mathcal{S}(X_0, \Omega)$, we have*

$$\mathcal{I}(g_L) = 4(n-1)\,\text{vol}[g_L] \, . \tag{5.7}$$

Thus the Einstein–Hilbert action restricted to the space $\mathcal{S}(X_0, \Omega)$ is simply the volume functional and depends only on the Reeb vector field ξ of the metric. This suggests we introduce

$$\mathcal{R}(X_0, \Omega) = \{\xi \in \mathcal{R}(X_0) \mid \mathcal{L}_\xi \Omega = in\Omega\} \, . \tag{5.8}$$

Since Sasaki–Einstein metrics are critical points of $\mathcal{I}$, we see that the Reeb vector field for a Sasaki–Einstein metric is determined by a *finite-dimensional extremal problem*, namely $d\mathcal{I} = 0$, where $\mathcal{I}$ is interpreted as a function on $\mathcal{R}(X_0, \Omega)$. For toric varieties, this is particularly simple:

Theorem 5.4. *Let X be an affine toric Gorenstein variety with fan (or Reeb polytope) $\mathcal{C} \subset \mathfrak{t}_n$, and generating vectors of the form $v_a = (1, w_a)$. Then the Reeb vector field ξ for a Ricci-flat Kähler cone metric on X_0 is uniquely determined as the critical point of the Euclidean volume of the polytope $\Delta(\xi)$*

$$\text{vol}[\Delta] : N_{int} \to \mathbb{R}_+ \tag{5.9}$$

where N is the $(n-1)$-dimensional polytope $N = \{\xi \in \mathcal{C} \mid \langle(1,0,\dots,0), \xi\rangle = n\}$.

Here we may take $\mathcal{R}(X_0) = \mathcal{C}_{int}$ and $\mathcal{R}(X_0, \Omega) = N_{int}$, following the classification of toric Sasakian metrics in Section 4. The first and second derivatives (5.3), (5.4) in Proposition 5.2 may be written [34]

$$\frac{\partial \, \text{vol}[\Delta]}{\partial \xi_i} = \frac{1}{2\xi_k \xi_k} \int_{H(\xi)} y^i \, d\sigma \tag{5.10}$$

$$\frac{\partial^2 \, \text{vol}[\Delta]}{\partial \xi_i \partial \xi_j} = \frac{2(n+1)}{\xi_k \xi_k} \int_{H(\xi)} y^i y^j \, d\sigma \, . \tag{5.11}$$

Here $d\sigma$ is the standard measure induced on the $(n-1)$-polytope $H(\xi) \subset \mathcal{C}^*$. Uniqueness and existence of the critical point follows from a standard convexity argument: $\text{vol}[\Delta]$ is a strictly convex (by (5.11)) positive function on the interior of a compact convex polytope N. Moreover, $\mathrm{vol}[\Delta]$ diverges to $+\infty$ at ∂N. It follows that $\mathrm{vol}[\Delta]$ must have precisely one critical point in the interior of N.

Given these results, it is natural to consider the general case, with $s \leq n$. We begin by recalling a classical result. Let $-\nabla_L^2$ be the scalar Laplacian on (L, g_L), with spectrum $\{E_\nu\}_{\nu=0}^\infty$. Then we may define the heat kernel trace

$$\Theta(t) = \sum_{\nu=0}^\infty \exp(-tE_\nu) \tag{5.12}$$

where $t \in (0, \infty)$. There is a holomorphic analogue of this. Let f be a holomorphic function on X_0 with

$$\mathcal{L}_\xi f = \lambda i f \tag{5.13}$$

where $\mathbb{R} \ni \lambda > 0$, and we refer to λ as the charge of f under ξ. Since f is holomorphic, this immediately implies that

$$f = r^\lambda \tilde{f} \tag{5.14}$$

where $\tilde{f}$ is homogeneous degree zero under $r\partial/\partial r$; that is, $\tilde{f}$ is the pull-back to X of a function on the link L. Moreover, since (X_0, g) is Kähler, one can show that f is *harmonic*, and that

$$-\nabla_L^2 \tilde{f} = E \tilde{f} \tag{5.15}$$

where

$$E = \lambda[\lambda + (2n - 2)] . \tag{5.16}$$

Thus any holomorphic function f of definite charge under ξ, or equivalently degree under $r\partial/\partial r$, corresponds to an eigenfunction of the Laplacian $-\nabla_L^2$ on the link. The charge λ is then related simply to the eigenvalue E by the above formula (5.16). We may thus in particular define the holomorphic spectral invariant

$$Z(t) = \sum_{i=0}^\infty \exp(-t\lambda_i) \tag{5.17}$$

where $\{\lambda_i\}_{i=0}^\infty$ is the holomorphic spectrum, in the above sense. This is also the trace of a kernel, namely the *Szegö kernel*.

Recall the following classical result:

Theorem 5.5 ([36]). *Let (L, g_L) be a compact Riemannian manifold with heat kernel trace $\Theta(t)$ given by (5.12). Then*

$$\mathrm{vol}[g_L] = \lim_{t \searrow 0} (4\pi t)^{n - \frac{1}{2}} \Theta(t) . \tag{5.18}$$

In the holomorphic setting we have:

Theorem 5.6 ([35]). *For $g_L \in \mathcal{S}(X_0, \Omega)$ with Reeb vector field ξ we have*

$$\frac{\text{vol}[g_L]}{\text{vol}[S^{2n-1}, g_{\text{can}}]} = \lim_{t \searrow 0} t^n \, Z(t) . \tag{5.19}$$

This result first appeared for regular Sasaki–Einstein manifolds in [2]. The proof of Theorem 5.6 is essentially the Riemann–Roch theorem. Suppose that $g_L \in \mathcal{S}(X_0, \Omega)$ is quasi-regular,[8] so that ξ generates a $U(1)$ action on L. The quotient is a *Fano* orbifold (V, ω_V):

Definition. A compact Kähler orbifold (V, ω_V) is *Fano* if the cohomology class of the Ricci-form in $H^{1,1}(V)$ is represented by a positive $(1,1)$-form.

Holomorphic functions on X_0 that are eigenstates under $\mathcal{L}_\xi$ correspond to holomorphic sections of a holomorphic orbifold line bundle $\mathcal{L}^{-k} \to V$ for some k; here $\mathcal{L}$ is the associated holomorphic line orbibundle to the $U(1)$ principal orbibundle $U(1) \hookrightarrow L \to V$. This is holomorphic since the curvature is proportional to the Kähler form on V, which is of Hodge type $(1,1)$. The number of holomorphic sections is given by an orbifold version of the Riemann–Roch theorem, involving characteristic classes on V. In general, this is rather more complicated than the smooth Riemman–Roch theorem, but the limit in Theorem 5.6 simplifies the formula considerably: only a leading term contributes. The essential point now is that the volume may also be written in terms of Chern classes:

Proposition 5.7. *For a quasi-regular Sasakian metric $g_L \in \mathcal{S}(X_0, \Omega)$, one has*

$$\frac{\text{vol}[g_L]}{\text{vol}[S^{2n-1}, g_{\text{can}}]} = \frac{\beta}{n^n} \int_V c_1(V)^{n-1} . \tag{5.20}$$

Here $(S^{2n-1}, g_{\text{can}})$ is the round sphere metric; $\beta \in \mathbb{Q}$ is defined by

$$c_1(\mathcal{L}) = -\frac{c_1(V)}{\beta} \in H^2_{\text{orb}}(V; \mathbb{Z}) \tag{5.21}$$

where $H^2_{\text{orb}}(V; \mathbb{Z})$ is the orbifold cohomology of Haefliger[9] [26], $c_1(V)$ is the first Chern class of the holomorphic tangent bundle of V, and $\mathcal{L}$ is the orbifold line bundle associated to the $U(1)$ principal orbibundle $U(1) \hookrightarrow L \to V$.

The cohomology group $H^2_{\text{orb}}(V; \mathbb{Z})$ classifies orbifold line bundles over V, in exactly the same way that $H^2(V; \mathbb{Z})$ classifies line bundles when V is smooth. The

---

[8] The reader who is uneasy with orbifolds may take a regular Sasakian manifold in what follows. However, the point here is that quasi-regular Reeb vector fields are dense in the space of Reeb vector fields, since the rationals are dense in the reals; regular Sasakian structures are considerably more special.

[9] One defines $H^*_{\text{orb}}(V; \mathbb{Z}) = H^*(BV; \mathbb{Z})$ where BV is the classifying space for V. For details, see for example [4].

proposition relates the volume of $g_L \in \mathcal{S}(X_0, \Omega)$, for quasi-regular ξ, to characteristic classes of V. Theorem 5.6 then follows from the fact that quasi-regular Reeb vector fields are dense in the space of all Reeb vector fields, and vol is continuous.

We end this section with a localization formula for the volume. We first require a definition.

Definition. Let (X, Ω) be a Gorenstein singularity with isolated singular point p, $X_0 = \mathbb{R}_+ \times L$, and let $\mathcal{S}(X_0, \Omega)$ be as above, with respect to an effective holomorphic action of $\mathbb{T}^s$. We say that an orbifold $\hat{X}$ is a *partial resolution* of X if

$$\pi : \hat{X} \to X \tag{5.22}$$

is a $\mathbb{T}^s$-equivariant map with $\pi : \hat{X} \setminus E \to X_0$ a $\mathbb{T}^s$-equivariant biholomorphism for some exceptional set E. If $\hat{X}$ is smooth, we say that it is a resolution of X.

First note that such a partial resolution $\hat{X}$ always exists: one can take any quasi-regular Reeb vector field $\xi \in \mathfrak{t}_s$ and blow up the corresponding orbifold V. In this case, the exceptional set $E = V$ and the partial resolution is clearly equivariant. We then have the following:

Theorem 5.8 ([35]). *Let $g_L \in \mathcal{S}(X_0, \Omega)$ and pick a partial resolution $\hat{X}$ of X. Suppose that the Kähler form of X_0 extends to a suitable smooth family of Kähler forms on $\hat{X}$ (see [35] for details). Then*

$$\frac{\text{vol}[g_L]}{\text{vol}[S^{2n-1}, g_{\text{can}}]} = \sum_{\{F\}} \frac{1}{d_F} \int_F \prod_{m=1}^R \frac{1}{\langle \xi, u_m \rangle^{n_m}} \left[\sum_{a \geq 0} \frac{c_a(\mathcal{E}_m)}{\langle \xi, u_m \rangle^a} \right]^{-1}. \tag{5.23}$$

Here

- $E \supset \{F\} = $ *set of connected components of the fixed point set, where ξ is a generic vector $\xi \in \mathfrak{t}_s$; that is, the orbits of ξ are dense in the torus $\mathbb{T}^s$.*
- *For fixed connected component F, the linearized $\mathbb{T}^s$ action on the normal bundle $\mathcal{E}$ of F in $\hat{X}$ is determined by a set of weights $u_1, \ldots, u_R \in \mathbb{Q}^s \subset \mathfrak{t}_s^*$. $\mathcal{E}$ then splits $\mathcal{E} = \bigoplus_{m=1}^R \mathcal{E}_m$ where $\text{rank}_{\mathbb{C}} \mathcal{E}_m = n_m$ and $\sum_{m=1}^R n_m = \text{rank}_{\mathbb{C}}(\mathcal{E})$.*
- $c_a(\mathcal{E}_m)$ *are the Chern classes of $\mathcal{E}_m$.*
- *When $\hat{X}$ has orbifold singularities, the normal fiber to a generic point on F is not a complex vector space, but rather an orbifold $\mathbb{C}^l/\Gamma$. Then $\mathcal{E}$ is more generally an orbibundle, and $d_F = |\Gamma|$ denotes the order of Γ.*

This theorem is proved as follows. One notes that the volume $\text{vol}[g_L]$ may be written [35] as

$$\text{vol}[g_L] = \frac{1}{2^{n-1}(n-1)!} \int_{X_0} e^{-r^2/2} \frac{\omega^n}{n!}. \tag{5.24}$$

The function $r^2/2$ is precisely the *Hamiltonian function* for the Reeb vector field ξ. One may then naively apply the theorem of Duistermaat–Heckman [15, 16], which easily extends to noncompact manifolds and orbifolds. This theorem localizes such an integral to the fixed point set of ξ. However, since $\|\xi\|_g^2 = r^2$, the integral formerly localizes at the *singular point* $p = \{r = 0\}$ of X. To obtain a sensible answer, one must first resolve the singularity, as in Theorem 5.8. Such a proof requires that the Kähler form of X_0 extends to a suitable smooth family of Kähler forms on $\hat{X}$. One then applies the Duistermaat–Heckman theorem to this family and takes the cone limit (X, g). The limit is independent of the choice of resolving family of Kähler metrics on $\hat{X}$.

Remark. The author believes that the technical condition requiring the ability to extend the Kähler form ω on X_0 to a smooth family of Kähler forms on $\hat{X}$ is probably redundant. In fact, one can also formerly apply the *equivariant* Riemann–Roch theorem to $\hat{X}$, with respect to the holomorphic action of $\mathbb{T}^s$, and Proposition 5.7 to obtain the same result. This has the advantage of not requiring existence of any Kähler metrics.

Note the theorem guarantees that vol : $\mathcal{R}(X_0) \to \mathbb{R}_+$, relative to the volume of the round sphere, is a rational function of ξ with rational coefficients. Since vol is strictly convex, its critical points are isolated. By Theorem 5.8, one thus sees that the volume of a Sasaki–Einstein manifold, relative to that of the round sphere, is an *algebraic number*.

Since the expression in Theorem 5.8 is rather formidable, we end with an example.

Example. Let X be an affine toric variety with moment polytope C^*. Let $\pi : \hat{X} \to X$ be a toric resolution of X, with some choice of Kähler metric, and denote the moment polytope by $\hat{\mu}(\hat{X}) = P \subset \mathfrak{t}_n^*$. It is standard that such a resolution and Kähler metric always exist. Let $\text{Vert}(P)$ denote the set of vertices of P. For each vertex $A \in \text{Vert}(P)$, there are precisely n outward-pointing edge vectors; these may be taken to be primitive vectors $u_A^i \in \mathbb{Z}^n \subset \mathfrak{t}_n^*$, $i = 1, \ldots, n$. This follows since the resolution is smooth. Then

$$\frac{\text{vol}[g_L]}{\text{vol}[S^{2n-1}, g_{\text{can}}]} = \sum_{A \in \text{Vert}(P)} \prod_{i=1}^n \frac{1}{\langle \xi, u_A^i \rangle} . \qquad (5.25)$$

Note that this gives an unusual way of computing the Euclidean volume of the polytope $\Delta(\xi)$, which, by (4.16), is also essentially the left-hand side of (5.25).

6 Obstructions to the Existence of Sasaki–Einstein Metrics

In this final section we examine the following:

Problem 6.1. Let (X, Ω) be a Gorenstein singularity with isolated singular point p, $X \setminus \{p\} = X_0 = \mathbb{R}_+ \times L$, and let $g_L \in \mathcal{S}(X_0, \Omega)$ be a Sasakian metric with Reeb vector field satisfying (5.5). When does X_0 admit a Ricci-flat Kähler cone metric with this Reeb vector field?

We shall describe three obstructions. The first is a natural corollary of the previous section, whereas the remaining two obstructions are based on classical theorems in differential geometry, and in particular lead to new obstructions to the existence of Kähler–Einstein orbifold metrics.

The previous section implies that a Sasaki–Einstein metric is a critical point of the volume functional, thought of as a function on the space $\mathcal{R}(X_0, \Omega)$ of Reeb vector fields satisfying (5.5). Thus the Reeb vector field in Problem 6.1 must be a critical point of this function, or equivalently of the Einstein–Hilbert action on L. We have already given this condition, in Proposition 5.2:

$$d\,\mathrm{vol}(Y) = -n \int_L \eta(Y) \,. \tag{6.1}$$

Here $Y \in \mathfrak{t}_s$ and recall that η is the contact one-form on L. For a Sasaki–Einstein metric (6.1) must be zero for all holomorphic vector fields $Y \in \mathfrak{t}_s$ satisfying $\mathcal{L}_Y \Omega = 0$. Suppose that the Sasakian metric $g_L \in \mathcal{S}(X_0, \Omega)$ is quasi-regular. Then there is a quotient $V = L/U(1)$ that is a Fano orbifold. In this case, we have the following:

Theorem 6.2. *Let $g_L \in \mathcal{S}(X_0, \Omega)$ be quasi-regular and let $Y \in \mathfrak{t}_s$ with $\mathcal{L}_Y \Omega = 0$. Then*

$$d\,\mathrm{vol}(Y) = -\frac{\ell}{2} F\left(\mathcal{J}_V(Y_V)\right) \,. \tag{6.2}$$

Here $\ell = 2\pi\beta/n$ is the length of the generic Reeb S^1 fiber, $\mathcal{J}_V$ is the complex struture tensor on V, Y_V is the push-forward of Y to V, and $F : \mathrm{aut}(V) \to \mathbb{R}$ is the Futaki *invariant of V.*

The Futaki invariant is a well-known obstruction to the existence of a Kähler–Einstein metric on V [19]. It is conventionally defined as follows. Let (V, ω_V) be a compact Kähler orbifold with Kähler form satisfying $[\rho_V] = 2n[\omega_V] \in H^{1,1}(V)$, where $\rho_V = \mathrm{Ric}(\mathcal{J}_V \cdot, \cdot)$ is the Ricci-form of (V, ω_V). Then there exists a real function f on V, unique up to an additive constant, satisfying

$$\rho_V - 2n\omega_V = i\partial\bar{\partial} f \,. \tag{6.3}$$

Given any real holomorphic vector field $\zeta \in \mathrm{aut}(V)$, we define

$$F(\zeta) = \int_V \mathcal{L}_\zeta f \, \frac{\omega_V^{n-1}}{(n-1)!} \,. \tag{6.4}$$

Clearly, if (V, ω_V) is Kähler–Einstein of scalar curvature $4n(n-1)$, then f is constant, and the function $F : \mathrm{aut}(V) \to \mathbb{R}$ vanishes. However, more generally F also satisfies the following rather remarkable properties [9, 19]:

- F is independent of the choice of Kähler metric representing $[\omega_V]$.
- The complexification $F_{\mathbb{C}}$ is a Lie algebra homomorphism $F_{\mathbb{C}} : \mathrm{aut}_{\mathbb{C}}(V) \to \mathbb{C}$.

Note that the first point is implied by Proposition 5.1 and Theorem 6.2. The second is due to Calabi [9]. The Sasakian setting thus gives a *dynamical* interpretation of the Futaki invariant.

We now turn to two further simple obstructions to the existence of solutions to Problem 6.1. These are based on the classical theorems of Lichnerowicz [30] and Bishop [3], respectively. We begin with the latter. Recall from Proposition 5.1 that the volume of a Sasakian metric $g_L \in \mathcal{S}(X_0, \Omega)$ is determined by its Reeb vector field ξ. In particular, one can compute this volume $\mathrm{vol}(\xi)$ using Theorem 5.6. In some cases, one can compute the trace over holomorphic functions on X directly. Now, Bishop's theorem [3] implies that for any $(2n-1)$-dimensional Einstein manifold (L, g_L) with $\mathrm{Ric}(g_L) = 2(n-1)g_L$, we have

$$\mathrm{vol}(L, g_L) \le \mathrm{vol}(S^{2n-1}, g_{\mathrm{can}}) . \tag{6.5}$$

Combining these two results, we have

Theorem 6.3 (Bishop obstruction [25]). *Let* (X, Ω) *be as in Problem 6.1, with Reeb vector field* ξ. *If* $\mathrm{vol}(\xi) > \mathrm{vol}(S^{2n-1}, g_{\mathrm{can}})$, *then* X_0 *admits no Ricci-flat Kähler cone metric with Reeb vector field* ξ.

A priori, it is not clear this condition can ever obstruct existence. We shall provide examples below. However, in [25] we conjectured that for *regular* Reeb vector fields ξ, Theorem 6.3 never obstructs. This is equivalent to the following conjecture about smooth Fano manifolds:

Conjecture 6.4 ([25]). *Let V be a smooth Fano manifold of complex dimension* $n-1$ *with Fano index* $I(V) \in \mathbb{N}$. *Then*

$$I(V) \int_V c_1(V)^{n-1} \le n \int_{\mathbb{CP}^{n-1}} c_1(\mathbb{CP}^{n-1})^{n-1} = n^n \tag{6.6}$$

with equality if and only if $V = \mathbb{CP}^{n-1}$.

This is related to, although slightly different from, a standard conjecture about Fano manifolds. For further details, see [25].

We turn now to the Lichnerowicz obstruction. Suppose that (L, g_L) is Einstein with $\mathrm{Ric}(g_L) = 2(n-1)g_L$. The first non-zero eigenvalue $E_1 > 0$ of $-\nabla_L^2$ is bounded from below:

$$E_1 \ge 2n - 1. \tag{6.7}$$

This is Lichnerowicz's theorem [30]. Moreover, equality holds if and only if (L, g_L) is isometric to the round sphere $(S^{2n-1}, g_{\mathrm{can}})$ [38]. From (5.16), we immediately see that for holomorphic functions f on X_0 of charge λ under ξ, Lichnerowicz's bound becomes $\lambda \ge 1$. This leads to a potential holomorphic obstruction to the existence of Sasaki–Einstein metrics:

Theorem 6.5 (Lichnerowicz obstruction [25]). *Let (X, Ω) be as in Problem 6.1, with Reeb vector field ξ. Suppose that f is a holomorphic function on X of positive charge $\lambda < 1$ under ξ. Then X_0 admits no Ricci-flat Kähler cone metric with Reeb vector field ξ.*

Again, it is not immediately clear that this can ever obstruct existence. Indeed, for *regular* ξ, one can prove [25] that this never obstructs. This follows from the fact that $I(V) \leq n$ for any smooth Fano V of complex dimension $n - 1$. However, there exist plenty of obstructed quasi-regular examples.

Notice then that the Lichnerowicz obstruction involves holomorphic functions on X of small charge with respect to ξ, whereas the Bishop obstruction is a statement about the volume, which is determined by the asymptotic growth of holomorphic functions on X, analogously to Weyl's asymptotic formula.

Example. Our main set of examples of Theorems 6.3 and 6.5 is provided by isolated quasi-homogeneous hypersurface singularities. Let $\mathbf{w} \in \mathbb{N}^{n+1}$ be a vector of positive weights. This defines an action of $\mathbb{C}^*$ on $\mathbb{C}^{n+1}$ via

$$(z_1, \dots, z_{n+1}) \mapsto (q^{w_1} z_1, \dots, q^{w_{n+1}} z_{n+1}) \tag{6.8}$$

where $q \in \mathbb{C}^*$. Without loss of generality, one can take the set $\{w_i\}$ of components of $\mathbf{w}$ to have no common factor. This ensures that the above $\mathbb{C}^*$ action is effective. Let

$$F : \mathbb{C}^{n+1} \to \mathbb{C} \tag{6.9}$$

be a quasi–homogeneous polynomial on $\mathbb{C}^{n+1}$ with respect to $\mathbf{w}$. This means that F has definite degree d under the above $\mathbb{C}^*$ action:

$$F(q^{w_1} z_1, \dots, q^{w_{n+1}} z_{n+1}) = q^d F(z_1, \dots, z_{n+1}) . \tag{6.10}$$

Moreover, we assume that the affine algebraic variety

$$X = \{F = 0\} \subset \mathbb{C}^{n+1} \tag{6.11}$$

is smooth everywhere except at the origin $(0, 0, \dots, 0)$. For obvious reasons, such X are called isolated quasi-homogeneous hypersurface singularities. The corresponding link L is the intersection of X with the unit sphere in $\mathbb{C}^{n+1}$:

$$\sum_{i=1}^{n+1} |z_i|^2 = 1 . \tag{6.12}$$

We define a nowhere zero holomorphic $(n, 0)$-form Ω on the smooth part of X by

$$\Omega = \frac{dz_1 \wedge \dots \wedge dz_n}{\partial F / \partial z_{n+1}} . \tag{6.13}$$

This defines Ω on the patch where $\partial F/\partial z_{n+1} \neq 0$. One has similar expressions on patches where $\partial F/\partial z_i \neq 0$ for each i, and it is simple to check that these glue together into a nowhere zero form Ω on X_0. Thus all such X are Gorenstein, and moreover they come equipped with a holomorphic $\mathbb{C}^*$ action by construction. The orbit space of this $\mathbb{C}^*$ action, or equivalently the orbit space of $U(1) \subset \mathbb{C}^*$ on the link, is a complex orbifold V. In fact, V is the weighted variety defined by $\{F = 0\}$ in the weighted projective space $\mathbb{WCP}^n_{[w_1, w_2, \ldots, w_{n+1}]}$. It is not difficult to show that V is a Fano orbifold if and only if

$$|\mathbf{w}| - d > 0 \tag{6.14}$$

where $|\mathbf{w}| = \sum_{i=1}^{n+1} w_i$. To see this, first notice that $|\mathbf{w}| - d$ is the charge of Ω under $U(1) \subset \mathbb{C}^*$. To be precise, if ζ denotes the holomorphic vector field on X with

$$\mathcal{L}_\zeta z_j = w_j i z_j \tag{6.15}$$

for each $j = 1, \ldots, n+1$, then

$$\mathcal{L}_\zeta \Omega = (|\mathbf{w}| - d) i \Omega . \tag{6.16}$$

Positivity of this charge then implies [35] that the cohomology class of the natural Ricci-form induced on V is represented by a positive $(1,1)$-form, which is the definition that V is Fano. If there exists a Ricci-flat Kähler metric on X_0 that is a cone under $\mathbb{R}_+ \subset \mathbb{C}^*$, then the correctly normalized Reeb vector field $\xi \in \mathcal{R}(X_0, \Omega)$ is thus

$$\xi = \frac{n}{|\mathbf{w}| - d} \zeta . \tag{6.17}$$

Bishop's theorem then requires, for existence of a Sasaki–Einstein metric on L with Reeb vector field ξ,

$$d(|\mathbf{w}| - d)^n \leq wn^n \tag{6.18}$$

where $w = \prod_{i=1}^{n+1} w_i$ is the product of the weights. The computation of the volume that gives this inequality may be found in [25]. It is simple to write down infinitely many examples of isolated quasi-homogeneous hypersurface singularities that violate this inequality and are thus obstructed by Theorem 6.3.

Lichnerowicz's theorem requires, on the other hand, that

$$|\mathbf{w}| - d \leq nw_{\min} \tag{6.19}$$

where $w_{\min}$ is the smallest weight. Moreover, this bound can be saturated if and only if (X_0, g) is $\mathbb{C}^n \setminus \{0\}$ with its flat metric. It is again clearly trivial to construct many examples of isolated hypersurface singularities that violate this bound and are hence obstructed by Theorem 6.5.

We conclude by making some remarks on the possible solution to Problem 6.1. Let us first comment on the case of toric varieties. The Reeb vector field for a critical point of the Einstein–Hilbert action, considered as a function on $\mathcal{R}(X_0, \Omega)$, exists and is unique, by Theorem 5.4. The remaining condition for a Ricci-flat Kähler cone metric may be written as a real Monge–Ampère equation on the polytope $\mathcal{C}^*$ [34]. This has recently been shown to always admit a solution in [20]. We state this as:

Theorem 6.6 ([20]). *Let (X, Ω) be an affine toric Gorenstein singularity. Then X_0 admits a $\mathbb{T}^n$-invariant Ricci-flat Kähler cone metric, with Reeb vector field determined by Theorem 5.4.*

Thus toric varieties are unobstructed. However, this still leaves us with the general non-toric case. Problem 6.1 is in fact closely related to a major open conjecture in the Kähler category. If ξ is *regular*, then existence of a Ricci-flat Kähler cone metric on X_0 with this Reeb vector field is equivalent to existence of a Kähler–Einstein metric on the Fano V, and we then have the following conjecture due to Yau:

Conjecture 6.7 ([45]). *A Fano manifold V admits a Kähler–Einstein metric if and only if it is stable in the sense of geometric invariant theory.*

A considerable amount of progress has been made on this conjecture, notably by Tian and Donaldson. However, it is still open in general. The conjecture is closely related to Problem 6.1, and it is clearly of interest to extend the work in the Kähler setting to the Sasakian setting.

Physics might also provide a different viewpoint. Suppose that X admits a *crepant resolution* $\pi : \hat{X} \to X$. This means that the holomorphic volume form Ω on X_0 extends smoothly[10] as a holomorphic volume form onto the resolution $\hat{X}$. Then one expects the derived category of coherent sheaves $\mathcal{D}^b(\mathrm{coh}(\hat{X}))$ on $\hat{X}$ to be equivalent to the derived category of representations $\mathcal{D}^b(\mathrm{Reps}\,(\mathcal{Q}, \mathcal{R}))$ of a quiver $\mathcal{Q}$ with relations $\mathcal{R}$. In fact, this could be stated formerly as a conjecture; it is known to be true for various sets of examples. A quiver $\mathcal{Q}$ is simply a directed graph. $\mathcal{R}$ is a set of relations on the path algebra $\mathbb{C}\mathcal{Q}$ of the quiver. The above correspondence between derived categories, if correct, would allow for a more precise mathematical statement of what the AdS/CFT map is. Physically, one is placing D3-branes at the singular point p of X; mathematically, a D3-brane at a point on X corresponds to the structure sheaf of that point. The corresponding representation of $\mathcal{Q}$ then defines an $\mathcal{N} = 1$ supersymmetric quantum field theory in four dimensions; this is precisely the quantum field theory on the D3-brane. The quiver representation determines the gauge group and matter content of the quantum field theory, and the relations specify the superpotential, which determines the interactions. The choice of $(\mathcal{Q}, \mathcal{R})$ is known to be nonunique, and this nonuniqueness is related to a duality known as Seiberg duality [40]. The AdS/CFT correspondence then implies that X admits a Ricci-flat Kähler cone metric (in dimension $n = 3$) if and only if this supersymmetric quantum field theory flows to a dual infrared fixed point under renormalization

---

[10] This is certainly an extra constraint. For example, the singularities $w^{2k+1} + x^2 + y^2 + z^2 = 0$ admit no crepant resolution by an argument in [25].

group flow; this infrared fixed point is precisely the superconformal field theory in the AdS/CFT correspondence. Thus the solution to Problem 6.1, in complex dimension $n = 3$, is related to the low-energy behavior of certain *supersymmetric quiver gauge theories* in four dimensions.

Acknowledgments I would very much like to thank the organizers of the conference for inviting me to speak. I also thank my collaborators: J. Gauntlett, D. Martelli, D. Waldram, and S.-T. Yau. This work was supported by NSF grants DMS-0244464, DMS-0074329, and DMS-9803347.

References

1. V. Apostolov, D. M. Calderbank, P. Gauduchon, "The geometry of weakly selfdual Kähler surfaces," Compositio Math. **135** (2003), 279–322.
2. A. Bergman, C. P. Herzog, "The volume of some non-spherical horizons and the AdS/CFT correspondence," JHEP **0201**, 030 (2002) [arXiv:hep-th/0108020].
3. R. L. Bishop, R. J. Crittenden, "Geometry of Manifolds," Academic Press, New York, 1964.
4. C. P. Boyer, K. Galicki, "On Sasakian–Einstein geometry," Int. J. Math. **11** (2000), no. 7, 873–909 [arXiv:math.DG/9811098].
5. C. P. Boyer, K. Galicki, "A note on toric contact geometry," J. Geom. Phys. **35** No. 4 (2000), 288–298, [arXiv:math.DG/9907043].
6. C. P. Boyer, K. Galicki, "Sasakian Geometry, Hypersurface Singularities, and Einstein Metrics," Supplemento ai Rendiconti del Circolo Matematico di Palermo Serie II. Suppl 75 (2005), 57–87 [arXiv:math.DG/0405256].
7. C. P. Boyer, K. Galicki, "Sasakian Geometry," Oxford Mathematical Monographs, Oxford University Press, ISBN-13: 978-0-19-856495-9.
8. C. P. Boyer, K. Galicki, S. R. Simanca, "Canonical Sasakian Metrics," Commun. Math. Phys. **279** (2008), 705–733.
9. E. Calabi, "Extremal Kähler metrics II," in "Differential Geometry and Complex Analysis" (ed. I. Chavel and H. M. Farkas), Springer-Verlag, Berlin, 1985.
10. J. Cheeger, G. Tian, "On the cone structure at infinity of Ricci flat manifolds with Euclidean volume growth and quadratic curvature decay," Invent. Math. **118** (1994), no. 3, 493–571.
11. W. Chen, H. Lu, C. N. Pope, J. F. Vazquez-Poritz, "A note on Einstein–Sasaki metrics in $D \geq 7$," Class. Quant. Grav. **22** (2005), 3421–3430 [arXiv:hep-th/0411218].
12. D. Conti, "Cohomogeneity one Einstein-Sasaki 5-manifolds," arXiv:math.DG/0606323.
13. M. Cvetic, H. Lu, D. N. Page, C. N. Pope, "New Einstein-Sasaki spaces in five and higher dimensions," Phys. Rev. Lett. **95**, 071101 (2005) [arXiv:hep-th/0504225].
14. M. Cvetic, H. Lu, D. N. Page, C. N. Pope, "New Einstein-Sasaki and Einstein spaces from Kerr-de Sitter," arXiv:hep-th/0505223.
15. J. J. Duistermaat, G. Heckman, "On the variation in the cohomology of the symplectic form of the reduced space," Inv. Math. **69**, 259–268 (1982).
16. J. J. Duistermaat, G. Heckman, Addendum, Inv. Math. **72**, 153–158 (1983).
17. S. Falcao de Moraes, C. Tomei, "Moment maps on symplectic cones," Pacific J. Math. 181 (2) (1997), 357–375.
18. Th. Friedrich, I. Kath, "Einstein manifolds of dimension five with small first eigenvalue of the Dirac operator," J. Differential Geom. **29** (1989), 263–279.
19. A. Futaki, "An obstruction to the existence of Einstein Kähler metrics," Invent. Math., **73** (1983), 437–443.
20. A. Futaki, H. Ono, G. Wang, "Transverse Kähler geometry of Sasaki manifolds and toric Sasaki–Einstein manifolds," arXiv:math.DG/0607586.
21. J. P. Gauntlett, D. Martelli, J. Sparks, D. Waldram, "Supersymmetric AdS$_5$ solutions of M-theory," Class. Quant. Grav. **21**, 4335 (2004) [arXiv:hep-th/0402153].

22. J. P. Gauntlett, D. Martelli, J. Sparks, D. Waldram, "Sasaki–Einstein metrics on $S^2 \times S^3$," Adv. Theor. Math. Phys. **8**, 711 (2004) [arXiv:hep-th/0403002].

23. J. P. Gauntlett, D. Martelli, J. F. Sparks, D. Waldram, "A new infinite class of Sasaki–Einstein manifolds," Adv. Theor. Math. Phys. **8**, 987 (2004) [arXiv:hep-th/0403038].

24. J. P. Gauntlett, D. Martelli, J. Sparks, D. Waldram, "Supersymmetric AdS Backgrounds in String and M-theory," IRMA Lectures in Mathematics and Theoretical Physics, vol. 8, published by the European Mathematical Society [arXiv:hep-th/0411194].

25. J. P. Gauntlett, D. Martelli, J. Sparks, S.-T. Yau, "Obstructions to the existence of Sasaki–Einstein metrics," Commun. Math. Phys. **273**, 803–827 (2007) [arXiv:hep-th/0607080].

26. A. Haefliger, "Groupoides d'holonomie et classifiants," Astérisque **116** (1984), 70–97.

27. E. Kähler, "Über eine bemerkenswerte Hermitesche Metrik," Abh. Math. Sem. Hamburg Univ. **9** (1933), 173–186.

28. J. Kollár, Y. Miyaoka, S. Mori, "Rational connectedness and boundedness of Fano manifolds," J. Diff. Geom. **36**, 765–769 (1992).

29. E. Lerman, "Contact toric manifolds," J. Symplectic Geom. 1 (2003), no. 4, 785–828 [arXiv:math.SG/0107201].

30. A. Lichnerowicz, "Géometrie des groupes de transformations," III, Dunod, Paris, 1958.

31. J. M. Maldacena, "The large N limit of superconformal field theories and supergravity," Adv. Theor. Math. Phys. **2**, 231 (1998) [Int. J. Theor. Phys. **38**, 1113 (1999)] [arXiv:hep-th/9711200].

32. D. Martelli, J. Sparks, "Toric geometry, Sasaki–Einstein manifolds and a new infinite class of AdS/CFT duals," Commun. Math. Phys. **262**, 51 (2006) [arXiv:hep-th/0411238].

33. D. Martelli, J. Sparks, "Toric Sasaki–Einstein metrics on $S^2 \times S^3$," Phys. Lett. B **621**, 208 (2005) [arXiv:hep-th/0505027].

34. D. Martelli, J. Sparks, S.-T. Yau, "The geometric dual of a-maximisation for toric Sasaki–Einstein manifolds," Commun. Math. Phys. **268**, 39–65 (2006) [arXiv:hep-th/0503183].

35. D. Martelli, J. Sparks, S.-T. Yau, "Sasaki–Einstein Manifolds and Volume Minimisation," Commun. Math. Phys. **280** (2008), 611–673 arXiv:hep-th/0603021.

36. S. Minakshisundaram, A. Pleijel, "Some properties of the eigenfunctions of the Laplace–operator on Riemannian manifolds," Canadian J. Math. **1** (1949), 242–256.

37. S. B. Myers, "Riemannian manifolds with positive mean curvature," Duke Math. J. **8** (1941), 401–404.

38. M. Obata, "Certain conditions for a Riemannian manifold to be isometric to a sphere," J. Math. Soc. Japan **14** (1962), 333–340.

39. S. Sasaki, "On differentiable manifolds with certain structures which are closely related to almost contact structure," Tôhoku Math. J. **2** (1960), 459–476.

40. N. Seiberg, "Electric - magnetic duality in supersymmetric nonAbelian gauge theories," Nucl. Phys. B **435**, 129 (1995) [arXiv:hep-th/9411149].

41. S. Smale, "On the structure of 5-manifolds," Ann. Math. **75** (1962), 38–46.

42. S. Tanno, "Geodesic flows on C_L-manifolds and Einstein metrics on $S^3 \times S^2$," in "Minimal Submanifolds and Geodesics" (Proc. Japan-United States Sem., Tokyo, 1977), pp. 283–292, North Holland, Amsterdam and New York, 1979.

43. G. Tian, "On Kähler–Einstein metrics on certain Kähler manifolds with $c_1(M) > 0$," Invent. Math. **89** (1987) 225–246.

44. G. Tian, S.-T. Yau, "On Kähler–Einstein metrics on complex surfaces with $C_1 > 0$," Commun. Math. Phys. **112** (1987) 175–203.

45. S.-T. Yau, "Open problems in geometry," Proc. Symp. Pure Math. **54** (1993), 1–28.

Some Examples of Toric Sasaki–Einstein Manifolds

Craig van Coevering

Abstract A series of examples of toric Sasaki–Einstein 5-manifolds is constructed, which first appeared in the author's Ph.D. thesis [40]. These are submanifolds of the toric 3-Sasakian 7-manifolds of C. Boyer and K. Galicki. And there is a unique toric quasi-regular Sasaki–Einstein 5-manifold associated to every simply connected toric 3-Sasakian 7-manifold. Using 3-Sasakian reduction as in [7, 8], an infinite series of examples is constructed of each odd second Betti number. They are all diffeomorphic to $\#kM_\infty$, where $M_\infty \cong S^2 \times S^3$, for k odd. We then make use of the same framework to construct positive Ricci curvature toric Sasakian metrics on the manifolds $X_\infty \# kM_\infty$ appearing in the classification of simply connected smooth 5-manifolds due to Smale and Barden. These manifolds are not spin, thus do not admit Sasaki–Einstein metrics. They are already known to admit toric Sasakian metrics (cf. [9]) that are not of positive Ricci curvature. We then make use of the join construction of C. Boyer and K. Galicki first appearing in [6], see also [9], to construct infinitely many toric Sasaki–Einstein manifolds with arbitrarily high second Betti number of every dimension $2m + 1 \geq 5$. This is in stark contrast with the analogous case of Fano manifolds in even dimensions.

1 Introduction

A new series of quasi-regular Sasaki–Einstein 5-manifolds is constructed. These examples first appeared in the author's Ph.D. thesis [40]. They are toric and arise as submanifolds of toric 3-Sasakian 7-manifolds. Applying 3-Sasakian reduction to torus actions on spheres, C. Boyer, K. Galicki, et al. [8] produced infinitely many toric 3-Sasakian 7-manifolds. More precisely, one has a 3-Sasakian 7-manifold $\mathcal{S}_\Omega$ for each integral weight matrix Ω satisfying some conditions to ensure smoothness.

C. van Coevering
Department of Mathematics, Massachusetts Institute of Technology, Cambridge, Massachusetts, USA

K. Galicki and S.R. Simanca (eds.), *Riemannian Topology and Geometric Structures on Manifolds,* Progress in Mathematics 271, DOI 10.1007/978-0-8176-4743-8,
© Springer Science+Business Media LLC 2009

This produces infinitely many examples of each $b_2(S_\Omega) \geq 1$. A result of D. Calderbank and M. Singer [12] shows that, up to finite coverings, this produces all examples of toric 3-Sasakian 7-manifolds. Associated to each S_Ω is its twistor space $\mathcal{Z}$, a complex contact Fano 3-fold with orbifold singularities. The action of T^2 complexifies to $T_{\mathbb{C}}^2 = \mathbb{C}^* \times \mathbb{C}^*$ acting on $\mathcal{Z}$. Furthermore, if $\mathbf{L}$ is the line bundle of the complex contact structure, the action defines a pencil

$$E = \mathbb{P}(\mathfrak{t}_{\mathbb{C}}) \subseteq |\mathbf{L}|,$$

where $\mathfrak{t}_{\mathbb{C}}$ is the Lie algebra of $T_{\mathbb{C}}^2$. We determine the structure of the divisors in E. The generic $X \in E$ is a toric variety with orbifold structure whose orbifold anticanonical bundle $\mathbf{K}_X^{-1}$ is positive. The total space M of the associated S^1 orbifold bundle to $\mathbf{K}_X$ has a natural Sasaki–Einstein structure. Associated to any toric 3-Sasakian 7-manifold S with $\pi_1(S) = e$, we have the following diagram.

$$
\begin{array}{ccc}
M & \longrightarrow & S \\
\downarrow & & \downarrow \\
X & \longrightarrow & \mathcal{Z} \\
& & \downarrow \\
& & \mathcal{M}
\end{array}
\qquad (1.1)
$$

The horizontal maps are inclusions, and the vertical are orbifold fibrations. And $\mathcal{M}$ is the 4-dimensional anti–self-dual Einstein orbifold over which S is an $Sp(1)$ or $SO(3)$ orbifold bundle.

It follows from the Smale/Barden classification of smooth 5-manifolds (cf. [38] and [3]) that M is diffeomorphic to $\#k(S^2 \times S^3)$ where $b_2(M) = k$. We have the following:

Theorem 1.1. *Associated to every simply connected toric 3-Sasakian 7-manifold S is a toric quasi-regular Sasaki–Einstein 5-manifold M. If $b_2(S) = k$, then $b_2(M) = 2k + 1$ and*

$$M \underset{diff}{\cong} \#m(S^2 \times S^3), \text{ where } m = b_2(M).$$

This gives an invertible correspondence. That is, given either X or M in diagram 1.1, one can recover the other spaces with their respective geometries.

This gives an infinite series of quasi-regular Sasaki–Einstein structures on $\#m(S^2 \times S^3)$ for every odd $m \geq 3$.

In Section 2, we review the basics of Sasakian geometry. In Section 3, we cover the toric geometry used in the proof of 1.1. The reader can find a complete proof of the solution to the Einstein equations giving Theorem 1.1 in [40]. But the existence problem of Sasaki–Einstein structures on toric Sasakian manifolds is completely solved in [41] and [18]. So this result is summarized in Section 4. The basics of 3-Sasakian manifolds and 3-Sasakian reduction are covered in Section 5. This section also contains some results on anti–self-dual Einstein orbifolds such as the classification result of D. Calderbank and M. Singer used in the correspondence in Theorem 1.1. Section 6 contains the proof of Theorem 1.1 and completes diagram 1.1. The only other possible topological types for simply connected toric Sasakian 5-manifolds are the non-spin manifolds $X_\infty \# kM_\infty$. In Section 7, we use the framework already constructed to construct positive Ricci curvature Sasakian metrics on these manifolds. They are already shown to admit Sasakian structures in [9]. In Section 8, we use the join construction of C. Boyer and K. Galicki [6,9] to construct higher dimensional examples of toric quasi-regular Sasaki–Einstein and positive Ricci curvature manifolds. In particular, in every possible dimension $n \geq 5$ there exist infinitely many toric quasi-regular Sasaki–Einstein manifolds with arbitrarily high second Betti number.

This article concentrates on the quasi-regular case. This is not for lack of interest in irregular Sasaki–Einstein manifolds, but that case is covered well elsewhere, such as in [18, 31, 32].

This article makes copious use of orbifolds, V-bundles on them, and orbifold invariants such as the orbifold fundamental group π_1^{orb} and orbifold cohomology H_{orb}^*. We will use the terminology "V-bundle" to denote an orbifold bundle. The characteristic classes of V-bundles will be elements of orbifold cohomology. The reader unfamiliar with these concepts can consult [24] or the appendices of [5] or [40].

2 Sasakian Manifolds

We summarize the basics of Sasakian geometry in this section. See the survey article of C. Boyer and K. Galicki [5] for more details.

Definition 2.1. *Let (M, g) be a Riemannian manifold of dimension $n = 2m + 1$, and ∇ the Levi–Civita connection. Then (M, g) is Sasakian if either of the following equivalent conditions hold:*
(i) There exists a unit length Killing vector field ξ on M so that the $(1,1)$ tensor $\Phi(X) = \nabla_X \xi$ satisfies the condition

$$(\nabla_X \Phi)(Y) = g(\xi, Y)X - g(X, Y)\xi$$

for vector fields X and Y on M.
(ii) The metric cone $C(M) = \mathbb{R}_+ \times M, \bar{g} = dr^2 + r^2 g$ is Kähler.

Define η to be the one form dual to ξ, i.e., $\eta(X) = g(X, \xi)$. We say that $\{g, \Phi, \xi, \eta\}$ defines a Sasakian structure on M. Note that $D = \ker \eta$ defines a contact structure on M, and Φ defines a CR-structure on D. Also, the integral curves of ξ are geodesics. The one form η extends to a one form on $C(M)$ as $\eta(X) = \frac{1}{r^2}\bar{g}(\xi, X)$. In (ii) of the definition, M is identified with the subset $r = 1$ of $C(M)$, and $\xi = Jr\partial_r$. And the complex structure arises as

$$Jr\partial_r = \xi \quad JY = \Phi(Y) - \eta(Y)r\partial_r \quad \text{for } Y \in TM. \tag{2.1}$$

The Kähler form of $(C(M), \bar{g})$ is given by

$$\frac{1}{2}d(r^2\eta) = \frac{1}{2}dd^c r^2, \tag{2.2}$$

where $d^c = \frac{i}{2}(\bar{\partial} - \partial)$.

Besides the Kähler structure on $C(M)$, there is *transverse Kähler structure* on M. The Killing vector field ξ generates the Reeb foliation $\mathscr{F}_\xi$ on M. The vector field $\xi - iJ\xi = \xi + ir\partial_r$ on $C(M)$ is holomorphic and generates a local $\mathbb{C}^*$-action extending that of ξ on M. The local orbits of this action define a transverse holomorphic structure on $\mathscr{F}_\xi$. One can choose an open covering $\{U_\alpha\}_{\alpha \in A}$ of M such that we have the projection onto the local leaf space $\pi_\alpha : U_\alpha \to V_\alpha$. Then when $U_\alpha \cap U_\beta \neq \emptyset$, the transition

$$\pi_\beta \circ \pi_\alpha^{-1} : \pi_\alpha(U_\alpha \cap U_\beta) \to \pi_\beta(U_\alpha \cap U_\beta)$$

is holomorphic. There is an isomorphism

$$d\pi_\alpha : D_p \to T_{\pi_\alpha(p)}V_\alpha,$$

for each $p \in U_\alpha$, which allows one to define a metric g^T with Kähler form $\omega = \frac{1}{2}d\eta$ as restrictions of g and $\frac{1}{2}d\eta = g(\Phi\cdot, \cdot)$ to D_p. These are easily seen to be invariant under coordinate changes. Straightforward calculation gives the following:

Proposition 2.2. *Let* (M, g) *be a Sasakian manifold, and* $\pi_\alpha : U_\alpha \to V_\alpha$ *as above. If* $Y, Z \in \Gamma(D)$ *are* π_α-*related to* $\tilde{Y}, \tilde{Z} \in \Gamma(TV_\alpha)$ *then*

$$\mathrm{Ric}^T(\tilde{Y}, \tilde{Z}) = \mathrm{Ric}(Y, Z) + 2g(X, Y), \tag{2.3}$$

$$s^T = s + 2m, \tag{2.4}$$

where Ric, s, *resp.* Ric^T, s^T, *are the Ricci and scalar curvatures of* g, *resp.* g^T.

Definition 2.3. *A* Sasaki–Einstein *manifold is a Sasakian manifold* (M, g) *with* $\mathrm{Ric} = 2mg$.

Note that one always has $\mathrm{Ric}(\xi, \xi) = 2m$ which fixes the Einstein constant at $2m$. Simple calculation shows that (M, g) is Einstein if, only if, $(C(M), \bar{g})$ is Ricci flat.

If ξ on M induces a free S^1-action, then $\{g, \Phi, \xi, \eta\}$ is a *regular* Sasakian structure. Another possibility is that all the orbits close but the action is not free, then the structure is *quasi-regular*. The third possibility is that not all the orbits close, in which case the generic orbit does not close. In this case, the Sasakian structure is *irregular*. In the regular, resp. quasi-regular, cases the leaf space, along with its transverse Kähler structure, is a manifold, resp. orbifold, X. And M is the total space of a principal S^1-bundle, resp. S^1 V-bundle, $\pi : M \to X$. A V-bundle is the orbifold analogue of a fiber bundle. See the appendices of [5] or [40] for details.

For a quasi-regular Sasakian manifold (M, g, Φ, ξ, η), the leaf space X of $\mathscr{F}_\xi$ is a normal projective, algebraic variety with an orbifold structure and a Kähler form ω with $[\omega] \in H^2_{orb}(X, \mathbb{Z})$. We will make use of the following well-known converse [6].

Proposition 2.4. *Let (X, ω) be a Kähler orbifold, with $[\omega] \in H^2_{orb}(X, \mathbb{Z})$. There is a holomorphic line V-bundle $\mathbf{L}$ with $c_1(\mathbf{L}) = -[\omega] \in H^2_{orb}(X, \mathbb{Z})$ and an S^1-principal subbundle $M \subset \mathbf{L}$ such that M has a family of Sasakian structures $\{g_a, \Phi, \xi_a, \eta_a\}, a \in \mathbb{R}_+$,*

$$g_a = a^2 \eta \otimes \eta + a\pi^* h, \text{ where } h \text{ is a Kähler metric on } X.$$

Furthermore, if $\mathbf{L}^\times$ is $\mathbf{L}$ minus the zero section, then $\mathbf{L}^\times$ is biholomorphic to $C(M)$.

If (X, ω) is Kähler–Einstein with positive scalar curvature, and $q[\omega] \in c_1^{orb}(X)$, for $q \in \mathbb{Z}_+$, then exactly one of the above Sasakian structures is Sasaki–Einstein, and one may take $\mathbf{L} = K_X^{\frac{1}{q}}$.

Note that if $\pi_1^{orb}(X) = e$ and $[\omega]$ is indivisible in $H^2_{orb}(X, \mathbb{Z})$, then $\pi_1^{orb}(M) = e$. Of course, we are interested in the case where M is smooth. This happens when the action of the local uniformizing groups of X inject into the group of the fibers of the V-bundle M.

We will mainly be interested in *toric* Sasaki–Einstein manifolds.

Definition 2.5. *A* toric Sasakian manifold *is a Sasakian manifold M of dimension $2m + 1$ whose Sasakian structure $\{g, \Phi, \xi, \eta\}$ is preserved by an effective action of an $(m + 1)$-dimensional torus T such that ξ is an element of the Lie algebra $\mathfrak{t}$ of T.*

Let $T_{\mathbb{C}} \cong (\mathbb{C}^*)^{m+1}$ be the complexification of T, then $T_{\mathbb{C}}$ acts on $C(M)$ by holomorphic automorphisms. One sees that this definition is equivalent to $C(M)$ being a toric Kähler manifold.

3 Toric Geometry

We give some basic definitions in the theory of toric varieties that we will need. See [16, 33, 34] for more details. In addition, we will consider the notion of a compatible orbifold structure on a toric variety and holomorphic V-bundles. We are interested in Kähler toric orbifolds and will give a description of the Kähler structure due to V. Guillemin [22].

3.1 Toric Varieties

Let $N \cong \mathbb{Z}^r$ be the free $\mathbb{Z}$-module of rank r and $M = \text{Hom}_{\mathbb{Z}}(N, \mathbb{Z})$ its dual. We denote $N_{\mathbb{Q}} = N \otimes \mathbb{Q}$ and $M_{\mathbb{Q}} = M \otimes \mathbb{Q}$ with the natural pairing

$$\langle \, , \, \rangle : M_{\mathbb{Q}} \times N_{\mathbb{Q}} \to \mathbb{Q}.$$

Similarly we denote $N_{\mathbb{R}} = N \otimes \mathbb{R}$ and $M_{\mathbb{R}} = M \otimes \mathbb{R}$.

Let $T_{\mathbb{C}} := N \otimes_{\mathbb{Z}} \mathbb{C}^* \cong \mathbb{C}^* \times \cdots \times \mathbb{C}^*$ be the algebraic torus. Each $m \in M$ defines a character $\chi^m : T_{\mathbb{C}} \to \mathbb{C}^*$ and each $n \in N$ defines a one-parameter subgroup $\lambda_n : \mathbb{C}^* \to T_{\mathbb{C}}$. In fact, this gives an isomorphism between M (resp. N) and the multiplicative group $\mathrm{Hom}_{\mathrm{alg.}}(T_{\mathbb{C}}, \mathbb{C}^*)$ (resp. $\mathrm{Hom}_{\mathrm{alg.}}(\mathbb{C}^*, T_{\mathbb{C}})$).

Definition 3.1. *A subset σ of $N_{\mathbb{R}}$ is a* strongly convex rational polyhedral cone *if there are $n_1, \ldots, n_r$ so that*

$$\sigma = \mathbb{R}_{\geq 0} n_1 + \cdots + \mathbb{R}_{\geq 0} n_r,$$

and one has $\sigma \cap -\sigma = \{o\}$, where $o \in N$ is the origin.

The dimension $\dim \sigma$ is the dimension of the $\mathbb{R}$-subspace $\sigma + (-\sigma)$ of $N_{\mathbb{R}}$. The dual cone to σ is

$$\sigma^{\vee} = \{x \in M_{\mathbb{R}} : \langle x, y \rangle \geq 0 \text{ for all } y \in \sigma\},$$

which is also a convex rational polyhedral cone. A subset τ of σ is a face, $\tau < \sigma$, if

$$\tau = \sigma \cap m^{\perp} = \{y \in \sigma : \langle m, y \rangle = 0\} \text{ for } m \in \sigma^{\vee}.$$

And τ is a strongly convex rational polyhedral cone.

Definition 3.2. *A* fan *in N is a collection Δ of strongly convex rational polyhedral cones such that:*

(i) For $\sigma \in \Delta$ every face of σ is contained in Δ.
(ii) For any $\sigma, \tau \in \Delta$, the intersection $\sigma \cap \tau$ is a face of both σ and τ.

We will consider *complete* fans for which the support $\bigcup_{\sigma \in \Delta} \sigma$ is $N_{\mathbb{R}}$. We will denote

$$\Delta(i) := \{\sigma \in \Delta : \dim \sigma = i\}, \quad 0 \leq i \leq n.$$

Definition 3.3. *A fan in N is* nonsingular *if each $\sigma \in \Delta(r)$ is generated by r elements of N that can be completed to a $\mathbb{Z}$-basis of N. A fan in N is* simplicial *if each $\sigma \in \Delta(r)$ is generated by r elements of N that can be completed to a $\mathbb{Q}$-basis of $N_{\mathbb{Q}}$.*

If σ is a strongly convex rational polyhedral cone, $S_{\sigma} = \sigma^{\vee} \cap M$ is a finitely generated semigroup. We denote by $\mathbb{C}[S_{\sigma}]$ the semigroup algebra. We will denote the generators of $\mathbb{C}[S_{\sigma}]$ by x^m for $m \in S_{\sigma}$. Then $U_{\sigma} := \mathrm{Spec}\, \mathbb{C}[S_{\sigma}]$ is a normal affine variety on which $T_{\mathbb{C}}$ acts algebraically with a (Zariski) open orbit isomorphic to $T_{\mathbb{C}}$. If σ is nonsingular, then $U_{\sigma} \cong \mathbb{C}^n$.

Theorem 3.4 ([16, 33, 34]). *For a fan Δ in N the affine varieties U_{σ} for $\sigma \in \Delta$ glue together to form an irreducible normal algebraic variety*

$$X_{\Delta} = \bigcup_{\sigma \in \Delta} U_{\sigma}.$$

Furthermore, X_{Δ} is nonsingular if, and only if, Δ is nonsingular. And X_{Δ} is compact if, and only if, Δ is complete.

Proposition 3.5. *The variety X_Δ has an algebraic action of $T_\mathbb{C}$ with the following properties.*

(i) To each $\sigma \in \Delta(i), 0 \le i \le n$, there corresponds a unique $(n-i)$-dimensional $T_\mathbb{C}$-orbit $\mathrm{Orb}(\sigma)$ so that X_Δ decomposes into the disjoint union

$$X_\Delta = \bigcup_{\sigma \in \Delta} \mathrm{Orb}(\sigma),$$

where $\mathrm{Orb}(o)$ is the unique n-dimensional orbit and is isomorphic to $T_\mathbb{C}$.

(ii) The closure $V(\sigma)$ of $\mathrm{Orb}(\sigma)$ in X_Δ is an irreducible $(n-i)$-dimensional $T_\mathbb{C}$-stable subvariety and

$$V(\sigma) = \bigcup_{\tau \ge \sigma} \mathrm{Orb}(\tau).$$

We will consider toric varieties with an orbifold structure.

Definition 3.6. *We will denote by Δ^* an augmented fan by which we mean a fan Δ with elements $n(\rho) \in N \cap \rho$ for every $\rho \in \Delta(1)$.*

Proposition 3.7. *For a complete simplicial augmented fan Δ^*, we have a natural orbifold structure compatible with the action of $T_\mathbb{C}$ on X_Δ. We denote X_Δ with this orbifold structure by X_{Δ^*}.*

Proof. Let $\sigma \in \Delta^*(n)$ have generators $p_1, p_2, \dots, p_n$ as in the definition. Let $N' \subseteq N$ be the sublattice $N' = \mathbb{Z}\{p_1, p_2, \dots, p_n\}$, and σ' the equivalent cone in N'. Denote by M' the dual lattice of N' and $T'_\mathbb{C}$ the torus. Then $U_{\sigma'} \cong \mathbb{C}^n$. It is easy to see that

$$N/N' = \mathrm{Hom}_\mathbb{Z}(M'/M, \mathbb{C}^*).$$

And N/N' is the kernel of the homomorphism

$$T'_\mathbb{C} = \mathrm{Hom}_\mathbb{Z}(M', \mathbb{C}^*) \to T_\mathbb{C} = \mathrm{Hom}_\mathbb{Z}(M, \mathbb{C}^*).$$

Let $\Gamma = N/N'$. An element $t \in \Gamma$ is a homomorphism $t : M' \to \mathbb{C}^*$ equal to 1 on M. The regular functions on $U_{\sigma'}$ consist of $\mathbb{C}$-linear combinations of x^m for $m \in \sigma'^\vee \cap M'$. And $t \cdot x^m = t(m)x^m$. Thus the invariant functions are the $\mathbb{C}$-linear combinations of x^m for $m \in \sigma^\vee \cap M$, the regular functions of U_σ. Thus $U_{\sigma'}/\Gamma = U_\sigma$. And the charts are easily seen to be compatible on intersections.   □

Conversely, we have the following.

Proposition 3.8. *Let Δ be a complete simplicial fan. Suppose for simplicity that the local uniformizing groups are abelian. Then every orbifold structure on X_Δ compatible with the action of $T_\mathbb{C}$ arises from an augmented fan Δ^*.*

See [40] for a proof.

Let Δ^* be an augmented fan in N. We will assume from now on that the fan Δ is simplicial and complete.

Definition 3.9. *A real function* $h : N_\mathbb{R} \to \mathbb{R}$ *is a* Δ^* *-linear support function if for each* $\sigma \in \Delta^*$ *with given* $\mathbb{Q}$*-generators* $p_1, \ldots, p_r$ *in* N*, there is an* $l_\sigma \in M_\mathbb{Q}$ *with* $h(s) = \langle l_\sigma, s \rangle$ *and* l_σ *is* $\mathbb{Z}$*-valued on the sublattice* $\mathbb{Z}\{p_1, \ldots, p_r\}$*. And we require that* $\langle l_\sigma, s \rangle = \langle l_\tau, s \rangle$ *whenever* $s \in \sigma \cap \tau$*. The additive group of* Δ^**-linear support functions will be denoted by* $\mathrm{SF}(\Delta^*)$*.*

Note that $h \in \mathrm{SF}(\Delta^*)$ is completely determined by the integers $h(n(\rho))$ for all $\rho \in \Delta(1)$. And conversely, an assignment of an integer to $h(n(\rho))$ for all $\rho \in \Delta(1)$ defines h. Thus

$$\mathrm{SF}(\Delta^*) \cong \mathbb{Z}^{\Delta(1)}.$$

Definition 3.10. *Let* Δ^* *be a complete augmented fan. For* $h \in \mathrm{SF}(\Delta^*)$*,*

$$\Sigma_h := \{m \in M_\mathbb{R} : \langle m, n \rangle \geq h(n), \text{ for all } n \in N_\mathbb{R}\},$$

is a, possibly empty, convex polytope in $M_\mathbb{R}$*.*

We will consider the holomorphic line V-bundles on $X = X_{\Delta^*}$. All V-bundles will be *proper* in this section. The set of isomorphism classes of holomorphic line V-bundles is denoted by $\mathrm{Pic}^{\mathrm{orb}}(X)$, which is a group under the tensor product.

Definition 3.11. *A Baily divisor is a* $\mathbb{Q}$*-Weil divisor* $D \in \mathrm{Weil}(X) \otimes \mathbb{Q}$ *whose inverse image* $D_{\tilde{U}} \in \mathrm{Weil}(\tilde{U})$ *in every local uniformizing chart* $\pi : \tilde{U} \to U$ *is Cartier. The additive group of Baily divisors is denoted* $\mathrm{Div}^{\mathrm{orb}}(X)$*.*

A Baily divisor D defines a holomorphic line V-bundle $[D] \in \mathrm{Pic}^{\mathrm{orb}}(X)$ in a way completely analogous to Cartier divisors. Given a nonzero meromorphic function $f \in \mathcal{M}$, we have the *principal divisor*

$$\mathrm{div}(f) := \sum_V v_V(f) V,$$

where $v_V(f) V$ is the order of the zero, or negative the order of the pole, of f along each irreducible subvariety of codimension one. We have the exact sequence

$$1 \to \mathbb{C}^* \to \mathcal{M}^* \to \mathrm{Div}^{\mathrm{orb}}(X) \xrightarrow{[\,]} \mathrm{Pic}^{\mathrm{orb}}(X). \tag{3.1}$$

A holomorphic line V-bundle $\pi : \mathbf{L} \to X$ is equivariant if there is an action of $T_\mathbb{C}$ on $\mathbf{L}$ such that π is equivariant, $\pi(tw) = t\pi(w)$ for $w \in \mathbf{L}$ and $t \in T_\mathbb{C}$ and the action lifts to a holomorphic action, linear on the fibers, over each uniformizing neighborhood. The group of isomorphism classes of equivariant holomorphic line V-bundles is denoted $\mathrm{Pic}^{\mathrm{orb}}{}_{T_\mathbb{C}}(X)$. Similarly, we have invariant Baily divisors, denoted $\mathrm{Div}^{\mathrm{orb}}{}_{T_\mathbb{C}}(X)$, and $[D] \in \mathrm{Pic}^{\mathrm{orb}}{}_{T_\mathbb{C}}(X)$ whenever $D \in \mathrm{Div}^{\mathrm{orb}}{}_{T_\mathbb{C}}(X)$.

Proposition 3.12. *Let* $X = X_{\Delta^*}$ *be compact with the standard orbifold structure, i.e.,* Δ^* *is simplicial and complete.*

(i) There is an isomorphism $\mathrm{SF}(\Delta^*) \cong \mathrm{Div}^{\mathrm{orb}}{}_{T_\mathbb{C}}(X)$ *obtained by sending* $h \in \mathrm{SF}(\Delta^*)$ *to*

$$D_h := - \sum_{\rho \in \Delta(1)} h(n(\rho)) V(\rho).$$

(ii) There is a natural homomorphism $\mathrm{SF}(\Delta^*) \to \mathrm{Pic}^{orb}{}_{T_{\mathbb{C}}}(X)$ *that associates an equivariant line V-bundle* $\mathbf{L}_h$ *to each* $h \in \mathrm{SF}(\Delta^*)$.

(iii) Suppose $h \in \mathrm{SF}(\Delta^*)$ *and* $m \in M$ *satisfies*

$$\langle m, n \rangle \geq h(n) \text{ for all } n \in N_{\mathbb{R}},$$

then m defines a section $\psi : X \to \mathbf{L}_h$ *which has the equivariance property* $\psi(tx) = \chi^m(t)(t\psi(x))$.

(iv) The set of sections $H^0(X, \mathcal{O}(\mathbf{L}_h))$ *is the finite dimensional* $\mathbb{C}$-*vector space with basis* $\{x^m : m \in \Sigma_h \cap M\}$.

(v) Every Baily divisor is linearly equivalent to a $T_{\mathbb{C}}$-*invariant Baily divisor. Thus for* $D \in \mathrm{Pic}^{orb}(X)$, $[D] \cong [D_h]$ *for some* $h \in \mathrm{SF}(\Delta^*)$.

(vi) If $\mathbf{L}$ *is any holomorphic line V-bundle, then* $\mathbf{L} \cong \mathbf{L}_h$ *for some* $h \in \mathrm{SF}(\Delta^*)$. *The homomorphism in part i. induces an isomorphism* $\mathrm{SF}(\Delta^*) \cong \mathrm{Pic}^{orb}{}_{T_{\mathbb{C}}}(X)$ *and we have the exact sequence*

$$0 \to M \to \mathrm{SF}(\Delta^*) \to \mathrm{Pic}^{orb}(X) \to 1.$$

Proof. (i) For each $\sigma \in \Delta(n)$ with uniformizing neighborhood $\pi : U_{\sigma'} \to U_\sigma$ as above, the map $h \to D_h$ assigns the principal divisor

$$\mathrm{div}(x^{-l_\sigma}) = - \sum_{\rho \in \Delta(1), \rho < \sigma} h(n(\rho)) V'(\rho),$$

where $V'(\rho)$ is the closure of the orbit $\mathrm{Orb}(\rho)$ in $U_{\sigma'}$. An element $\mathrm{Div}^{orb}{}_{T_{\mathbb{C}}}(X)$ must be a sum of closures of codimension one orbits $V(\rho)$ in Proposition 3.5, and by above remarks the map is an isomorphism.

(ii) One defines $\mathbf{L}_h := [D_h]$, where $[D_h]$ is constructed as follows. Consider a uniformizing chart $\pi : U_{\sigma'} \to U_\sigma$ as in Proposition 3.7. Define $\mathbf{L}_h|_{U_{\sigma'}}$ to be the invertible sheaf $\mathcal{O}_{U_{\sigma'}}(D_h)$, with D_h defined on $U_{\sigma'}$ by x^{-l_σ}. So $\mathbf{L}_h|_{U_{\sigma'}} \cong U_{\sigma'} \times \mathbb{C}$ with an action of $T'_{\mathbb{C}}$,

$$t(x, v) = (tx, \chi^{-l_\sigma}(t)v) \text{ where } t \in T'_{\mathbb{C}}, \ (x, v) \in U_{\sigma'} \times \mathbb{C}.$$

Then $\mathbf{L}_h|_{U_\sigma}$ is the quotient by the subgroup $N/N' \subset T'_{\mathbb{C}}$, so it has an action of $T_{\mathbb{C}}$. And the $\mathbf{L}_h|_{U_\sigma}$ glue together equivariantly with respect to the action.

(iii) For $\sigma \in \Delta$, we have $\langle m, n \rangle \geq \langle l_\sigma, n \rangle$ for all $n \in \sigma$. Then $m - l_\sigma \in M' \cap \sigma'^\vee$ and x^{m-l_σ} is a section of the invertible sheaf $\mathcal{O}_{U_{\sigma'}}(D_h)$ and is equivariant with respect to N/N' so it defines a section of $\mathbf{L}_h|_{U_\sigma}$. And these sections are compatible.

(iv) We will make use of the GAGA theorems of A. Grothendieck [20, 21]. As with any holomorphic V-bundle, the sheaf of sections $\mathcal{O}(\mathbf{L}_h)$ is a coherent sheaf. It follows from GAGA that we may consider $\mathcal{O}(\mathbf{L}_h)$ as a coherent algebraic sheaf, and all global sections are algebraic. If ϕ is a global section, then $\phi \in H^0(T_{\mathbb{C}}, \mathcal{O}(\mathbf{L}_h)) \subset \mathbb{C}[M]$. And in the uniformizing chart $\pi : U_{\sigma'} \to U_\sigma$, ϕ lifts to an element of the module $\mathcal{O}_{U_{\sigma'}} \cdot x^{l_\sigma}$ that has a basis $\{x^m : m \in l_\sigma + M' \cap \sigma'^\vee\}$. So $\phi|_{U_\sigma}$ is a $\mathbb{C}$-linear combination of x^m with $m \in M$ and $\langle m, n \rangle \geq h(n)$ for all $n \in \sigma$. Thus $m \in \Sigma_h$.

(v) The divisor $T_{\mathbb{C}} \cap D$ is a Cartier divisor on $T_{\mathbb{C}}$ which is also principal since $\mathbb{C}[M]$ is a unique factorization domain. Thus there is a nonzero rational function f so that $D' = D - \mathrm{div}(f)$ satisfies $D' \cap T_{\mathbb{C}} = \emptyset$. Then $D' \in \mathrm{Div}^{\mathrm{orb}}{}_{T_{\mathbb{C}}}(X)$, and the result follows from i.

(vi) Consider $\mathbf{L}_{U_{\sigma'}}$ on a uniformizing neighborhood $U_{\sigma'}$ as above. For each $\rho \in \Delta(1), \rho < \sigma$ the subgroup $H_\rho \subseteq N/N'$ fixing $V'(\rho)$ is cyclic and generated by $n' \in N$ where n' is the primitive element with $a_\rho n' = n(\rho)$. Now H_ρ acts linearly on the fibers of $\mathbf{L}_{U_{\sigma'}}$ over $V'(\rho)$. Suppose n' acts with weight $e^{2\pi i \frac{k}{a}}$, then let $D_\rho := kV(\rho)$. If $D' := \sum_{\rho \in \Delta(1)} D_\rho$, then $\mathbf{L}' := \mathbf{L} \otimes [-D']$ is Cartier on $X_0 := X \setminus \mathrm{Sing}(X)$, where $\mathrm{Sing}(X)$ has codimension at least two. The sheaf $\mathcal{O}(\mathbf{L}')$ is not only coherent but is a rank-1 reflexive sheaf. By GAGA $\mathcal{O}(\mathbf{L}') \cong E \otimes \mathcal{O}'$, where E is an algebraic reflexive rank-1 sheaf and $\mathcal{O}'$ is the sheaf of analytic functions. It is well-known that $E = \mathcal{O}(D)$ for $D \in \mathrm{Weil}(X)$. And as a Baily divisor, we have $\mathbf{L}' \cong [D]$. So $\mathbf{L} \cong [D + D']$, and by v. we have $\mathbf{L} \cong \mathbf{L}_h$ for some $h \in \mathrm{SF}(\Delta^*)$. $\square$

The sign convention in the proposition is adopted to make subsequent discussions involving Σ_h consistent with the existing literature, although having $D_{-m} = \mathrm{div}(x^m)$ may be bothersome. Note also that we denote a Baily divisor by a formal $\mathbb{Z}$-linear sum the coefficient giving the multiplicity of the irreducible component in the *uniformizing chart*. This is different from its expression as a Weil divisor when irreducible components are contained in codimension-1 components of the singular set of the orbifold.

For $X = X_{\Delta^*}$, there is a unique $k \in \mathrm{SF}(\Delta^*)$ such that $k(n(\rho)) = 1$ for all $\rho \in \Delta(1)$. The corresponding Baily divisor

$$D_k := - \sum_{\rho \in \Delta(1)} V(\rho) \tag{3.2}$$

is the *(orbifold) canonical divisor*. The corresponding V-bundle is $\mathbf{K}_X$, the V-bundle of holomorphic n-forms. This will in general be different from the canonical sheaf in the algebraic geometric sense.

Definition 3.13. *Consider support functions as above but that are only required to be $\mathbb{Q}$-valued on $N_{\mathbb{Q}}$, denoted $\mathrm{SF}(\Delta, \mathbb{Q})$. h is strictly upper convex if $h(n + n') \geq h(n) + h(n')$ for all $n, n' \in N_{\mathbb{Q}}$ and for any two $\sigma, \sigma' \in \Delta(n), l_\sigma$ and $l_{\sigma'}$ are different linear functions.*

Given a strictly upper convex support function h, the polytope Σ_h is the convex hull in $M_{\mathbb{R}}$ of the vertices $\{l_\sigma : \sigma \in \Delta(n)\}$. Each $\rho \in \Delta(1)$ defines a facet by

$$\langle m, n(\rho) \rangle \geq h(n(\rho)).$$

If $n(\rho) = a_\rho n'$ with $n' \in N$ primitive and $a_\rho \in \mathbb{Z}^+$ we may label the face with a_ρ to get the labeled polytope Σ_h^*, which encodes the orbifold structure. Conversely, from a rational convex polytope Σ^*, we associate a fan Δ^* and a support function h as follows. For an l-dimensional face $\theta \subset \Sigma^*$, define the rational n-dimensional cone $\sigma^\vee(\theta) \subset M_{\mathbb{R}}$ consisting of all vectors $\lambda(p - p')$, where $\lambda \in \mathbb{R}_{\geq 0}, p \in \Sigma$, and $p' \in \theta$.

Then $\sigma(\theta) \subset N_{\mathbb{R}}$ is the $(n-l)$-dimensional cone dual to $\sigma^{\vee}(\theta)$. The set of all $\sigma(\theta)$ defines the complete fan Δ^*, where one assigns $n(\rho)$ to $\rho \in \Delta(1)$ if $n(\rho) = an'$ with n' primitive and a is the label on the corresponding $(n-1)$-dimensional face of Σ^*. The corresponding rational support function is then

$$h(n) = \inf\{\langle m, n \rangle : m \in \Sigma^*\} \text{ for } n \in N_{\mathbb{R}}.$$

Proposition 3.14 ([16, 34]). *There is a one-to-one correspondence between the set of pairs (Δ^*, h) with $h \in \mathrm{SF}(\Delta, \mathbb{Q})$ strictly upper convex, and rational convex marked polytopes Σ_h^*.*

We will be interested in toric orbifolds X_{Δ^*} with such a support function and polytope, Σ_h^*. More precisely, we will be concerned with the following.

Definition 3.15. *Let $X = X_{\Delta^*}$ be a compact toric orbifold. We say that X is Fano if $-k \in \mathrm{SF}(\Delta^*)$, which defines the anticanonical V-bundle $\mathbf{K}_X^{-1}$, is strictly upper convex.*

These toric varieties are not necessarily Fano in the usual sense, since $\mathbf{K}_X^{-1}$ is the *orbifold* anticanonical class. This condition is equivalent to $\{n \in N_{\mathbb{R}} : k(n) \leq 1\} \subset N_{\mathbb{R}}$ being a convex polytope with vertices $n(\rho), \rho \in \Delta(1)$. We will use Δ^* to denote both the augmented fan and this polytope in this case.

If $\mathbf{L}_h$ is a line V-bundle, then for certain $s > 0$, $\mathbf{L}_h^s \cong \mathbf{L}_{sh}$ will be a holomorphic line bundle. For example, $s = \mathrm{Ord}(X)$, the least common multiple of the orders of the uniformizing groups, will do. So suppose $\mathbf{L}_h$ is a holomorphic line bundle. If the global holomorphic sections generate $\mathbf{L}_h$, by Proposition 3.12 $M \cap \Sigma_h = \{m_0, m_1, \ldots, m_r\}$ and we have a holomorphic map $\psi_h : X \to \mathbb{C}P^r$ where

$$\psi_h(w) := [x^{m_0}(w) : x^{m_1}(w) : \cdots : x^{m_r}(w)]. \tag{3.3}$$

Proposition 3.16 ([34]). *Suppose $\mathbf{L}_h$ is a line bundle, so $h \in \mathrm{SF}(\Delta^*)$ is integral, and suppose h is strictly upper convex. Then $\mathbf{L}_h$ is ample, meaning that for large enough $v > 0$*

$$\psi_{vh} : X \to \mathbb{C}P^N,$$

is an embedding, where $M \cap \Sigma_{vh} = \{m_0, m_1, \ldots, m_N\}$.

Corollary 3.17. *Let X be a Fano toric orbifold. If $v > 0$ is sufficiently large with $-vk$ integral, $\mathbf{K}^{-v}$ will be very ample and $\psi_{-vk} : X \to \mathbb{C}P^N$ an embedding.*

Let X_{Δ^*} be an orbifold surface with $h \in \mathrm{SF}(\Delta^*)$. Then the total spaces of $\mathbf{L}_h$ and $\mathbf{L}_h^{\times}$ are toric varieties. The fan of $\mathbf{L}_h^{\times}$ is as follows. If $\sigma \in \Delta^*$ is spanned by $n(\rho_1), \ldots, n(\rho_k) \in \mathbb{Z}^n$ as in Definition 3.6, let $\bar{\sigma}$ be the cone in $\mathbb{R}^{n+1}$ spanned by $(n(\rho_1), h(n(\rho_1))), \ldots, (n(\rho_k), h(n(\rho_k))) \in \mathbb{Z}^{n+1}$. The collection of $\bar{\sigma}, \sigma \in \Delta^*$ defines a fan $\mathcal{C} = \mathcal{C}_h$, which is the fan of $\mathbf{L}_h^{\times}$. Furthermore, if h, or $-h$, is strictly upper convex, then one can add an additional $(n+1)$-cone to $\mathcal{C}_h$ to get an affine variety $Y = \mathbf{L}_h^{\times} \cup \{p\}$. We will make use of the smoothness condition on $\mathbf{L}_h^{\times}$. The toric variety $\mathbf{L}_h^{\times}$ is smooth if for every n-cone $\bar{\sigma}$ of $\mathcal{C}$ as above spanned by $\tau_1, \ldots, \tau_n \in \mathbb{Z}^{n+1}$ we have

$$(\mathbb{R}_{\geq 0}\tau_1 + \cdots + \mathbb{R}_{\geq 0}\tau_n) \cap \mathbb{Z}^{n+1} = \mathbb{Z}_{\geq 0}\tau_1 + \cdots + \mathbb{Z}_{\geq 0}\tau_n. \tag{3.4}$$

Suppose $\mathbf{L}_h^q \cong \mathbf{K}_X$ for some $q \in \mathbb{Z}_{>0}$. Then $\mathbf{K}_Y$ is trivial. That is, Y has a Gorenstein singularity at the apex.

3.2 Kähler Structures

We review the construction of toric Kähler metrics on toric varieties. Any compact toric orbifold associated to a polytope admits a Kähler metric (see [29]). Due to T. Delzant [13] and E. Lerman and S. Tolman [29] in the orbifold case, the symplectic structure is uniquely determined up to symplectomorphism by the polytope, which is the image of the moment map. This polytope is Σ_h^* of the previous section with h generalized to be real valued. There are infinitely many Kähler structures on a toric orbifold with fixed polytope Σ_h^*, but there is a canonical Kähler metric obtained by reduction. V. Guillemin gave an explicit formula [11, 22] for this Kähler metric. In particular, we show that every toric Fano orbifold admits a Kähler metric $\omega \in c_1(X)$.

Let Σ^* be a convex polytope in $M_\mathbb{R} \cong \mathbb{R}^{n*}$ defined by the inequalities

$$\langle x, u_i \rangle \geq \lambda_i, \quad i = 1, \ldots, d, \tag{3.5}$$

where $u_i \in N \subset N_\mathbb{R} \cong \mathbb{R}^n$ and $\lambda_i \in \mathbb{R}$. If Σ_h^* is associated to (Δ^*, h), then the u_i and λ_i are the set of pairs $n(\rho)$ and $h(n(\rho))$ for $\rho \in \Delta(1)$. We allow the λ_i to be real but require any set $u_{i_1}, \ldots, u_{i_n}$ corresponding to a vertex to form a $\mathbb{Q}$-basis of $N_\mathbb{Q}$.

Let $(e_1, \ldots, e_d)$ be the standard basis of $\mathbb{R}^d$ and $\beta : \mathbb{R}^d \to \mathbb{R}^n$ be the map that takes e_i to u_i. Let $\mathfrak{n}$ be the kernel of β, so we have the exact sequence

$$0 \to \mathfrak{n} \xrightarrow{\iota} \mathbb{R}^d \xrightarrow{\beta} \mathbb{R}^n \to 0, \tag{3.6}$$

and the dual exact sequence

$$0 \to \mathbb{R}^{n*} \xrightarrow{\beta^*} \mathbb{R}^{d*} \xrightarrow{\iota^*} \mathfrak{n}^* \to 0. \tag{3.7}$$

Since (3.6) induces an exact sequence of lattices, we have an exact sequence

$$1 \to N \to T^d \to T^n \to 1, \tag{3.8}$$

where the connected component of the identity of N is an $(d-n)$-dimensional torus. The standard representation of T^d on $\mathbb{C}^d$ preserves the Kähler form

$$\frac{i}{2} \sum_{k=1}^d dz_k \wedge d\bar{z}_k, \tag{3.9}$$

and is Hamiltonian with moment map

$$\mu(z) = \frac{1}{2}\sum_{k=1}^{d}|z_k|^2 e_k + c, \qquad (3.10)$$

unique up to a constant c. We will set $c = \sum_{k=1}^{d}\lambda_k e_k$. Restricting to $\mathfrak{n}^*$, we get the moment map for the action of N on $\mathbb{C}^d$

$$\mu_N(z) = \frac{1}{2}\sum_{k=1}^{d}|z_k|^2 \alpha_k + \lambda, \qquad (3.11)$$

with $\alpha_k = \iota^* e_k$ and $\lambda = \sum\lambda_k\alpha_k$. Let $Z = \mu_N^{-1}(0)$ be the zero set. By the exactness of (3.7) $z \in \mu_N^{-1}(0)$ if and only if there is a $v \in \mathbb{R}^{n*}$ with $\mu(z) = \beta^* v$. Since β^* is injective, we have a map

$$v : Z \to \mathbb{R}^{n*}, \qquad (3.12)$$

where $\beta^* v(z) = \mu(z)$ for all $z \in Z$. For $z \in Z$

$$\begin{aligned}
\langle v(z), u_i\rangle &= \langle \beta^* v(z), e_i\rangle \\
&= \langle \mu(z), e_i\rangle \\
&= \frac{1}{2}|z_i|^2 + \lambda_i,
\end{aligned} \qquad (3.13)$$

thus $v(z) \in \Sigma^*$. Conversely, if $v \in \Sigma^*$, then $v = v(z)$ for some $z \in Z$ and in fact a T^d orbit in Z. Thus Z is compact. The following is not difficult to show.

Theorem 3.18. *The action of N on Z is locally free. Thus the quotient*

$$X_{\Sigma^*} = Z/N$$

is a compact orbifold. Let

$$\pi : Z \to X$$

be the projection and

$$\iota : Z \to \mathbb{C}^d$$

the inclusion. Then X_{Σ^*} has a canonical Kähler structure with Kähler form ω uniquely defined by

$$\pi^*\omega = \iota^*\left(\frac{i}{2}\sum_{k=1}^{d}dz_k \wedge d\bar{z}_k\right).$$

We have an action of $T^n = T^d/N$ on X_{Σ^*}, which is Hamiltonian for ω. The map v is T^d invariant, and it descends to a map, which we also call v,

$$v : X_{\Sigma^*} \to \mathbb{R}^{n*}, \qquad (3.14)$$

which is the moment map for this action. The above comments show that $\text{Im}(v) = \Sigma^*$. The action T^n extends to the complex torus $T_{\mathbb{C}}^n$, and one can show that as an analytic variety and orbifold, X_{Σ^*} is the toric variety constructed from Σ^* in the previous section. See [23] for more details.

Let $\sigma : \mathbb{C}^d \to \mathbb{C}^d$ be the involution $\sigma(z) = \bar{z}$. The set Z is stable under σ, and σ descends to an involution on X. We denote the fixed point sets by Z_r and X_r. And we have the projection

$$\pi : Z_r \to X_r. \tag{3.15}$$

We equip Z_r and X_r with Riemannian metrics by restricting the Kähler metrics on $\mathbb{C}^d$ and X, respectively.

Proposition 3.19. *The map (3.15) is a locally finite covering and is an isometry with respect to these metrics*

Note that Z_r is a subset of $\mathbb{R}^d$ defined by

$$\frac{1}{2} \sum_{k=1}^{d} x_k^2 \alpha_k = -\lambda. \tag{3.16}$$

Restrict to the orthant $x_k > 0$ $k = 1, \ldots, d$ of $\mathbb{R}^d$. Let Z_r' be the component of Z_r in this orthant. Under the coordinates

$$s_k = \frac{x_k^2}{2}, \quad k = 1, \ldots, d. \tag{3.17}$$

The flat metric on $\mathbb{R}^d$ becomes

$$\frac{1}{2} \sum_{k=1}^{d} \frac{(ds_k)^2}{s_k}. \tag{3.18}$$

Consider the moment map v restricted to Z_r'. The above arguments show that v maps Z_r' diffeomorphically onto the interior Σ° of Σ. In particular, we have

$$\langle v(x), u_k \rangle = \lambda_k + s_k, \ k = 1, \ldots, d, \text{ for } x \in Z_r'. \tag{3.19}$$

Let $l_k : \mathbb{R}^{n*} \to \mathbb{R}$ be the affine function

$$l_k(x) = \langle x, u_k \rangle - \lambda_k, \quad k = 1, \ldots, d.$$

Then by equation (3.19), we have

$$l_k \circ v = s_k. \tag{3.20}$$

Thus the moment map v pulls back the metric

$$\frac{1}{2} \sum_{k=1}^{d} \frac{(dl_k)^2}{l_k}, \tag{3.21}$$

on Σ° to the metric (3.18) on Z_r'. We obtain the following.

Proposition 3.20. *The moment map $v : X_r' \to \Sigma^\circ$ is an isometry when Σ° is given the metric (3.21).*

Let $W \subset X$ be the orbit of $T_{\mathbb{C}}^n$ isomorphic to $T_{\mathbb{C}}^n$. Then by restriction, W has a T^n-invariant Kähler form ω. Identify $T_{\mathbb{C}}^n = \mathbb{C}^n / 2\pi i \mathbb{Z}^n$, so there is an inclusion $\iota : \mathbb{R}^n \to T_{\mathbb{C}}^n$.

Proposition 3.21. *Let ω be a T^n-invariant Kähler form on W. Then the action of T^n is Hamiltonian if and only if ω has a T^n-invariant potential function, that is, a function $F \in C^\infty(\mathbb{R}^n)$ such that*

$$\omega = 2i\partial\bar{\partial}F.$$

Proof. Suppose the action is Hamiltonian. Any T^n-orbit is Lagrangian, so ω restricts to zero. The inclusion $T^n \subset T_{\mathbb{C}}^n$ is a homotopy equivalence. Thus ω is exact. Let γ be a T^n-invariant 1-form with $\omega = d\gamma$. Let $\gamma = \beta + \bar{\beta}$ where $\beta \in \Omega^{0,1}$. Then

$$\omega = d\gamma = \partial\beta + \bar{\partial}\bar{\beta},$$

since $\bar{\partial}\beta = \partial\bar{\beta} = 0$. Since $H^{0,k}(W)_{T^n} = 0$ for $k > 0$, there exists a T^n-invariant function f with $\beta = \bar{\partial}f$. Then

$$\omega = \partial\bar{\partial}f + \bar{\partial}\partial\bar{f} = 2i\partial\bar{\partial}\operatorname{Im}f.$$

The converse is a standard result. $\square$

Suppose the T^n action on W is Hamiltonian with moment map $v : W \to \mathbb{R}^{n*}$. Denote by $x + iy$ the coordinates given by the identification $W = \mathbb{C}^n / 2\pi i \mathbb{Z}^n$.

Proposition 3.22 ([22]). *Up to a constant, v is the Legendre transform of F, i.e.,*

$$v(x+iy) = \frac{\partial F}{\partial x} + c, \quad c \in \mathbb{R}^{n*}$$

Proof. By definition

$$dv_k = -\iota\left(\frac{\partial}{\partial y_k}\right)\omega.$$

But by Proposition 3.21,

$$\omega = \sum_{j,k=1}^{n} \frac{\partial^2 F}{\partial x_j \partial x_k} dx_j \wedge dy_k,$$

so

$$dv_k = -\iota\left(\frac{\partial}{\partial y_k}\right)\omega = d\left(\frac{\partial F}{\partial x_k}\right).$$

Therefore $v_k = \frac{\partial F}{\partial x_k} + c_k$. $\square$

We can eliminate c by replacing F with $F - \sum_{k=1}^n c_k x_k$.

Notice that the metric (3.21) on Σ° can be written

$$\sum_{j,k} \frac{\partial^2 G}{\partial y_j \partial y_k} dy_j dy_k, \tag{3.22}$$

with

$$G = \frac{1}{2} \sum_{k=1}^{d} l_k(y) \log l_k(y). \tag{3.23}$$

V. Guillemin [22] showed that the Legendre transform of G is the inverse Legendre transform of F, i.e.,

$$\frac{\partial F}{\partial x} = y \text{ and } \frac{\partial G}{\partial y} = x. \tag{3.24}$$

From this it follows that

$$F(x) = \sum_{i=1}^{n} x_i y_i - G(y), \text{ where } y = \frac{\partial F}{\partial x}. \tag{3.25}$$

Define

$$l_\infty(x) = \sum_{i=1}^{d} \langle x, u_i \rangle.$$

From equations (3.23) and (3.25), it follows that F has the expression

$$F = \frac{1}{2} v^* \left(\sum_{k=1}^{d} \lambda_k \log l_k + l_\infty \right), \tag{3.26}$$

which gives us the following.

Theorem 3.23 ([11,22]). *On the open $T_{\mathbb{C}}^n$ orbit of X_{Σ^*} the Kähler form ω is given by*

$$i \partial \bar{\partial} v^* \left(\sum_{k=1}^{d} \lambda_k \log l_k + l_\infty \right).$$

Suppose we have an embedding as in Proposition 3.16,

$$\psi_h : X_{\Sigma^*} \to \mathbb{C}P^N.$$

So Σ_h is an integral polytope and $M \cap \Sigma_h = \{m_0, m_1, \dots, m_N\}$. Let ω_{FS} be the Fubini–Study metric on $\mathbb{C}P^N$. Note that $\psi_h^* \omega_{FS}$ is degenerate along the singular set of X, so does not define a Kähler form.

Consider the restriction of ψ_h to the open $T_{\mathbb{C}}^n$ orbit $W \subset X$. Let $\iota = \psi_h|_W$. It is induced by a representation

$$\tau : T_{\mathbb{C}}^n \to GL(N+1, \mathbb{C}), \tag{3.27}$$

with weights $m_0, m_1, \dots, m_N$. If $z = x + iy \in \mathbb{C}^n / 2\pi i \mathbb{Z}^n = T_{\mathbb{C}}^n$, and $w = (w_0, \dots, w_N)$, then

$$\tau(\exp z) w = (e^{\langle m_0, x+iy \rangle} w_0, \dots, e^{\langle m_N, x+iy \rangle} w_N). \tag{3.28}$$

Recall the Fubini–Study metric is

$$\omega_{FS} = i \partial \bar{\partial} \log |w|^2. \tag{3.29}$$

Let $[w_0 : \cdots : w_N]$ be homogeneous coordinates of a point in the image of W, then

$$\iota^* \omega_{FS} = i\partial\bar{\partial} \log\left(\sum_{k=0}^{N} |w_k|^2 e^{2\langle m_k, x\rangle} \right). \tag{3.30}$$

From equation (3.23) we have

$$x = \frac{\partial G}{\partial y} = \frac{1}{2}\left(\sum_{j=1}^{d} u_j \log l_j + u \right),$$

where $u = \sum u_j$. Then

$$2\langle m_i, x\rangle = 2\langle m_i, \frac{\partial G}{\partial y}\rangle = \sum_{j=1}^{d} \langle m_i, u_j\rangle \log l_j + \langle m_i, u\rangle.$$

So setting $d_i = e^{\langle m_i, u\rangle}$, gives

$$e^{2\langle m_i, x\rangle} = v^*\left(d_i \prod_{j=1}^{d} l_j^{\langle m_i, u_j\rangle} \right).$$

But from (3.26),

$$e^{2F} = v^*\left(e^{l_\infty} \prod_{j=1}^{d} l_j^{\lambda_j} \right).$$

Combining these,

$$e^{2\langle m_i, x\rangle} = e^{2F} v^*\left(d_i e^{-l_\infty} \prod_{j=1}^{d} l_j^{l_j(m_i)} \right).$$

Let $k_i = |w_i|^2 d_i$, then summing gives

$$\sum_{i=1}^{N} |w_i|^2 e^{2\langle m_i, x\rangle} = e^{2F} v^*(e^{-l_\infty} Q),$$

where

$$Q = \sum_{i=1}^{N} k_i \prod_{j=1}^{d} l_j^{l_j(m_i)}.$$

Thus we have

$$\psi_h^* \omega_{FS} = \omega + i\partial\bar{\partial} v^*(-l_\infty + \log Q). \tag{3.31}$$

Using that Σ_h is integral, and $k_i \neq 0$ for m_i a vertex of Σ_h, it is not difficult to show that Q is a positive function on Σ_h. Thus equation (3.31) is valid on all of X.

Theorem 3.24. *Suppose $\mathbf{L}_h$ is very ample for some $h \in \mathrm{SF}(\Delta^*)$ strictly upper convex and integral, and let ω be the canonical Kähler metric for the polytope Σ_h. Then*

$$[\omega] = 2\pi c_1(\mathbf{L}) = [\psi_h^* \omega_{FS}].$$

Corollary 3.25. *Suppose* $X = X_{\Delta^*}$ *is Fano. Let* ω *be the canonical metric of the integral polytope* Σ^*_{-k}. *Then*

$$[\omega] = 2\pi c_1(\mathbf{K}^{-1}) = 2\pi c_1(X).$$

Thus $c_1(X) > 0$. *Conversely, if* $c_1(X) > 0$, *then* $\mathbf{K}^{-p}$ *is very ample for some* $p > 0$ *and* X *is Fano as defined in definition (3.15).*

Proof. For some $p \in \mathbb{Z}^+$, $-pk \in \mathrm{SF}(\Delta^*)$ is integral and $\mathbf{L}_{-pk} = \mathbf{K}^{-p}$ is very ample. Let $\tilde{\omega}$ be the canonical metric of the integral polytope Σ^*_{-pk}. From the theorem we have

$$[\tilde{\omega}] = 2\pi c_1(\mathbf{K}^{-p}) = 2\pi p c_1(X).$$

Let ω be the canonical metric for Σ^*_{-k}. Theorem (3.23) implies that $[\tilde{\omega}] = p[\omega]$

For the converse, it follows from the extension to orbifolds of the Kodaira embedding theorem of W. Baily [2] that $\mathbf{K}^{-p}$ is very ample for some $p > 0$ sufficiently large. It follows from standard results on toric varieties that $-k$ is strictly upper convex (see [34]). $\square$

The next result will have interesting applications to the Einstein manifolds constructed later.

Proposition 3.26. *With the canonical metric, the volume of* X_{Σ^*} *is* $(2\pi)^n$ *times the Euclidean volume of* Σ.

Proof. Let $W \subset X$ be the open $T^n_{\mathbb{C}}$ orbit. We identify W with $\mathbb{C}^n/2\pi i \mathbb{Z}^n$ with coordinates $x + iy$. The restriction of ω to W is

$$\omega|_W = \sum_{j,k=1}^{n} \frac{\partial^2 F}{\partial x_j \partial x_k} dx_j \wedge dy_k.$$

Thus

$$\frac{\omega^n}{n!} = \det\left(\frac{\partial^2 F}{\partial x_j \partial x_k}\right) dx \wedge dy.$$

Integrating over dy gives

$$\mathrm{Vol}(X, \omega) = (2\pi)^n \int_{\mathbb{R}^n} \det\left(\frac{\partial^2 F}{\partial x_j \partial x_k}\right) dx.$$

By Proposition 3.22, $x \to z = v(x+iy) = \frac{\partial F}{\partial x}$ is a diffeomorphism from $\mathbb{R}^n$ to Σ°. By the change of variables,

$$\mathrm{Vol}(\Sigma) = \int_\Sigma dz = \int_{\mathbb{R}^n} \det\left(\frac{\partial^2 F}{\partial x_j \partial x_k}\right) dx.$$

$\square$

Corollary 3.27. *Let $X = X_{\Delta^*}$ be a toric Fano orbifold. And let ω be any Kähler form with $\omega \in c_1(X)$. Then*

$$\text{Vol}(X, \omega) = \frac{1}{n!} c_1(X)^n[X] = \text{Vol}(\Sigma_{-k}).$$

Proof. Let ω_c be the canonical metric associated to Σ^*_{-k}, then $\frac{1}{2\pi}\omega_c \in c_1(X)$ by Corollary 3.25. Then

$$\text{Vol}(X, \omega) = \frac{1}{(2\pi)^n} \text{Vol}(X, \omega_c) = \text{Vol}(\Sigma_{-k}).$$

$\square$

3.3 Moment Map and Futaki Invariant

A closer analysis of the moment map in the Fano case, originally due to T. Mabuchi [30], will be useful. Suppose in this section that $X = X_{\Delta^*}$ is toric Fano with $\dim_{\mathbb{C}} X = n$. As above, $z_k = x_k + iy_k$, $1 \le k \le n$ are logarithmic coordinates, $(z_1, \ldots, z_n) \in \mathbb{C}^n/2\pi i\mathbb{Z}^n \cong T_{\mathbb{C}}$. For $(t_1, \ldots, t_n) \in T_{\mathbb{C}}$, $x_k = \log|t_k|$. Then a T-invariant function $u \in C^{\infty}(T_{\mathbb{C}})$ is considered as a C^{∞} function $u = u(x_1, \ldots, x_n)$ on $\mathbb{R}^n$. There exists a T-invariant fiber metric Ω on $\mathbf{K}_X^{-1}$ with positive Chern form. Thus there exists a C^{∞} function $u = u(x_1, \ldots, x_n)$ so that

$$e^{-u} \prod_{k=1}^{n} (dx_k \wedge dy_k), \tag{3.32}$$

extends to a volume form Ω on all of X and $i\partial\bar{\partial}u$ extends to a Kähler form ω. The moment map $v_u : X \to M_{\mathbb{R}}$ can be given, without an ambiguous constant, as

$$v_u(t) = \left(\frac{\partial u}{\partial x_1}(t), \ldots, \frac{\partial u}{\partial x_n}(t) \right), \quad \text{for } t \in T_{\mathbb{C}}. \tag{3.33}$$

Theorem 3.28. *The closure of the image $v_u(T_{\mathbb{C}})$ in $M_{\mathbb{R}}$ is Σ_{-k}. Furthermore, v_u extends to a C^{∞} map $v_u : X \to M_{\mathbb{R}}$, which is the usual moment map.*

Only the first statement remains to be proved. This is a slight generalization of a similar result in [30].

We define the Futaki invariant. Let $\text{Aut}^o(X) \subseteq \text{Aut}(X)$ be the subgroup of the homomorphic automorphism group preserving the orbifold structure, and let $\mathfrak{g}$ be its Lie algebra. Let $\omega \in 2\pi c_1(X)$ be a Kähler form. There exists $f \in C^{\infty}(X)$ with $\text{Ricci}(\omega) - \omega = i\partial\bar{\partial}f$. Set $c = -2^{n+1}((2\pi c_1(X))^n[X])^{-1}$. Then the *Futaki invariant* $F : \mathfrak{g} \to \mathbb{C}$ is defined by

$$F(V) = c \int_X V f \omega^n, \quad \text{for } V \in \mathfrak{g}. \tag{3.34}$$

Note that, as proved in [17], F is zero on $[\mathfrak{g}, \mathfrak{g}]$. We have the Cartan decomposition

$$\mathfrak{g} = \mathfrak{t}_{\mathbb{C}} \oplus \sum_i \mathbb{C}v_i,$$

where the v_i are eigenvalues for the adjoint action of $\mathfrak{t}_{\mathbb{C}}$. Since $\mathfrak{t}$ is the Lie algebra of a maximal torus, one sees that the v_i are contained in $[\mathfrak{g}, \mathfrak{g}]$. Thus we may restrict F to $\mathfrak{t}_{\mathbb{C}}$.

Suppose $\mathbf{L}$ is an equivariant holomorphic line V-bundle on X. In our case of interest, $\mathbf{L} = \mathbf{K}_X^{-1}$ with the usual action. Let H be the space of Hermitian metrics on $\mathbf{L}$. For $h \in H$, denote by $c_1(\mathbf{L}, h)$ the Chern form, $\frac{i}{2\pi} \bar{\partial} \partial \log(h)$ in local holomorphic coordinates. For a pair $(h', h'') \in H \times H$, we define

$$R_{\mathbf{L}}(h', h'') := \int_{t_0}^{t_1} \left(\int_X h_t^{-1} \dot{h}_t (2\pi c_1(\mathbf{L}, h_t))^n \right) dt,$$

where $h_t, t_0 \le t_1$ is any piecewise smooth path with $h_{t_0} = h'$ and $h_{t_1} = h''$. One has that $R_{\mathbf{L}}(h', h'')$ is independent of the path h_t and satisfies

$$R_{\mathbf{L}}(g^* h', g^* h'') = R_{\mathbf{L}}(h', h''), \quad \text{for } g \in \mathrm{Aut}^o(X),$$

and the cocycle conditions

$$R_{\mathbf{L}}(h', h'') + R_{\mathbf{L}}(h'', h') = 0, \text{and} \tag{3.35}$$

$$R_{\mathbf{L}}(h, h') + R_{\mathbf{L}}(h', h'') + R_{\mathbf{L}}(h'', h) = 0, \tag{3.36}$$

for any $h, h', h'' \in H$. These identities imply that

$$r_{\mathbf{L}}(g) := \exp(R_{\mathbf{L}}(h, g^* h)), \quad \text{for } g \in \mathrm{Aut}^o(X), \tag{3.37}$$

is in independent of $h \in H$ and is a Lie group homomorphism into $\mathbb{R}_+$. It has associated Lie algebra character $r_{\mathbf{L}*} : \mathfrak{g} \to \mathbb{R}$.

Let σ be a $T_{\mathbb{C}}$-invariant section of $\mathbf{L}^*$. Then $h \in H$ is $h = e^{-u_h} \sigma \otimes \bar{\sigma}$ on $W \cong T_{\mathbb{C}}$ for some $u_h \in C^\infty(W)$. We denote by $V_{\mathbb{R}}$ the real component of a homomorphic vector field. Then differentiating (3.37) gives

$$r_{\mathbf{L}*}(V) = - \int_W V_{\mathbb{R}}(u_h)(i\partial \bar{\partial} u_h)^n, \tag{3.38}$$

for $V \in \mathfrak{t}_{\mathbb{C}}$ independent of $h \in H$. We have the following.

Proposition 3.29. *Suppose X_{Δ^*} is a toric Fano orbifold. Then with $\mathbf{L} = \mathbf{K}_X^{-1}$ we have*

$$F_X = -2^{n+1}((2\pi c_1(X))^n[X])^{-1} r_{\mathbf{L}*},$$

where both sides are restricted to $\mathfrak{t}_{\mathbb{C}}$.

Proof. By assumption there is an $h \in H$ with positive Chern form. Let β be a $T_{\mathbb{C}}$-invariant section of $\mathbf{K}_X^{-1}$ over W. Then h may be written as the volume form on W

$\Omega = i^n (-1)^{n(n-1)/2} e^{-u_h} \beta \wedge \bar{\beta}$, and $i\partial\bar{\partial}u_h$ extends to a Kähler form $\omega \in 2\pi c_1(X)$. Then if $f = \log(\frac{\Omega}{\omega^n})$, we have

$$\mathrm{Ricci}(\omega) - \omega = i\partial\bar{\partial}f.$$

Then

$$\begin{aligned}
0 &= \int_X \mathscr{L}_{V_\mathbb{R}}\left(e^{-f}\Omega\right) \\
&= -\int_X V_\mathbb{R}(f)\omega^n + \int_X e^{-f}\mathscr{L}_{V_\mathbb{R}}\Omega \\
&= -\int_X V_\mathbb{R}(f)\omega^n - \int_W V_\mathbb{R}(u_h)\omega^n
\end{aligned}$$

And the result follows from (3.38). □

The $t_k \frac{\partial}{\partial t_k}, k = 1, \dots, n$ from a basis of $\mathfrak{t}_\mathbb{C}$. We may assume that the Kähler form ω is T-invariant. Then we have

$$\begin{aligned}
F_X\left(t_k \frac{\partial}{\partial t_k}\right) &= -2^{n+1}((2\pi c_1(X))^n[X])^{-1} r_{\mathbf{L}*}\left(t_k \frac{\partial}{\partial t_k}\right) \\
&= 2^n((2\pi c_1(X))^n[X])^{-1} \int_W \frac{\partial u_h}{\partial x_k}(i\partial\bar{\partial}u_h)^n \\
&= 2^n((2\pi c_1(X))^n[X])^{-1} \int_W \frac{\partial u_h}{\partial x_k}\left(\frac{1}{2}\right)^n n! \det\left(\frac{\partial^2 u_h}{\partial x_j \partial x_k}\right) \prod_{l=1}^n dx_l \wedge dy_l \\
&= (2\pi)^n((2\pi c_1(X))^n[X])^{-1} \int_{\mathbb{R}^n} \frac{\partial u_h}{\partial x_k} n! \det\left(\frac{\partial^2 u_h}{\partial x_j \partial x_k}\right) dx \\
&= (2\pi)^n n!((2\pi c_1(X))^n[X])^{-1} \int_{\Sigma_{-k}} y_k dy \\
&= \mathrm{Vol}(\Sigma_{-k})^{-1} \int_{\Sigma_{-k}} y_k dy
\end{aligned}$$

where $y_k = \frac{\partial u_h}{\partial x_k}$.

We have the following simple interpretation of the Futaki invariant in this case.

Proposition 3.30. *Suppose $X = X_{\Delta^*}$ is a toric Fano orbifold. Then the Futaki invariant F_X is the barycenter of the polytope Σ_{-k}.*

3.4 Symmetric Toric Orbifolds

Let X_Δ be an n-dimensional toric variety. Let $\mathcal{N}(T_\mathbb{C}) \subset \mathrm{Aut}(X)$ be the normalizer of $T_\mathbb{C}$. Then $\mathcal{W}(X) := \mathcal{N}(T_\mathbb{C})/T_\mathbb{C}$ is isomorphic to the finite group of all symmetries of

Δ, i.e., the subgroup of $GL(n,\mathbb{Z})$ of all $\gamma \in GL(n,\mathbb{Z})$ with $\gamma(\Delta) = \Delta$. Then we have the exact sequence.

$$1 \to T_{\mathbb{C}} \to \mathcal{N}(T_{\mathbb{C}}) \to \mathcal{W}(X) \to 1. \tag{3.39}$$

Choosing a point $x \in X$ in the open orbit, defines an inclusion $T_{\mathbb{C}} \subset X$. This also provides a splitting of (3.39). Let $\mathcal{W}_0(X) \subseteq \mathcal{W}(X)$ be the subgroup that are also automorphisms of Δ^*; $\gamma \in \mathcal{W}_0(X)$ is an element of $\mathcal{N}(T_{\mathbb{C}}) \subset \text{Aut}(X)$, which preserves the orbifold structure. Let $G \subset \mathcal{N}(T_{\mathbb{C}})$ be the compact subgroup generated by T^n, the maximal compact subgroup of $T_{\mathbb{C}}$, and $\mathcal{W}_0(X)$. Then we have the, split, exact sequence

$$1 \to T^n \to G \to \mathcal{W}_0(X) \to 1. \tag{3.40}$$

Definition 3.31. *A symmetric Fano toric orbifold X is a Fano toric orbifold with $\mathcal{W}_0$ acting on N with the origin as the only fixed point. Such a variety and its orbifold structure is characterized by the convex polytope Δ^* invariant under $\mathcal{W}_0$. We call a toric orbifold special symmetric if $\mathcal{W}_0(X)$ contains the involution $\sigma : N \to N$, where $\sigma(n) = -n$.*

The following is immediate from Proposition 3.30.

Proposition 3.31. *For a symmetric Fano toric orbifold X, one has $F_X \equiv 0$.*

Definition 3.33. *The index of a Fano orbifold X is the largest positive integer m such that there is a holomorphic V-bundle $\mathbf{L}$ with $\mathbf{L}^m \cong \mathbf{K}_X^{-1}$. The index of X is denoted $\text{ind}(X)$.*

Note that $c_1(X) \in H_{orb}^2(X,\mathbb{Z})$, and $\text{ind}(X)$ is the greatest positive integer m such that $\frac{1}{m}c_1(X) \in H_{orb}^2(X,\mathbb{Z})$.

Proposition 3.34. *Let X_{Δ^*} be a special symmetric toric Fano orbifold. Then $\text{ind}(X) = 1$ or 2.*

Proof. We have $\mathbf{K}^{-1} \cong \mathbf{L}_{-k}$ with $-k \in \text{SF}(\Delta^*)$ where $-k(n_\rho) = -1$ for all $\rho \in \Delta(1)$. Suppose we have $\mathbf{L}^m \cong \mathbf{K}^{-1}$. By Proposition 3.12 there is an $h \in \text{SF}(\Delta^*)$ and $f \in M$ so that $mh = -k + f$. For some $\rho \in \Delta(1)$,

$$mh(n_\rho) = -1 + f(n_\rho)$$
$$mh(-n_\rho) = -1 - f(n_\rho).$$

Thus $m(h(n_\rho) + h(-n_\rho)) = -2$, and $m = 1$ or 2. $\square$

We will now restrict to dimension two. In the smooth case, every Fano surface, called a *Del Pezzo surface*, is either $\mathbb{C}P^1 \times \mathbb{C}P^1$ or $\mathbb{C}P^2$ blown up at r points in general position $0 \leq r \leq 8$. The smooth toric Fano surfaces are $\mathbb{C}P^1 \times \mathbb{C}P^1$, $\mathbb{C}P^2$, the Hirzebruch surface F_1, the equivariant blow-up of $\mathbb{C}P^2$ at two $T_{\mathbb{C}}$-fixed points, and the equivariant blow-up of $\mathbb{C}P^2$ at three $T_{\mathbb{C}}$-fixed points. There are only three examples of smooth symmetric toric Fano surfaces, which are $\mathbb{C}P^1 \times \mathbb{C}P^1$, $\mathbb{C}P^2$, and the equivariant blow up of $\mathbb{C}P^2$ at three $T_{\mathbb{C}}$-fixed points. Their marked fans are shown in Figure 1. The smooth toric Fano surfaces admitting a Kähler–Einstein metric are precisely the symmetric cases.

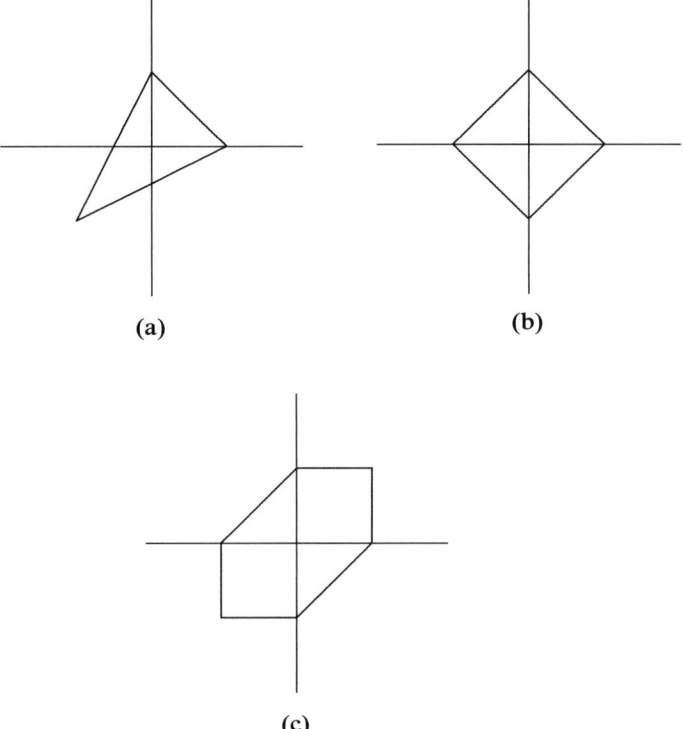

(a)   (b)

(c)

Fig. 1 The three smooth examples.

4 Einstein Equation

We consider the existence of Sasaki–Einstein metrics on toric Sasakian manifolds. This problem is completely solved in [18] where the more generally the existence of Sasaki–Ricci solitons is proved extending the existence of Kähler–Ricci solitons on toric Fano manifolds proved in [41]. For the examples of Sasaki–Einstein manifolds considered in this article, a complete proof can be found in [40].

Given a Sasakian manifold (M, g, Φ, ξ, η), we consider deformations of the transversal Kähler structure. That is, for a basic function $\phi \in C_B^\infty(M)$, set

$$\tilde{\eta} = \eta + 2d_B^c\phi.$$

Then for ϕ small enough,

$$\tilde{\omega}^T = \frac{1}{2}d\tilde{\eta} = \frac{1}{2}d\eta + d_Bd_B^c\phi = \omega^T + d_Bd_B^c\phi$$

is a transversal Kähler metric and $\tilde{\eta} \wedge \tilde{\eta}^m$ is nowhere zero. Then there is a Sasakian structure $(M, \tilde{g}, \tilde{\Phi}, \xi, \tilde{\eta})$ with the same Reeb vector field, transverse holomorphic structure, and basic Kähler class $[\tilde{\omega}^T]_B = [\omega^T]_B$. The existence of such a deformation of the transverse Kähler structure to a Sasakian Einstein structure requires the following.

Proposition 4.1. *The following three conditions are equivalent.*

(i) $(2m+2)\omega \in 2\pi c_1^B(M)$.

(ii) $\mathbf{K}_{C(M)}^q$ *is holomorphically trivial for some* $q \in \mathbb{Z}_+$, *and there is a nowhere zero section* Ω *of* $\mathbf{K}_{C(M)}$, *which will be multivalued if* $\mathbf{K}_{C(M)}^q$ *is not trivial, with* $\mathscr{L}_{r\partial_r}\Omega = (m+1)\Omega$.

(iii) $\mathbf{K}_{C(M)}^q$ *is holomorphically trivial for some* $q \in \mathbb{Z}_+$, *and there is a section* Ω *of* $\mathbf{K}_{C(M)}$, *which will be multivalued if* $\mathbf{K}_{C(M)}^q$ *is not trivial, such that*

$$\frac{i^{m+1}}{2^{m+1}}(-1)^{\frac{m(m+1)}{2}}\Omega \wedge \bar{\Omega} = e^f \frac{1}{n!}\bar{\omega}^{m+1},$$

for $f \in C_B^\infty(M)$ *pulled back to an element of* $C^\infty(C(M))$. *Here* $\bar{\omega}$ *is the Kähler form on* $C(M)$.

In Proposition 2.4 with $\mathbf{L} = \mathbf{K}_X^{\frac{1}{q}}$, the conditions of Proposition 4.1 are satisfied. We will need the following definition.

Definition 4.2. *A* Hamiltonian holomorphic vector field *on M is a complex vector field Y invariant by ξ so that*

(i) For any local leaf space projection $\pi_\alpha : U_\alpha \to V_\alpha$, $\pi_\alpha(Y)$ *is a holomorphic vector field,*

(ii) the complex function $\theta_Y = \sqrt{-1}\eta(Y)$ *satisfies*

$$\bar{\partial}_B\theta_Y = -\frac{\sqrt{-1}}{2}Y \neg d\eta.$$

The Lie algebra of Hamiltonian holomorphic vector fields on M is denoted $\mathfrak{h}$.

Note that $\tilde{Y} = Y + i\eta(Y)r\partial_r$ is a holomorphic vector field on $C(M)$. These correspond exactly to transversely holomorphic vector fields, i.e., those satisfying (i), which have a potential function (cf. [10]). In other words if $Y \in \Gamma(D \otimes \mathbb{C})$ satisfies (i) and $\mathscr{L}_\xi Y = 0$, and there exits a complex function $\theta_Y \in C_B^\infty(M)$ with

$$\bar{\partial}_B\theta_Y = -\frac{\sqrt{-1}}{2}Y \neg d\eta,$$

then $Y - \sqrt{-1}\theta_Y\xi$ is a Hamiltonian holomorphic vector field. Furthermore, in the case $c_1^B(M) > 0$ by the transverse Calabi–Yau theorem (cf. [28]), there exits a transversal Kähler deformation $(M, \tilde{g}, \tilde{\Phi}, \xi, \tilde{\eta})$ with $\tilde{\eta} = \eta + 2d_B^c\phi$ for some basic $\phi \in C_B^\infty(M)$ with $\mathrm{Ric}_{\tilde{g}^T}^T$ positive. The usual Weitzenböck formula shows that the space of basic harmonic 1-forms $\mathcal{H}_B^1$ is zero. It follows that if $Y \in \Gamma(D \otimes \mathbb{C})$ satisfies (i) and $\mathscr{L}_\xi Y = 0$, there exits a potential function $\theta_Y \in C_B^\infty(M)$ so that $Y - \sqrt{-1}\theta_Y \xi$ is a Hamiltonian holomorphic vector field. Thus $\mathfrak{h}$ is isomorphic to the space of transversely holomorphic vector fields commuting with ξ, which is isomorphic to the space of holomorphic vector fields on $C(M)$ commuting with $\xi + \sqrt{-1}r\partial_r$.

We now suppose that the conditions of Proposition 4.1 hold for (M, g, Φ, ξ, η). Let Y be a Hamiltonian holomorphic vector field. Then (M, g, Φ, ξ, η) is a *Sasakian–Ricci soliton* if

$$\operatorname{Ric}^T - (2m+2)g^T = \mathscr{L}_Y g^T. \tag{4.1}$$

Let $h \in C_B^\infty(M)$ be a basic function with

$$\operatorname{Ricci}(\omega^T) - (2m+2)\omega^T = i\partial_B\bar{\partial}_B h, \tag{4.2}$$

where $\omega^T = \frac{1}{2}d\eta$ is the transverse Kähler form. In [39], Tian and Zhu defined a modified Futaki invariant F_Y

$$F_Y(v) = \int_M v(h - \theta_Y)e^{\theta_Y}\eta \wedge (\frac{1}{2}d\eta)^m, \quad v \in \mathfrak{h}. \tag{4.3}$$

One can show as in [39] that F_Y is unchanged under transversal deformations $\eta \to \tilde{\eta} + 2d_B^c\phi$ of the Sasakian structure; and $F_Y(v) = 0$, for all $v \in \mathfrak{h}$, is a necessary condition for a solution to (4.1). If θ_Y is a constant, i.e., $Y = c\xi$ for some $c \in \mathbb{C}$, then (4.3) defines the usual Futaki invariant. And if M is quasi-regular this is, up to a constant, the same invariant defined in (3.34).

Let H be the transversal holomorphic automorphism group generated by $\mathfrak{h}$. Let $K \subset H$ be a compact group with Lie algebra $\mathfrak{k}$, and let $\mathfrak{k}^\mathbb{C}$ be its complexification. Note that we may choose a K-invariant Sasakian structure $(M, \tilde{g}, \tilde{\Phi}, \xi, \tilde{\eta})$. Then for $Y \in \mathfrak{k}$, one can take θ_Y to be imaginary. So one has $\mathfrak{k}^\mathbb{C} \subset \mathfrak{h}$. As in [39], we have the following.

Proposition 4.3. *There exists a* $Y \in \mathfrak{k}^\mathbb{C}$ *with* $\operatorname{Im}Y \in \mathfrak{k}$ *so that*

$$F_Y(v) = 0, \quad \text{for all } v \in \mathfrak{k}^\mathbb{C}.$$

Furthermore, Y *is unique up to addition of* $c\xi$, *for* $c \in \mathbb{C}$, *and*

$$F_Y([v, w]) = 0, \quad \text{for all } v \in \mathfrak{k}^\mathbb{C} \text{ and } w \in \mathfrak{h}.$$

Suppose now that (M, g, Φ, ξ, η) is a toric Sasakian manifold. If $\mathfrak{t}$ is the Lie algebra of the $m+1$-torus T acting on M, then $\mathfrak{t} \subset \mathfrak{t}_\mathbb{C} \subseteq \mathfrak{h}$. Using the same argument as after (3.34), we have the following.

Corollary 4.4. *If* M *is toric, then there exists a unique* $Y \in \mathfrak{t}_\mathbb{C}$ *with* $\operatorname{Im}Y \in \mathfrak{t}$ *so that*

$$F_Y(v) = 0, \quad \text{for all } v \in \mathfrak{h}.$$

For $\phi \in C_B^\infty(M)$, $\tilde{\eta} = \eta + 2d_B^c \phi$ defines a transversally deformed Sasakian structure, with transverse Kähler form $\tilde{\omega}^T = \omega + i\partial\bar{\partial}\phi$. And the Hamiltonian function for $Y \in \mathfrak{h}$ becomes $\tilde{\theta}_Y = \theta_Y + Y\phi$ (see [39]). In transverse holomorphic coordinates (4.1) becomes the Monge–Ampère equation

$$\frac{\det(g_{i\bar{j}}^T + \phi_{i\bar{j}})}{\det(g_{i\bar{j}}^T)} = e^{-(2m+2)\phi - \theta_Y - Y\phi + h}. \tag{4.4}$$

In the toric case A. Futaki, H. Ono, and G. Wang [18] prove the necessary C^0 estimate on ϕ to solve (4.4) using the continuity method.

Theorem 4.5. *Let (M, g, Φ, ξ, η) be a compact toric Sasakian manifold satisfying the conditions in (4.1). Then there exists a unique transversal deformation $(M, \tilde{g}, \tilde{\Phi}, \xi, \tilde{\eta})$ which is a Sasakian–Ricci soliton.*

If the Futaki invariant vanishes, then one has $Y = 0$ in Corollary 4.4. Therefore we have the following.

Corollary 4.6. *The solution in Theorem 4.5 is Einstein if, and only if, the Futaki invariant vanishes.*

If (M, g, Φ, ξ, η) is quasi-regular, then the leaf space of $\mathscr{F}_\xi$ is a toric orbifold $X = X_{\Sigma_{-k}^*}$ where Σ_{-k}^* is the marked polytope of the anticanonical bundle. Then (M, g, Φ, ξ, η) admits a transversal deformation to an Sasaki–Einstein structure if, and only if, the barycenter of Σ_{-k}^* is the origin. In particular, if M is quasi-regular and the leaf space X_{Δ^*} is symmetric, then solution is Einstein.

5 Three-Sasakian Manifolds

In this section, we define 3-Sasakian manifolds, the closely related quaternionic-Kähler spaces, and their twistor spaces. For more details, see [5]. These are sister geometries where one is able to pass from one to the other two by considering the appropriate orbifold fibration. Given a 3-Sasakian manifold $\mathcal{S}$, there is the associated twistor space $\mathcal{Z}$, quaternionic-Kähler orbifold $\mathcal{M}$, and hyperkähler cone $C(\mathcal{S})$. This is characterized by the *diamond*:

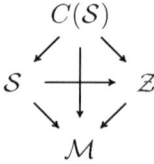

The equivalent 3-Sasakian and quaternionic-Kähler reduction procedures provide an elementary method for constructing 3-Sasakian and quaternionic-Kähler orbifolds (cf. [8,19]). This method is effective in producing smooth 3-Sasakian manifolds, though the quaternionic-Kähler spaces obtained are rarely smooth. In particular, we are interested in toric 3-Sasakian 7-manifolds $\mathcal{S}$ and their associated four-dimensional quaternionic-Kähler orbifolds $\mathcal{M}$. Here toric means that the structure is preserved by an action of the real two torus T^2. In four dimensions, quaternionic-Kähler means that $\mathcal{M}$ is Einstein and anti–self-dual, i.e., the self-dual half of the Weyl curvature vanishes $W_+ \equiv 0$. These examples are well-known and they are all obtained by reduction (cf. [8] and [12]). In this case, we will associate two more Einstein spaces with the four Einstein spaces in the diamond. To each diamond of a toric 3-Sasakian manifold, we have a special symmetric toric Fano surface X and a Sasaki–Einstein manifold M, which complete Diagram 1.1.

The motivation is twofold. First, it adds two more Einstein spaces to the examples on the right considered by C. Boyer, K. Galicki, and others in [5,8] and also by D. Calderbank and M. Singer [12]. Second, M is smooth when the 3-Sasakian space $\mathcal{S}$ is. And the smoothness of $\mathcal{S}$ is ensured by a relatively mild condition on the moment map. Thus we get infinitely many quasi-regular Sasaki–Einstein 5-manifolds with arbitrarily high second Betti numbers paralleling the 3-Sasakian manifolds constructed in [8].

5.1 Definitions and Basic Properties

We cover some of the basics of 3-Sasakian manifolds and 3-Sasakian reduction. See [5] for more details.

Definition 5.1. *Let $(\mathcal{S}, g)$ be a Riemannian manifold of dimension $n = 4m + 3$. Then $\mathcal{S}$ is 3-Sasakian if it admits three Killing vector fields $\{\xi^1, \xi^2, \xi^3\}$ each satisfying definition (2.1) such that $g(\xi^i, \xi^k) = \delta_{ij}$ and $[\xi^i, \xi^j] = 2\varepsilon_{ijk}\xi^k$.*

We have a triple of Sasakian structures on $\mathcal{S}$. For $i = 1, 2, 3$ we have $\eta^i(X) = g(\xi^i, X)$ and $\Phi^i(X) = \nabla_X \xi^i$. We say that $\{g, \Phi^i, \xi^i, \eta^i : i = 1, 2, 3\}$ defines a 3-Sasakian structure on $\mathcal{S}$.

Proposition 5.2. *The tensors $\Phi^i, i = 1, 2, 3$ satisfy the following identities.*

(i) $\Phi^i(\xi^j) = -\varepsilon_{ijk}\xi^k$,
(ii) $\Phi^i \circ \Phi^j = -\varepsilon_{ijk}\Phi^k + \xi^i \otimes \eta^j - \delta_{ij}Id$ □

Notice that if $\alpha = (a_1, a_2, a_3) \in S^2 \subset \mathbb{R}^3$, then $\xi(\alpha) = a_1\xi^1 + a_2\xi^2 + a_3\xi^3$ is a Sasakian structure. Thus a 3-Sasakian manifold comes equipped with an S^2 of Sasakian structures.

As in Definition 2.1, 3-Sasakian manifolds can be characterized by the holonomy of the cone $C(\mathcal{S})$.

Proposition 5.3. *Let $(\mathcal{S}, g)$ be a Riemannian manifold of dimension $n = 4m + 3$. Then $(\mathcal{S}, g)$ is 3-Sasakian if, and only if, the holonomy of the metric cone $(C(\mathcal{S}), \bar{g})$ is a subgroup of $Sp(m + 1)$. In other words, $(C(\mathcal{S}), \bar{g})$ is hyperkähler.*

Proof. Define almost complex structures $I_i, i = 1, 2, 3$ by

$$I_i X = -\Phi^i(X) + \eta^i(X) r \partial_r, \text{ and } I_i r \partial_r = \xi^i.$$

It is straightforward to verify that they satisfy $I_i \circ I_j = \varepsilon_{ijk} I_k - \delta_{ij} Id$. And from the integrability condition on each Φ^i in Definition 2.1 each $I_i, i = 1, 2, 3$ is parallel. $\square$

Since a hyperkähler manifold is Ricci flat, we have the following.

Corollary 5.4. *A 3-Sasakian manifold $(\mathcal{S}, g)$ of dimension $n = 4m + 3$ is Einstein with positive scalar curvature $s = 2(2m + 1)(4m + 3)$. Furthermore, if $(\mathcal{S}, g)$ is complete, then it is compact with finite fundamental group.*

The structure group of a 3-Sasakian manifold reduces to $Sp(m) \times \mathbb{I}_3$ where $\mathbb{I}_3$ is the 3×3 identity matrix. Thus we have

Corollary 5.5. *A 3-Sasakian manifold (M, g) is spin.*

Suppose $(\mathcal{S}, g)$ is compact. This will be the case in all examples considered here. Then the vector fields $\{\xi^1, \xi^2, \xi^3\}$ are complete and define a locally free action of $Sp(1)$ on $(\mathcal{S}, g)$. This defines a foliation $\mathscr{F}_3$, the *3-Sasakian foliation*. The generic leaf is either $SO(3)$ or $Sp(1)$, and all the leaves are compact. So $\mathscr{F}_3$ is quasi-regular, and the space of leaves is a compact orbifold, denoted $\mathcal{M}$. The projection $\varpi : \mathcal{S} \to \mathcal{M}$ exhibits $\mathcal{S}$ as an $SO(3)$ or $Sp(1)$ V-bundle over $\mathcal{M}$. The leaves of $\mathscr{F}_3$ are constant curvature 3-Sasakian 3-manifolds that must be homogeneous spherical space forms. Thus a leaf is $\Gamma \backslash S^3$ with $\Gamma \subset Sp(1)$.

For $\beta \in S^2$, we also have the characteristic vector field ξ_β with the associated 1-dimensional foliation $\mathscr{F}_\beta \subset \mathscr{F}_3$. In this case, $\mathscr{F}_\beta$ is automatically quasi-regular. Denote the leaf space of $\mathscr{F}_\beta$ as $\mathcal{Z}_\beta$ or just $\mathcal{Z}$. Then the natural projection $\pi : \mathcal{S} \to \mathcal{Z}$ is an S^1 V-bundle. And $\mathcal{Z}$ is a $(2m + 1)$-dimensional projective, normal algebraic variety with orbifold singularities and a Kähler form $\omega \in c_1^{orb}(\mathcal{Z})$, i.e., is Fano.

Fix a Sasakian structure $\{\Phi^1, \xi^1, \eta^1\}$ on $\mathcal{S}$. The horizontal subbundle $\mathcal{H} = \ker \eta^1$ to the foliation $\mathcal{F}$ of ξ^1 with the almost complex structure $I = -\Phi^1|_{\mathcal{H}}$ define a CR structure on $\mathcal{S}$. The form $\eta = \eta^2 + i\eta^3$ is of type $(1, 0)$ with respect to I. And $d\eta|_{\mathcal{H} \cap \ker(\eta)} \in \Omega^{2,0}(\mathcal{H} \cap \ker(\eta))$ is nondegenerate as a complex 2-form on $\mathcal{H} \cap \ker(\eta)$. Consider the complex 1-dimensional subspace $P \subset \Lambda^{1,0}\mathcal{H}$ spanned by η. Letting $\exp(it\xi^1)$ denote an element of the circle subgroup $U(1) \subset Sp(1)$ generated by ξ^1, one sees that $\exp(it\xi^1)$ acts on P with character e^{-2it}. Then $\mathbf{L} \cong \mathcal{S} \times_{U(1)} P$ defines a holomorphic line V-bundle over $\mathcal{Z}$. And we have a holomorphic section θ of $\Lambda^{1,0}(\mathcal{Z}) \otimes \mathbf{L}$ such that

$$\theta(X) = \eta(\tilde{X}),$$

where $\tilde{X}$ is the horizontal lift of a vector field X on $\mathcal{Z}$. Let $D = \ker(\theta)$ be the complex distribution defined by θ. Then $d\theta|_D \in \Gamma(\Lambda^2 D \otimes \mathbf{L})$ is nondegenerate. Thus $D = \ker(\theta)$ is complex contact structure on $\mathcal{Z}$, that is, a maximally non-integrable holomorphic subbundle of $T^{1,0}\mathcal{Z}$. Also, $\theta \wedge (d\theta)^m$ is a nowhere zero section of

$\mathbf{K}_{\mathcal{Z}} \otimes \mathbf{L}^{m+1}$. Thus $\mathbf{L} \cong \mathbf{K}_{\mathcal{Z}}^{-\frac{1}{m+1}}$ as holomorphic line V-bundles. We have the following for 3-Sasakian manifolds.

Theorem 5.6. *Let* $(\mathcal{S}, g)$ *be a compact 3-Sasakian manifold of dimension* $n = 4m + 3$, *and let* $\mathcal{Z}_\beta$ *be the leaf space of the foliation* $\mathcal{F}_\beta$ *for* $\beta \in S^2$. *Then* $\mathcal{Z}_\beta$ *is a compact* $\mathbb{Q}$-*factorial contact Fano variety with a Kähler–Einstein metric* h *with scalar curvature* $s = 8(2m+1)(m+1)$. *The projection* $\pi : \mathcal{S} \to \mathcal{Z}$ *is an orbifold Riemannian submersion with respect to the metrics* g *on* $\mathcal{S}$ *and* h *on* $\mathcal{Z}$.

The space $\mathcal{Z} = \mathcal{Z}_\beta$ is, up to isomorphism of all structures, independent of $\beta \in S^2$. We call $\mathcal{Z}$ the *twistor space* of $\mathcal{S}$. Consider again the natural projection $\varpi : \mathcal{S} \to \mathcal{M}$ coming from the foliation $\mathcal{F}_3$. This factors into $\pi : \mathcal{S} \to \mathcal{Z}$ and $\rho : \mathcal{Z} \to \mathcal{M}$. The generic fibers of ρ is a $\mathbb{C}P^1$, and there are possible singular fibers $\Gamma \backslash \mathbb{C}P^1$ that are simply connected and for which $\Gamma \subset U(1)$ is a finite group. And restricting to a fiber $\mathbf{L}|_{\mathbb{C}P^1} = \mathcal{O}(2)$, which is a V-bundle on singular fibers. Consider $g = \exp(\frac{\pi}{2}\xi^2) \in Sp(1)$, which gives an isometry of $\mathcal{S}$ $\varsigma_g : \mathcal{S} \to \mathcal{S}$ for which $\varsigma_g(\xi^1) = -\xi^1$. And ς_g descends to an anti-holomorphic isometry $\sigma : \mathcal{Z} \to \mathcal{Z}$ preserving the fibers.

We now consider the orbifold $\mathcal{M}$ more closely. Let $(\mathcal{M}, g)$ be any $4m$ dimensional Riemannian orbifold. An *almost quaternionic* structure on $\mathcal{M}$ is a rank 3 V-subbundle $\mathcal{Q} \subset End(T\mathcal{M})$ that is locally spanned by almost complex structures $\{J_i\}_{i=1,2,3}$ satisfying the quaternionic identities $J_i^2 = -Id$ and $J_1 J_2 = -J_2 J_1 = J_3$. We say that $\mathcal{Q}$ is compatible with g if $J_i^* g = g$ for $i = 1, 2, 3$. Equivalently, each $J_i, i = 1, 2, 3$ is skew symmetric.

Definition 5.7. *A Riemannian orbifold* $(\mathcal{M}, g)$ *of dimension* $4m, m > 1$ *is quaternionic Kähler if there is an almost quaternionic structure* $\mathcal{Q}$ *compatible with* g *that is preserved by the Levi–Civita connection.*

This definition is equivalent to the holonomy of $(\mathcal{M}, g)$ being contained in $Sp(1)Sp(m)$. For orbifolds, this is the holonomy on $\mathcal{M} \setminus S_{\mathcal{M}}$ where $S_{\mathcal{M}}$ is the singular locus of $\mathcal{M}$. See [37] for more on quaternionic Kähler manifolds. Notice that this definition always holds on an oriented Riemannian 4-manifold ($m = 1$). This case requires a different definition. Consider the *curvature operator* $\mathcal{R} : \Lambda^2 \to \Lambda^2$ of an oriented Riemannian 4-manifold. With respect to the decomposition $\Lambda^2 = \Lambda_+^2 \oplus \Lambda_-^2$, we have

$$\mathcal{R} = \begin{pmatrix} W_+ + \frac{s}{12} & \overset{\circ}{r} \\ \overset{\circ}{r} & W_- + \frac{s}{12} \end{pmatrix}, \tag{5.1}$$

where W_+ and W_- are the selfdual and anti–self-dual pieces of the Weyl curvature and $\overset{\circ}{r} = \text{Ric} - \frac{s}{4}g$ is the trace-free Ricci curvature. An oriented 4-dimensional Riemannian orbifold $(\mathcal{M}, g)$ is quaternionic Kähler if it is Einstein and anti–self-dual, meaning that $\overset{\circ}{r} = 0$ and $W_+ = 0$.

Theorem 5.8. *Let* $(\mathcal{S}, g)$ *be a compact 3-Sasakian manifold of dimension* $n = 4m + 3$. *Then there is a natural quaternionic Kähler structure on the leaf space of* $\mathcal{F}_3$, $(\mathcal{M}, \check{g})$, *such that the V-bundle map* $\varpi : \mathcal{S} \to \mathcal{M}$ *is a Riemannian submersion. Furthermore,* $(\mathcal{M}, \check{g})$ *is Einstein with scalar curvature* $16m(m+2)$.

5.2 3-Sasakian Reduction

We now summarize 3-Sasakian reduction and its application to producing infinitely many 3-Sasakian 7-manifolds. In particular, we are interested in toric 3-Sasakian 7-manifolds that have a T^2 action preserving the 3-Sasakian structure. Up to coverings they are all obtainable by taking 3-Sasakian quotients of S^{4n-1} by a torus T^k, $k = n - 2$. See [5, 8] for more details.

Let $(\mathcal{S}, g)$ be a 3-Sasakian manifold. And let $I(\mathcal{S}, g)$ be the subgroup in the isometry group $\mathrm{Isom}(\mathcal{S}, g)$ of 3-Sasakian automorphisms.

Definition 5.9. *Let $(\mathcal{S}, g)$ be a 3-Sasakian 7-manifold. Then $(\mathcal{S}, g)$ is* toric *if there is a real 2-torus $T^2 \subseteq I(\mathcal{S}, g)$.*

Let $G \subseteq I(\mathcal{S}, g)$ be compact with Lie algebra $\mathfrak{g}$. One can define the *3-Sasakian moment map*

$$\mu_{\mathcal{S}} : \mathcal{S} \to \mathfrak{g}^* \otimes \mathbb{R}^3 \tag{5.2}$$

by

$$\langle \mu_{\mathcal{S}}^a, X \rangle = \frac{1}{2} \eta^a(\tilde{X}), \quad a = 1, 2, 3 \text{ for } X \in \mathfrak{g}, \tag{5.3}$$

where $\tilde{X}$ be the vector field on $\mathcal{S}$ induced by $X \in \mathfrak{g}$.

Proposition 5.10. *Let $(\mathcal{S}, g)$ be a 3-Sasakian manifold and $G \subset I(\mathcal{S}, g)$ a connected compact subgroup. Assume that G acts freely (locally freely) on $\mu_{\mathcal{S}}^{-1}(0)$. Then $\mathcal{S}/G = \mu_{\mathcal{S}}^{-1}(0)/G$ has the structure of a 3-Sasakian manifold (orbifold). Let $\iota : \mu_{\mathcal{S}}^{-1}(0) \to \mathcal{S}$ and $\pi : \mu_{\mathcal{S}}^{-1}(0) \to \mu_{\mathcal{S}}^{-1}(0)/G$ be the corresponding embedding and submersion. Then the metric $\check{g}$ and 3-Sasakian vector fields are defined by $\pi^* \check{g} = \iota^* g$ and $\pi_* \xi^i|_{\mu_{\mathcal{S}}^{-1}(0)} = \check{\xi}^i$.*

Consider the unit sphere $S^{4n-1} \subset \mathbb{H}^n$ with the metric g obtained by restricting the flat metric on $\mathbb{H}^n$. Give S^{4n-1} the standard 3-Sasakian structure induced by the right action of $Sp(1)$. Then $I(S^{4n-1}, g) = Sp(n)$ acting by the standard linear representation on the left. We have the maximal torus $T^n \subset Sp(n)$, and every representation of a subtorus T^k is conjugate to an inclusion $\iota_\Omega : T^k \to T^n$, which is represented by a matrix

$$\iota_\Omega(\tau_1, \ldots, \tau_k) = \begin{pmatrix} \prod_{i=1}^{k} \tau_1^{a_1^i} & \cdots & 0 \\ \vdots & \ddots & \vdots \\ 0 & \cdots & \prod_{i=1}^{k} \tau_n^{a_n^i} \end{pmatrix}, \tag{5.4}$$

where $(\tau_1, \ldots, \tau_k) \in T^k$. Every such representation is defined by the $k \times n$ integral *weight matrix*

$$\Omega = \begin{pmatrix} a_1^1 & \cdots & a_k^1 & \cdots & a_n^1 \\ a_1^2 & \cdots & a_k^2 & \cdots & a_n^2 \\ \vdots & \ddots & \vdots & \ddots & \vdots \\ a_1^k & \cdots & a_k^n & \cdots & a_n^k \end{pmatrix} \tag{5.5}$$

Let $\{e_i\}, i = 1, \ldots, k$ be a basis for the dual of the Lie algebra of T^k, $\mathfrak{t}_k^* \cong \mathbb{R}^k$. Then the moment map $\mu_\Omega : S^{4n-1} \to \mathfrak{t}_k^* \otimes \mathbb{R}^3$ can be written as $\mu_\Omega = \sum_j \mu_\Omega^j e_j$, where in terms of complex coordinates $u_l = z_l + w_l j$ on $\mathbb{H}^n$

$$\mu_\Omega^j(\mathbf{z}, \mathbf{w}) = i \sum_l a_l^j (|z_l|^2 - |w_l|^2) + 2k \sum_l a_l^j \bar{w}_l z_l. \tag{5.6}$$

Denote by $\Delta_{\alpha_1,\ldots,\alpha_k}$ the $\binom{n}{k}$ $k \times k$ minor determinants of Ω.

Definition 5.11. *Let $\Omega \in \mathscr{M}_{k,n}(\mathbb{Z})$ be a weight matrix.*
(i) Ω is nondegenerate *if $\Delta_{\alpha_1,\ldots,\alpha_k} \neq 0$, for all $1 \leq \alpha_1 < \cdots < \alpha_k \leq n$.*
Let Ω be nondegenerate, and let d be the gcd of all the $\Delta_{\alpha_1,\ldots,\alpha_k}$, the kth determinantal divisor. Then Ω is admissible *if*
(ii) $\gcd(\Delta_{\alpha_2,\ldots,\alpha_{k+1}}, \ldots, \Delta_{\alpha_1,\ldots,\hat{\alpha}_t,\ldots,\alpha_{k+1}}, \ldots, \Delta_{\alpha_1,\ldots,\alpha_k}) = d$ for all length $k+1$ sequences $1 \leq \alpha_1 < \cdots < \alpha_t < \cdots < \alpha_{k+1} \leq n+1$.

The quotient obtained in Proposition 5.10 $\mathcal{S}_\Omega = S^{4n-1}/T^k(\Omega)$ will depend on Ω only up to a certain equivalence. Choosing a different basis of $\mathfrak{t}_k$ results in an action on Ω by an element in $Gl(k, \mathbb{Z})$. We also have the normalizer of T^n in $Sp(n)$, the Weyl group $\mathscr{W}(Sp(n)) = \Sigma_n \times \mathbb{Z}_2^n$ where Σ_n is the permutation group. $\mathscr{W}(Sp(n))$ acts on S^{4n-1} preserving the 3-Sasakian structure, and it acts on weight matrices by permutations and sign changes of columns. The group $Gl(k, \mathbb{Z}) \times \mathscr{W}(Sp(n))$ acts on $\mathscr{M}_{k,n}(\mathbb{Z})$. We say Ω is *reduced* if $d = 1$ in Definition 5.11. It is a result in [8] that we may assume that a nondegenerate weight matrix Ω is *reduced, as this is the case precisely when (5.4) is an inclusion.*

Theorem 5.12 ([5,8]). *Let $\Omega \in \mathscr{M}_{k,n}(\mathbb{Z})$ be reduced.*
(i) If Ω is nondegenerate, then $\mathcal{S}_\Omega$ is an orbifold.
(ii) Supposing Ω is nondegenerate, $\mathcal{S}_\Omega$ is smooth if and only if Ω is admissible.

Notice that the automorphism group of $\mathcal{S}_\Omega$ contains $T^{n-k} \cong T^n/\iota_\Omega(T^k)$.

We now restrict to the case of 7-dimensional toric quotients, so $n = k+2$. We may take matrices of the form

$$\Omega = \begin{pmatrix} 1 & 0 & \cdots & 0 & a_1 & b_1 \\ 0 & 1 & \cdots & 0 & a_2 & b_2 \\ \vdots & \vdots & \ddots & \vdots & \vdots & \vdots \\ 0 & 0 & \cdots & 1 & a_k & b_k \end{pmatrix}. \tag{5.7}$$

Proposition 5.13 ([8]). *Let $\Omega \in \mathscr{M}_{k,k+2}(\mathbb{Z})$ be as above. Then Ω is admissible if and only if $a_i, b_j, i, j = 1, \ldots, k$ are all nonzero, $\gcd(a_i, b_i) = 1$ for $i = 1, \ldots, k$, and we do not have $a_i = a_j$ and $b_i = b_j$, or $a_i = -a_j$ and $b_i = -b_j$ for some $i \neq j$.*

This shows that for $n = k+2$ there are infinitely many reduced admissible weight matrices. One can, for example, choose $a_i, b_j, i, j = 1, \ldots k$ be all pairwise relatively prime. We will make use of the cohomology computation of R. Hepworth [26] to show that we have infinitely many smooth 3-Sasakian 7-manifolds of each second Betti number $b_2 \geq 1$. Let $\Delta_{p,q}$ denote the $k \times k$ minor determinant of Ω obtained by deleting the pth and q^{th} columns.

Theorem 5.14 ([26][7,8]). *Let* $\Omega \in \mathcal{M}_{k,k+2}(\mathbb{Z})$ *be a reduced admissible weight matrix. Then* $\pi_1(\mathcal{S}_\Omega) = e$. *And the cohomology of* $\mathcal{S}_\Omega$ *is*

p	0	1	2	3	4	5	6	7
H^p	$\mathbb{Z}$	0	$\mathbb{Z}^k$	0	G_Ω	$\mathbb{Z}^k$	0	$\mathbb{Z}$

where G_Ω *is a torsion group of order*

$$\sum |\Delta_{s_1,t_1}| \cdots |\Delta_{s_{k+1},t_{k+1}}|$$

with the summand with index $s_1, t_1, \ldots, s_{k+1}, t_{k+1}$ *included if and only if the graph on the vertices* $\{1, \ldots, k+2\}$ *with edges* $\{s_i, t_i\}$ *is a tree.*

If we consider weight matrices as in Proposition 5.13, then the order of G_Ω is greater than $|a_1 \cdots a_k| + |b_1 \cdots b_k|$. We have the following.

Corollary 5.15 ([8, 26]). *There are smooth toric 3-Sasakian 7-manifolds with second Betti number* $b_2 = k$ *for all* $k \geq 0$. *Furthermore, there are infinitely many possible homotopy types of examples* $\mathcal{S}_\Omega$ *for each* $k > 0$.

5.3 Anti–Self-Dual Einstein Spaces

We consider the anti–self-dual Einstein orbifolds $\mathcal{M} = \mathcal{M}_\Omega$ associated to the toric 3-Sasakian 7-manifolds $\mathcal{S}$ in greater detail. Since $\mathcal{M}$ is a 4-dimensional orbifold with an effective action of T^2, the techniques of [25] show that $\mathcal{M}$ is characterized by the polygon $\mathcal{Q}_\Omega = \mathcal{M}/T^2$ with $k+2$ edges, $b_2(\mathcal{M}) = k$, labeled in cyclic order with $(m_0, n_0), (m_1, n_1), \ldots, (m_{k+2}, n_{k+2})$ in $\mathbb{Z}^2$, $(m_0, n_0) = -(m_{k+2}, n_{k+2})$, denoting the isotropy subgroups. For the quotients $\mathcal{M}_\Omega$, one can show the following (cf [40]):

a. The sequence m_i, $i = 0, \ldots k+2$ is strictly increasing.
b. The sequence $(n_i - n_{i-1})/(m_i - m_{i-1})$, $i = 1, \ldots k+2$ is strictly increasing.

We will make use of the following result of D. Calderbank and M. Singer [12], which classifies those compact orbifolds which admit toric anti–self-dual Einstein metrics. The case for which the associated 3-Sasakian space is smooth is originally due to R. Bielawski [4].

Theorem 5.16. *Let* $\mathcal{M}$ *be a compact toric 4-orbifold with* $\pi_1^{orb}(\mathcal{M}) = e$ *and* $k = b_2(\mathcal{M})$. *Then the following are equivalent.*

(i) One can arrange that the isotropy data of $\mathcal{M}$ *satisfy a. and b. above by cyclic permutations, changing signs, and acting by* $Gl(2, \mathbb{Z})$.

(ii) $\mathcal{M}$ *admits a toric anti–self-dual Einstein metric unique up to homothety and equivariant diffeomorphism. Furthermore,* $(\mathcal{M}, g)$ *is isometric to the quaternionic Kähler reduction of* $\mathbb{H}P^{k+1}$ *by a torus* $T^k \subset Sp(k+2)$.

It is well-known that the only possible smooth compact anti–self-dual Einstein spaces with positive scalar curvature are S^4 and $\overline{\mathbb{CP}}^2$, which are both toric.

Suppose $\mathcal{M}$ has isotropy data $v_0, v_1, \ldots, v_{k+2}$. Then it is immediate that $v_0, v_1, \ldots, v_{k+2}, -v_1, -v_2, \ldots, -v_{k+1}$ are the vertices of a convex polygon in $N_\mathbb{R} = \mathbb{R}^2$, which defines an augmented fan Δ^* defining a toric Fano surface X. A bit more thought gives the following.

Theorem 5.17. *There is a one-to-one correspondence between compact toric anti–self-dual Einstein orbifolds $\mathcal{M}$ with $\pi_1^{orb}(\mathcal{M}) = e$ and special symmetric toric Fano orbifold surfaces X with $\pi_1^{orb}(X) = e$. And X has a Kähler–Einstein metric of positive scalar curvature. Under the correspondence if $b_2(\mathcal{M}) = k$, then $b_2(X) = 2k+2$.*

This will be reproved in Section 5.4 by exhibiting X as a divisor in the twistor space.

Example. Consider the admissible weight matrix

$$\Omega = \begin{pmatrix} 1 & 0 & 1 & 1 \\ 0 & 1 & 1 & 2 \end{pmatrix}.$$

Then the 3-Sasakian space $\mathcal{S}_\Omega$ is smooth and $b_2(\mathcal{S}_\Omega) = b_2(\mathcal{M}_\Omega) = 2$. And the anti–self-dual orbifold $\mathcal{M}_\Omega$ has isotropy data

$$v_0 = (-7, -2), (-5, -2), (-1, -1), (5, 1), (7, 2) = v_4.$$

The singular set of $\mathcal{M}$ consists of two points with stablizer group $\mathbb{Z}_3$ and two with $\mathbb{Z}_4$. The associated toric Kähler–Einstein surface is that in Figure 2. $\diamondsuit$

The generic fiber of $\varpi : \mathcal{S} \to \mathcal{M}$ is either $SO(3)$ of $Sp(1)$. In general, the existence of a lifting to an $Sp(1)$ V-bundle is obstructed by the *Marchiafava–Romani class*, which when $\mathcal{M}$ is 4-dimensional is identical to $w_2(\mathcal{M}) \in H_{orb}^2(\mathcal{M}, \mathbb{Z}_2)$. In other words, the contact line bundle $\mathbf{L}$ has a square root $\mathbf{L}^{\frac{1}{2}}$ if, and only if, $w_2(\mathcal{M}) = 0$.

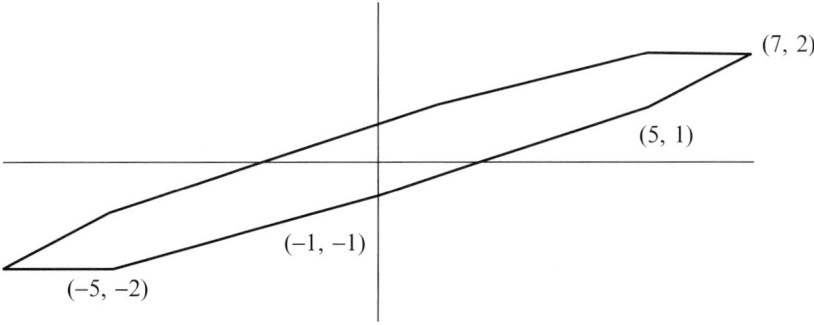

Fig. 2 Example with 8-point singular set and $\mathcal{W}_0 = \mathbb{Z}_2$.

Proposition 5.18. *Let X be the symmetric toric Fano surface associated to the anti–self-dual Einstein orbifold* $\mathcal{M}$. *Then* $\mathrm{ind}(X) = 2$ *if and only if* $w_2(\mathcal{M}) = 0$. *In other words,* $\mathbf{K}_X^{-1}$ *has a square root if and only if the contact line bundle* $\mathbf{L}$ *does.*

See [40] Proposition 5.22 for a proof.

5.4 Twistor Space and Divisors

We will consider the twistor space $\mathcal{Z}$ introduced in Theorem 5.6 more closely for the case when $\mathcal{M}$ is an anti–self-dual Einstein orbifold. For now suppose $(\mathcal{M}, [g])$ is an anti–self-dual, i.e., $W_+ \equiv 0$, conformal orbifold. There is a twistor space of $(\mathcal{M}, [g])$ that is originally due to R. Penrose [36]. See [1] for positive definite case.

The *twistor space* of $(\mathcal{M}, [g])$ is a complex 3-dimensional orbifold $\mathcal{Z}$ with the following properties:

a. There is a V-bundle fibration $\varpi : \mathcal{Z} \to \mathcal{M}$.
b. The general fiber of $P_x = \varpi^{-1}(x), x \in \mathcal{Z}$ is a projective line $\mathbb{C}P^1$ with normal bundle $N \cong \mathcal{O}(1) \oplus \mathcal{O}(1)$, which holds over singular fibers with N a V-bundle.
c. There exists an anti-holomorphic involution σ of $\mathcal{Z}$ leaving the fibers P_x invariant.

Let T be an oriented real 4-dimensional vector space with inner product g. Let $C(T)$ be set of orthogonal complex structures inducing the orientation, i.e., if $r, s \in T$ is a complex basis, then r, Jr, s, Js defines the orientation. One has $C(T) = S^2 \subset \Lambda_+^2(T)$, where S^2 is the sphere of radius $\sqrt{2}$. Now take T to be $\mathbb{H}$. Recall that $Sp(1)$ is the group of unit quaternions. Let

$$Sp(1)_+ \times Sp(1)_- \tag{5.8}$$

act on $\mathbb{H}$ by

$$w \to gwg'^{-1}, \text{ for } w \in \mathbb{H} \text{ and } (g, g') \in Sp(1)_+ \times Sp(1)_-. \tag{5.9}$$

Then we have

$$Sp(1)_+ \times_{\mathbb{Z}_2} Sp(1)_- \cong SO(4), \tag{5.10}$$

where $\mathbb{Z}_2$ is generated by $(-1, -1)$. Let

$$C = \{ai + bj + ck : a^2 + b^2 + c^2 = 1, a, b, c \in \mathbb{R}\}$$
$$= \{g \in Sp(1)_+ : g^2 = -1\} \cong S^2. \tag{5.11}$$

Then $g \in C$ defines an orthogonal complex structure by

$$w \to gw, \text{ for } w \in \mathbb{H},$$

giving an identification $C = C(\mathbb{H})$. Let $V_+ = \mathbb{H}$ considered as a representation of $Sp(1)_+$ and a right $\mathbb{C}$-vector space. Define $\pi : V_+ \setminus \{0\} \to C$ by $\pi(h) = -hih^{-1}$. Then the fiber of π over hih^{-1} is $h\mathbb{C}$. Then π is equivariant if $Sp(1)_+$ acts on C by $q \to gqg^{-1}, g \in Sp(1)_+$. We have a the identification

$$C = V_+ \setminus \{0\}/\mathbb{C}^* = \mathbb{P}(V_+). \tag{5.12}$$

Fix a Riemannian metric g in $[g]$. Let $\phi : \tilde{U} \to U \subset \mathcal{M}$ be a local uniformizing chart with group Γ. Let $F_{\tilde{U}}$ be the bundle of orthonormal frames on $\tilde{U}$. Then

$$F_{\tilde{U}} \times_{SO(4)} \mathbb{P}(V_+) = F_{\tilde{U}} \times_{SO(4)} C \tag{5.13}$$

defines a local uniformizing chart for $\mathcal{Z}$ mapping to

$$F_{\tilde{U}} \times_{SO(4)} \mathbb{P}(V_+)/\Gamma = F_{\tilde{U}}/\Gamma \times_{SO(4)} \mathbb{P}(V_+).$$

Right multiplication by j on $V_+ = \mathbb{H}$ defines the anti-holomorphic involution σ, which is fixed point free on (5.13). We will denote a neighborhood as in (5.13) by $\tilde{U}_{\mathcal{Z}}$.

An almost complex structure is defined as follows. At a point $z \in \tilde{U}_{\mathcal{Z}}$, the Levi–Civita connection defines a horizontal subspace H_z of the real tangent space T_z and we have a splitting

$$T_z = H_z \oplus T_z P_x = T_x \oplus T_z P_x, \tag{5.14}$$

where $\varpi(z) = x$ and T_x is the real tangent space of $\tilde{U}$. Let J_z be the complex structure on T_x given by $z \in P_x = C(T_x)$, and let J_z' be complex structure on $T_x \oplus T_z P_x$ arising from the natural complex structure on P_x. Then the almost complex structure on T_z is the direct sum of J_z and J_z'. This defines a natural almost complex structure on $\mathcal{Z}_{\tilde{U}}$ that is invariant under Γ. We get an almost complex structure on $\mathcal{Z}$ that is integrable precisely when $W_+ \equiv 0$.

Assume that $\mathcal{M}$ anti–self-dual Einstein with non-zero scalar curvature. Then $\mathcal{Z}$ has a complex contact structure $D \subset T^{1,0}\mathcal{Z}$ with holomorphic contact form $\theta \in \Gamma(\Lambda^{1,0}\mathcal{Z} \otimes \mathbf{L})$ where $\mathbf{L} = T^{1,0}\mathcal{Z}/D$.

The group of isometries $\mathrm{Isom}(\mathcal{M})$ lifts to an action on $\mathcal{Z}$ by real holomorphic transformations. Real means commuting with σ. This extends to a holomorphic action of the complexification $\mathrm{Isom}(\mathcal{M})_{\mathbb{C}}$. For $X \in \mathfrak{Isom}(\mathcal{M}) \otimes \mathbb{C}$, the Lie algebra of $\mathrm{Isom}(\mathcal{M})_{\mathbb{C}}$, we will also denote by X the holomorphic vector field induced on $\mathcal{Z}$. Then $\theta(X) \in H^0(\mathcal{Z}, \mathcal{O}(\mathbf{L}))$. By a well-known twistor correspondence, the map $X \to \theta(X)$ defines an isomorphism

$$\mathfrak{Isom}(\mathcal{M}) \otimes \mathbb{C} \cong H^0(\mathcal{Z}, \mathcal{O}(\mathbf{L})), \tag{5.15}$$

which maps real vector fields to real sections of $\mathbf{L}$.

Suppose for now on that $\mathcal{M}$ is a toric anti–self-dual Einstein orbifold with twistor space $\mathcal{Z}$. We will assume that $\pi_1^{orb}(\mathcal{M}) = e$, which can always be arranged by taking the orbifold cover. Then as above T^2 acts on $\mathcal{Z}$ by holomorphic transformations. And the action extends to $T_{\mathbb{C}}^2 = \mathbb{C}^* \times \mathbb{C}^*$, which in this case is an algebraic action.

Let $\mathfrak{t}$ be the Lie algebra of T^2 with $\mathfrak{t}_{\mathbb{C}}$ the Lie algebra of $T_{\mathbb{C}}^2$. Then we have from (5.15) the pencil

$$E = \mathbb{P}(\mathfrak{t}_{\mathbb{C}}) \subseteq |\mathbf{L}|, \tag{5.16}$$

where for $t \in E$ we denote $X_t = (\theta(t))$ the divisor of the section $\theta(t) \in H^0(\mathcal{Z}, \mathcal{O}(\mathbf{L}))$. Note that E has an equator of real divisors. Also, since $T_{\mathbb{C}}^2$ is Abelian, every $X_t, t \in E$ is $T_{\mathbb{C}}^2$ invariant.

Our goal is to determine the structure of the divisors in the pencil E. As before we will consider the one parameter groups $\rho_i \in N = \mathbb{Z} \times \mathbb{Z}$, where N is the lattice of one parameter $\mathbb{C}^*$-subgroups of $T_{\mathbb{C}}^2$. Also, we will identify the Lie algebra $\mathfrak{t}$ of T^2 with $N \otimes \mathbb{R}$ and the Lie algebra $\mathfrak{t}_{\mathbb{C}}$ of $T_{\mathbb{C}}^2$ with $N \otimes \mathbb{C}$. Since $\mathbf{L}|_{P_x} = \mathcal{O}(2)$, a divisor $X_t \in E$ intersects a generic twistor line P_x at two points.

Recall that the set of nontrivial stablizers of the T^2-action on $\mathcal{M}$ is $B = \cup_{i=1}^{k+2} B_i$ where B_i is topologically a 2-sphere. Denote by $x_i = B_i \cap B_{i+1}$ the $k+2$ fixed points of the action. We will denote $P_i := P_{x_i}, i = 1, \dots, k+2$. One can show there exist two irreducible rational curves $C_i^{\pm}, i = 1, \dots, k+2$ mapped diffeomorphically to B_i by ϖ. Furthermore, $\sigma(C_i^{\pm}) = C_i^{\mp}$. The singular set for the T^2-action on $\mathcal{Z}$ is the union of rational curves

$$\Sigma = \left(\cup_{i=1}^{k+2} P_i \right) \bigcup \left(\cup_{i+1}^{k+2} C_i^+ \cup C_i^- \right). \tag{5.17}$$

With a closer analysis of the action of $T_{\mathbb{C}}^2$ on $\mathcal{Z}$, one can prove the following. See [40] for the proof. The term *suborbifold* denotes a subvariety that is a submanifold in every local uniformizing neighborhood.

Theorem 5.19. *Let $\mathcal{M}$ be a compact anti–self-dual Einstein orbifold with $b_2(\mathcal{M}) = k$ and $\pi_1^{orb}(\mathcal{M}) = e$. Let $n = k+2$. Then there are distinct real points $t_1, t_2, \dots, t_n \in E$ so that for $t \in E \setminus \{t_1, t_2, \dots, t_n\}$, $X_t \subset \mathcal{Z}$ is a suborbifold. Furthermore X_t is a special symmetric toric Fano surface. The anticanonical cycle of X_t is $C_1, C_2, \dots, C_{2n}$, and the corresponding stabilizers are $\rho_1, \rho_2, \dots, \rho_{2n}$, which define the vertices in $N = \mathbb{Z} \times \mathbb{Z}$ of Δ^* with $X_t = X_{\Delta^*}$.*

For $t_i \in E$, $X_{t_i} = D + \bar{D}$, where $D, \bar{D}$ are irreducible degree one divisors with $\sigma(D) = \bar{D}$. The $D, \bar{D}$ are suborbifolds of $\mathcal{Z}$ and are toric Fano surfaces. We have $D \cap \bar{D} = P_i$, and the elements $\pm(\rho_1, \dots, \rho_i, -\rho_i + \rho_{i+1}, \rho_{n+i+1}, \dots, \rho_{2n})$ define the augmented fans for D and $\bar{D}$.

Note that a consequence of the theorem is that Σ given by equation (5.17) is the set with nontrivial stabilizers of the action of the complex torus $T_{\mathbb{C}}^2$.

6 Sasakian Submanifolds

In this section, we use the results on toric 3-Sasakian manifolds to produce a new infinite series of toric Sasaki–Einstein 5-manifolds corresponding to the toric 3-Sasakian 7-manifolds discussed above and completing the correspondence in

Diagram (1.1). But first we review the Smale/Barden classification of smooth 5-manifolds that will be used. The possible diffeotypes of toric examples in dimension 5 is very limited.

6.1 Classification of 5-Manifolds

Closed smooth simply connected spin 5-manifolds were classified by S. Smale [38]. Subsequently D. Barden extended the classification to the non-spin case [3]. Consider the primary decomposition

$$H_2(M,\mathbb{Z}) \cong \mathbb{Z}^r \oplus \mathbb{Z}_{k_1} \oplus \mathbb{Z}_{k_2} \oplus \cdots \oplus \mathbb{Z}_{k_s}, \qquad (6.1)$$

where k_j divides k_{j+1}. Of course the decomposition is not unique, but the $r, k_1, \ldots, k_s$ are. The second Stiefel–Whitney class defines a homomorphism $w_2 : H_2(M,\mathbb{Z}) \rightarrow \mathbb{Z}_2$. One can arrange the decomposition 6.1 so that w_2 is non-zero on only one component $\mathbb{Z}_{k_j}$, or $\mathbb{Z}$ of $\mathbb{Z}^r$. Then define $i(M)$ to be i if 2^i is the 2-primary component of $\mathbb{Z}_{k_j}$, or ∞. Alternatively, $i(M)$ is the minimum i so that w_2 is non-zero on a 2-primary component of order 2^i of $H_2(M,\mathbb{Z})$.

Theorem 6.1 ([3,38]). *Smooth simply connected closed 5-manifolds are classifiable up to diffeomorphism. Any such manifold is diffeomorphic to one of*

$$X_j \# M_{k_1} \# \cdots \# M_{k_s},$$

where $-1 \leq j \leq \infty, s \geq 0, 1 < k_1$ *and* K_i *divides* k_{i+1} *or* $k_{i+1} = \infty$. *A complete set of invariants is given by* $H_2(M,\mathbb{Z})$ *and* $i(M)$, *and the manifolds* $X_{-1}, X_0, X_j, X_\infty, M_k, M_\infty$ *are as follows*

M	$H_2(M,\mathbb{Z})$	$i(M)$
$M_0 = X_0 = S^5$	0	0
$M_k, 1 < k < \infty$	$\mathbb{Z}_k \oplus \mathbb{Z}_k$	0
$M_\infty = S^2 \times S^3$	$\mathbb{Z}$	0
X_{-1}	$\mathbb{Z}_2$	1
$X_j, 0 < j < \infty$	$\mathbb{Z}_{2^j} \oplus \mathbb{Z}_{2^j}$	j
X_∞	$\mathbb{Z}$	∞

where $X_{-1} = SU(3)/SO(3)$ *is the Wu manifold, and* X_∞ *is the nontrivial* S^3*-bundle over* S^2.

The existence of an effective T^3 action severely restricts the topology by the following theorem of Oh [35].

Theorem 6.2. *Let M be a compact simply connected 5-manifold with an effective T^3-action. Then $H_2(M,\mathbb{Z})$ has no torsion. Thus, M is diffeomorphic to S^5, $\#kM_\infty$, or $X_\infty\#(k-1)M_\infty$, where $k = b_2(M) \geq 1$. Conversely, these manifolds admit effective T^3-actions.*

By direct construction, C. Boyer, K. Galicki, and L. Ornea [9] showed that the manifolds in this theorem admit toric Sasakian structures and, in fact, admit regular Sasakian structures.

A simply connected Sasaki–Einstein manifold must have $w_2 = 0$, therefore we have the following:

Corollary 6.3. *Let M be a compact simply connected 5-manifolds with a toric Sasaki–Einstein structure. Then M is diffeomorphic to S^5, or $\#kM_\infty$, where $k = b_2(M) \geq 1$.*

6.2 Sasakian Submanifolds and Examples

Associated to each compact toric anti–self-dual Einstein orbifold $\mathcal{M}$ with $\pi_1^{orb}(\mathcal{M}) = e$ is the twistor space $\mathcal{Z}$ and a family of embeddings $X_t \subset \mathcal{Z}$ where $t \in E \setminus \{t_1, t_2, \ldots, t_{k+2}\}$ and $X = X_t$ is the symmetric toric Fano surface canonically associated to $\mathcal{M}$. We denote the family of embeddings by

$$\iota_t : X \to \mathcal{Z}. \tag{6.2}$$

Let M be the total space of the S^1 V-bundle associated to $\mathbf{K}_X$ or $\mathbf{K}_X^{\frac{1}{2}}$, depending on whether $\mathrm{ind}(X) = 1$ or 2.

Theorem 6.4. *Let $\mathcal{M}$ be a compact toric anti–self-dual Einstein orbifold with $\pi_1^{orb}(\mathcal{M}) = e$. There exists a Sasakian structure $\{\tilde{g}, \tilde{\Phi}, \tilde{\xi}, \tilde{\eta}\}$ on M, such that if $(X, \tilde{h})$ is the Kähler structure making $\pi : M \to X$ a Riemannian submersion, then we have the following diagram where the horizontal maps are isometric embeddings.*

$$
\begin{array}{ccc}
M & \xrightarrow{\bar{\iota}_t} & \mathcal{S} \\
\downarrow & & \downarrow \\
X & \xrightarrow{\iota_t} & \mathcal{Z} \\
& & \downarrow \\
& & \mathcal{M}
\end{array}
\tag{6.3}
$$

If the 3-Sasakian space $\mathcal{S}$ is smooth, then so is M. If M is smooth, then

$$M \underset{diff}{\cong} \#kM_\infty, \text{ where } k = 2b_2(\mathcal{S}) + 1.$$

Proof. Let $\{g, \Phi, \xi, \eta\}$ be the fixed Sasakian structure on $\mathcal{S}$ with Φ descending to the complex structure on $\mathcal{Z}$. The adjunction formula gives

$$\mathbf{K}_X \cong \mathbf{K}_{\mathcal{Z}} \otimes [X]|_X = \mathbf{K}_{\mathcal{Z}} \otimes \mathbf{K}_{\mathcal{Z}}^{-\frac{1}{2}}|_X = \mathbf{K}_{\mathcal{Z}}^{\frac{1}{2}}|_X. \tag{6.4}$$

Let h be the Kähler–Einstein metric on $\mathcal{Z}$ related to the 3-Sasakian metric g on $\mathcal{S}$ by Riemannian submersion. Recall that $\mathcal{S}$ is the total space of the S^1 V-bundle associated to $\mathbf{L}^{-1}$, (resp. $\mathbf{L}^{-\frac{1}{2}}$ if $w_2(\mathcal{M}) = 0$). Also M is the total space of the S^1 V-bundle associated to $\mathbf{K}_X$, (resp. $\mathbf{K}_X^{\frac{1}{2}}$ if $w_2(\mathcal{M}) = 0$). Using the isomorphism in (6.4), we lift ι_t to $\bar{\iota}_t$. The metric on $\mathcal{S}$ is

$$g = \eta \otimes \eta + \pi^* h.$$

Pull back h and η to $\tilde{h} = \iota_t^* h$ and $\tilde{\eta} = \bar{\iota}_t^* \eta$, respectively. It follows that $\frac{1}{2} \tilde{\eta} = \tilde{\omega}$, where $\tilde{\omega}$ is the Kähler form of $\tilde{h}$. Then $\tilde{\Phi} = \tilde{\nabla} \tilde{\xi}$ is a lift of the complex structure J on X. And the integrability condition in Definition 2.1 follows from the integrability of J. Then

$$\tilde{g} = \tilde{\eta} \otimes \tilde{\eta} + \pi^* \tilde{h}$$

is a Sasakian metric on M.

If $\mathcal{S}$ is smooth, then the orbifold uniformizing groups act on $\mathbf{L}^{-1}$ (or $\mathbf{L}^{\frac{1}{2}}$) minus the zero section without nontrivial stabilizers. By (6.4) this holds for the bundle $\mathbf{K}_X$ (or $\mathbf{K}_X^{\frac{1}{2}}$) on X.

It follows from $\pi_1^{orb}(\mathcal{M}) = e$ that $\pi_1^{orb}(X) = e$ (cf. [25]). And $\pi_1^{orb}(M) = e$ by arguments as after Proposition 2.4. If M is smooth, Corollary 6.3 gives the diffeomorphism.

We are more interested in M with the Sasaki–Einstein metric that exists by Corollary 4.6. In this case, the horizontal maps are not isometries.

Consider the reducible cases, $t_i \in E, i = 1, \ldots, k+2$, where $X_{t_i} = D + \bar{D} \subset \mathcal{Z}$. Then restricting $\mathbf{L}^{-1}$, (resp. $\mathbf{L}^{-\frac{1}{2}}$ if $w_2(\mathcal{M}) = 0$), and arguing as above we obtain Sasakian manifolds $N_i, \bar{N}_i \subset \mathcal{S}$, where smoothness follows from that of $\mathcal{S}$, whose Sasakian structures $(N_i, g_i, \Phi_i, \xi_i, \eta_i)$ and $(\bar{N}_i, \bar{g}_i, \bar{\Phi}_i, \bar{\xi}_i, \bar{\eta}_i)$ pull back from that of $\mathcal{S}$. And $N_{t_i} \cap \bar{N}_{t_i}$ is a lens space with the constant curvature metric. Note that ς restricts to an isometry $\varsigma : N_i \to \bar{N}_i$, which gives a conjugate isomorphism between Sasakian structures. These manifolds do not satisfy the conditions of Proposition 4.1, so cannot be transversally deformed to Sasaki–Einstein structures. But $c_1^B(N_i) > 0$, so they can be transversally deformed to positive Ricci curvature Sasakian by the transverse Calabi–Yau theorem [28]. By Theorem 6.2, N_i is diffeomorphic to $\#kM_\infty$, or $X_\infty \#(k-1)M_\infty$, where $k = b_2(N_i) = b_2(\mathcal{S})$. By the remarks after Proposition 7.1, $N_i, i = 1, \ldots, k+2$, is diffeomorphic to $X_\infty \#(k-1)M_\infty$ if $w_2(\mathcal{M}) \neq 0$.

The family of submanifolds $\bar{\iota}_t : M \to \mathcal{S}$ for $t \in E \setminus \{t_1, t_2, \ldots, t_{k+2}\}$ and $N_i, \bar{N}_i, i = 1, \ldots, k+2$ for the reducible cases have a simple description. Recall the 1-form $\eta = \eta^2 + i\eta^3$ of Section 5.1, which is $(1, 0)$ with respect to the CR structure $I = -\Phi^1$. For $t \in \mathfrak{t}$ let Y_t, denote the killing vector field on $\mathcal{Z}$ with lift $\bar{Y}_t \in I(\mathcal{S}, g)$. Then $\theta(Y_t) \in H^0(\mathcal{Z}, \mathcal{O}(\mathbf{L}))$, which defines a holomorphic function on $\mathbf{L}^{-1}$. Since $\mathcal{S}$ is the S^1 subbundle of $\mathbf{L}^{-1}$, we have $\theta(Y_t) = \eta(\bar{Y}_t)$. Complexifying gives the same equality for $t \in \mathfrak{t}_\mathbb{C}$. Thus for $t \in E \setminus \{t_1, t_2, \ldots, t_{k+2}\}$, we have $M_t := \bar{\iota}_t(M) = (\eta(\bar{Y}_t)) \subset \mathcal{S}$ and

$N_i \cup \bar{N}_i = (\eta(\bar{Y}_{t_i})) \subset \mathcal{S}$, where of course $(\eta(\bar{Y}_t))$ denotes the submanifold $\eta(\bar{Y}_t) = 0$. Note that here we are setting 2/3 s of the moment map to zero.

This gives us the new infinite families of Sasaki–Einstein manifolds and the diagram 1.1.

Theorem 6.5. *Let $(\mathcal{S}, g)$ be a toric 3-Sasakian 7-manifold with $\pi_1(\mathcal{S}) = e$. Canonically associated to $(\mathcal{S}, g)$ are a special symmetric toric Fano surface X and a toric quasi-regular Sasaki–Einstein 5-manifold (M, g, Φ, ξ, η) that fit in the commutative Diagram (6.3). We have $\pi_1^{orb}(X) = e$ and $\pi_1(M) = e$. And*

$$M \underset{diff}{\cong} \#kM_\infty, \text{ where } k = 2b_2(\mathcal{S}) + 1$$

Furthermore, $(\mathcal{S}, g)$ can be recovered from either X or M with their torus actions.

Proof. Note that the homotopy sequence

$$\cdots \to \pi_1(G) \to \pi_1(\mathcal{S}) \to \pi_1^{orb}(\mathcal{M}) \to e,$$

where $G = SO(3)$ or $Sp(1)$, shows that $\pi_1^{orb}(\mathcal{M}) = e$. The toric surface X_{Δ^*} and Sasakian 5-manifold $(M, \tilde{g}, \tilde{\Phi}, \tilde{\xi}, \tilde{\eta})$ are given in Theorem 6.4. By Corollary 4.6, this Sasakian structure has a transversal deformation to a Sasaki–Einstein structure.

Given either X or M, by the discussion in Section 5.3, the orbifold $\mathcal{M}$ can be recovered. By Theorem 5.16 $\mathcal{M}$ admits a unique anti–self-dual Einstein structure up to homothety compatible with the toric structure. The 3-Sasakian space $\mathcal{S}$ and its twistor space $\mathcal{Z}$ can be constructed from $\mathcal{M}$ and its anti–self-dual Einstein structure. (cf. [5]). □

Corollary 6.6. *For each odd $k \geq 3$, there is a countably infinite number of distinct toric quasi-regular Sasaki–Einstein structures on $\#kM_\infty$.*

We do not know the Sasaki–Einstein metrics explicitly. But if $c = \text{ind}(M)$, an application of Corollary 3.27 gives

$$\text{Vol}(M, g) = \frac{2\pi c}{3} \text{Vol}(X, \omega)$$
$$= 2c \left(\frac{\pi}{3}\right)^3 \text{Vol}(\Sigma_{-k}),$$

where ω is the transversal Kähler metric. We have $c = 1$ or 2. Let $(M, g_i), i \in \mathbb{Z}_+$, be any infinite sequence of metrics on $\#kM_\infty$ in Corollary 6.6. These Sasaki–Einstein structures have leaf spaces X_i, where $X_i = X_{\Delta_i^*}$. Observe that the polygons Δ_i^* get arbitrarily large, and the anticanonical polytopes $(\Sigma_{-k})_i$ satisfy

$$\text{Vol}((\Sigma_{-k})_i) \to 0, \text{ as } i \to \infty.$$

Thus we have $\text{Vol}(M, g_i) \to 0$, as $i \to \infty$.

6.3 Examples

We consider some of the examples obtained starting with the simplest. In particular, we can determine some of the spaces in Diagram (1.1) associated to a toric 3-Sasakian 7-manifold more explicitly in some cases.

6.3.1 Smooth Examples

It is well-known that there exists only two complete examples of positive scalar curvature anti–self-dual Einstein manifolds [27] [15], S^4 and $\mathbb{C}P^2$ with the round and Fubini–Study metrics, respectively. Note that we are considering $\mathbb{C}P^2$ with the opposite of the usual orientation.

$\mathcal{M} = S^4$

For the spaces in Diagram (1.1) we have: $\mathcal{M} = S^4$ with the round metric; its twistor space $\mathcal{Z} = \mathbb{C}P^3$ with the Fubini–Study metric; the quadratic divisor $X \subset \mathcal{Z}$ is $\mathbb{C}P^1 \times \mathbb{C}P^1$ with the homogeneous Kähler–Einstein metric; $M = S^2 \times S^3$ with the homogeneous Sasaki–Einstein structure; and $\mathcal{S} = S^7$ has the round metric. In this case Diagram (1.1) becomes the following.

$$
\begin{array}{ccc}
S^2 \times S^3 & \longrightarrow & S^7 \\
\downarrow & & \downarrow \\
\mathbb{C}P^1 \times \mathbb{C}P^1 & \longrightarrow & \mathbb{C}P^3 \\
& & \downarrow \\
& & S^4
\end{array}
\tag{6.5}
$$

This is the only example, I am aware of, for which the horizontal maps are isometric immersions when the toric surface and Sasakian space are equipped with the Einstein metrics.

$\mathcal{M} = \mathbb{C}P^2$

In this case, $\mathcal{M} = \mathbb{C}P^2$ with the Fubini–Study metric and the reverse of the usual orientation; its twistor space is $\mathcal{Z} = F_{1,2}$, the manifold of flags $V \subset W \subset \mathbb{C}^3$ with $\dim V = 1$ and $\dim W = 2$, with the homogeneous Kähler–Einstein metric. The projection $\pi : F_{1,2} \to \mathbb{C}P^2$ is as follows. If $(p, l) \in F_{1,2}$ so l is a line in $\mathbb{C}P^2$ and $p \in l$, then $\pi(p, l) = p^\perp \cap l$, where $p^\perp$ is the orthogonal compliment with respect to the standard Hermitian inner product. We can define $F_{1,2} \subset \mathbb{C}P^2 \times (\mathbb{C}P^2)^*$ by

$$
F_{1,2} = \{([p_0 : p_1 : p_2], [q^0 : q^1 : q^2]) \in \mathbb{C}P^2 \times (\mathbb{C}P^2)^* : \sum p_i q^i = 0\}.
$$

And the complex contact structure is given by $\theta = q^i dp_i - p_i dq^i$. Fix the action of T^2 on $\mathbb{C}P^2$ by

$$
(e^{i\theta}, e^{i\phi})[z_0 : z_1 : z_2] = [z_0 : e^{i\theta} z_1 : e^{i\phi} z_2].
$$

Then this induces the action on $F_{1,2}$

$$(e^{i\theta}, e^{i\phi})([p_0 : p_1 : p_2], [q^0 : q^1 : q^2]) = ([p_0 : e^{i\theta}p_1 : e^{i\phi}p_2], [q^0 : e^{-i\theta}q^1 : e^{-i\phi}q^2]).$$

Given $[a,b] \in \mathbb{C}P^1$ the one parameter group $(e^{ia\tau}, e^{ib\tau})$ induces the holomorphic vector field $W_\tau \in \Gamma(T^{1,0}F_{1,2})$ and the quadratic divisor $X_\tau = (\theta(W_\tau))$ given by

$$X_\tau = (ap_1q^1 + bp_2q^2 = 0, \quad p_iq^i = 0).$$

One can check directly that X_τ is smooth for $\tau \in \mathbb{C}P^1 \setminus \{[1,0], [0,1], [1,1]\}$ and $X_\tau = \mathbb{C}P^2_{(3)}$, the equivariant blow-up of $\mathbb{C}P^2$ at 3 points. For $\tau \in \{[1,0], [0,1], [1,1]\}$, $X_\tau = D_\tau + \bar{D}_\tau$ where both $D_\tau, \bar{D}_\tau$ are isomorphic to the Hirzebruch surface $F_1 = \mathbb{P}(\mathcal{O}_{\mathbb{C}P^1} \oplus \mathcal{O}_{\mathbb{C}P^1}(1))$.

The Sasaki–Einstein space is $M = \#3(S^2 \times S^3)$. And the Sasakian manifolds $N_\tau, \bar{N}_\tau$ are diffeomorphic to X_∞. $\mathcal{S} = \mathcal{S}(1,1,1) = SU(3)/U(1)$ with the homogeneous 3-Sasakian structure. This case has the following diagram.

$$
\begin{array}{ccc}
\#3(S^2 \times S^3) & \longrightarrow & SU(3)/U(1) \\
\downarrow & & \downarrow \\
\mathbb{C}P^2_{(3)} & \longrightarrow & F_{1,2} \\
& & \downarrow \\
& & \mathbb{C}P^2
\end{array}
\tag{6.6}
$$

6.3.2 Galicki–Lawson Quotients

The simplest examples of quaternionic-Kähler quotients are the Galicki–Lawson examples first appearing in [19] and further considered in [8]. These are circle quotients of $\mathbb{H}P^2$. In this case, the weight matrices are of the form $\Omega = \mathbf{p} = (p_1, p_2, p_3)$ with the admissible set

$$\{\mathcal{A}_{1,3}(Z) = \{\mathbf{p} \in \mathbb{Z}^3 | p_i \neq 0 \text{ for } i = 1, 2, 3 \text{ and } \gcd(p_i, p_j) = 1 \text{ for } i \neq j\}$$

We may take $p_i > 0$ for $i = 1, 2, 3$. The zero locus of the 3-Sasakian moment map $N(\mathbf{p}) \subset S^{11}$ is diffeomorphic to the Stiefel manifold $V^{\mathbb{C}}_{2,3}$ of complex 2-frames in $\mathbb{C}^3$, which can be identified as $V^{\mathbb{C}}_{2,3} \cong U(3)/U(1) \cong SU(3)$. Let $f_{\mathbf{p}} : U(1) \to U(3)$ be

$$
f_{\mathbf{p}}(\tau) = \begin{bmatrix} \tau^{p_1} & 0 & 0 \\ 0 & \tau^{p_2} & 0 \\ 0 & 0 & \tau^{p_3} \end{bmatrix}.
$$

Then the 3-Sasakian space $\mathcal{S}(\mathbf{p})$ is diffeomorphic to the quotient of $SU(3)$ by the action of $U(1)$

$$\tau \cdot W = f_{\mathbf{p}}(\tau) W f_{(0,0,-p_1-p_2-p_3)}(\tau) \text{ where } \tau \in U(1) \text{ and } W \in SU(3).$$

Thus $\mathcal{S}(\mathbf{p}) \cong SU(3)/U(1)$ is a biquotient similar to the examples considered by Eschenburg in [14].

The action of the group $SU(2)$ generated by $\{\xi^1, \xi^2, \xi^3\}$ on $N(\mathbf{p}) \cong SU(3)$ commutes with the action of $U(1)$. We have $N(\mathbf{p})/SU(2) \cong SU(3)/SU(2) \cong S^5$ with $U(1)$ acting by

$$\tau \cdot v = f_{(-p_2-p_3, -p_1-p_3, -p_1-p_2)} v \text{ for } v \in S^5 \subset \mathbb{C}^3.$$

We see that $\mathcal{M}_\Omega \cong \mathbb{C}P^2_{a_1,a_2,a_3}$ where $a_1 = p_2 + p_3, a_2 = p_1 + p_3, a_3 = p_1 + p_2$ and the quotient metric is anti–self-dual with the reverse of the usual orientation. If p_1, p_2, p_3 are all odd, then the generic leaf of the 3-Sasakian foliation $\mathscr{F}_3$ is $SO(3)$. If exactly one is even, then the generic leaf is $Sp(1)$. Denote by X_{p_1,p_2,p_3} the toric Fano divisor, which can be considered as a generalization of $\mathbb{C}P^2_{(3)}$. We have the following spaces and embeddings.

$$\begin{array}{ccc}
\#3(S^2 \times S^3) & \longrightarrow & \mathcal{S}(p_1, p_2, p_3) \\
\downarrow & & \downarrow \\
X_{p_1,p_2,p_3} & \longrightarrow & \mathcal{Z}(p_1, p_2, p_3) \\
& & \downarrow \\
& & \mathbb{C}P^2_{a_1,a_2,a_3}
\end{array} \qquad (6.7)$$

7 Positive Ricci Curvature Examples

In this section, we use the toric geometry developed to construct examples of positive Ricci curvature Sasakian structures on the manifolds $X_\infty \# (k-1)M_\infty$ in Theorem 6.1. By Theorem 6.2, these are the only simply connected non-spin 5-manifolds that can admit toric Sasakian structures. They are already known to admit Sasakian structures [9].

Define a marked fan $\Delta^* = \Delta^*_{k,p}, k \geq 2, p \geq 0$ as follows. Let $\sigma_0 = (-1,0)$, $\sigma_1 = (0,1)$; $\sigma_j = (j-1, \frac{j(j-1)}{2} - 1), j = 2, \ldots, k$; $\sigma_{k+1} = (k, \frac{(k+1)k}{2} - 1 + p)$, $\sigma_{k+2} = (0, \frac{(k+1)k}{2} + p)$. For $k = 1$ define $\Delta^*_{1,p}, p \geq 0$ by $\sigma_0 = (-1,0), \sigma_1 = (0,1), \sigma_2 = (1, 1+p), \sigma_3 = (0, 2+p)$. And define $l \in \mathrm{SF}(\Delta^*)$ by

$$l(\sigma_j) := \begin{cases} 0 & \text{if } j = 0, \\ -1 & \text{if } 1 \leq j \leq k+2. \end{cases}$$

It is easy to check that l is strictly upper convex, and we have $\pi_1^{orb}(X) = e$ for all of the above fans. The cone $\mathcal{C}_l$ corresponding to $(\mathbf{L}_l^{-1})^\times$ on X_{Δ^*} satisfies the smoothness condition of equation (3.4), and $(\mathbf{L}_l^{-1})^\times$ is simply connected. Also,

$-k \in SF(\Delta^*)$ is strictly upper convex. Thus by Corollary 3.25, $2\pi c_1(X_{\Delta^*}) = [\omega]$ for a Kähler form ω.

Proposition 7.1. *Each of the manifolds $M_{k,p}$, for $k \geq 1, p \geq 0$, is diffeomorphic to $X_\infty \#(k-1)M_\infty$. And for each p has a distinct positive Ricci curvature Sasakian structure.*

Proof. Let $X = X_{\Delta^*_{k,p}}$, and let $\mathbf{L}$ be the holomorphic line V-bundle associated with $l \in SF(\Delta^*)$. Since l is strictly upper convex, if ω is the canonical metric on X of Σ_l^*, then $[\omega] \in 2\pi c_1(\mathbf{L})$ by Theorem 3.24. By Proposition 2.4, there is a Sasakian structure (M, g, Φ, ξ, η) where M is the principle S^1 subbundle of $\mathbf{L}^{-1}$. The the toric cone $\mathcal{C}$ of $(\mathbf{L}^{-1})^\times \cong C(M)$ satisfies the smoothness condition. Also, by Theorem 3.24, $\mathbf{K}^{-1} > 0$. In other words, $c_1^B(M)$ has a positive representative. By the transverse Calabi–Yau theorem [28], there is a transversal deformation $(M, \tilde{g}, \tilde{\Phi}, \xi, \tilde{\eta})$ with $\tilde{\eta} = \eta + 2d^c\phi$, for $\phi \in C_B^\infty(M)$, with $\mathrm{Ric}_{\tilde{g}}^T$ positive. Then by (2.3), a homothetic deformation $\tilde{g}_a = a^2\tilde{\eta} \otimes \tilde{\eta} + a\tilde{g}^T$ for small enough $a \in \mathbb{R}_+$ has $\mathrm{Ric}_{\tilde{g}_a}$ positive on M.

Considering $\pi : M \to X$ as an S^1 V-bundle over X, it lifts to a genuine fiber bundle over $B(X) = M \times_{S^1} E(S^1)$, $\tilde{\pi} : \tilde{M} \to B(X)$. Here $E(S^1)$ is the universal S^1-principal bundle, and $B(X)$ is the orbifold classifying space (cf. [6]). Since M is smooth, we have a homotopy equivalence $\tilde{M} \simeq M$. And since $\pi_1^{org}(X) = e$, the first few terms of the Gysin sequence give

$$\mathbb{Z} \xrightarrow{\cup e} H_{orb}^2(X, \mathbb{Z}) \xrightarrow{\pi^*} H^2(M, \mathbb{Z}) \to 0,$$

where $e = c_1^{orb}(\mathbf{L}) \in H_{orb}^2(X, \mathbb{Z})$. We have $w_2(M) \equiv c_1(D) \mod 2 \equiv \pi^*(c_1^{orb}(X))$ mod 2, which is zero precisely when $x = c_1^{orb}(X)$ is divisible by 2 in $H_{orb}^2(X, \mathbb{Z})/\mathbb{Z}(e)$. If this is the case, then there is a $u \in H_{orb}^2(X, \mathbb{Z})$ with $2u = x + al$, with $a \in \mathbb{Z}$ odd. It is easy to see that $u \in H_{orb}^2(X, \mathbb{Z})$ is represented by a holomorphic line V-bundle, since $2u$ is. By Proposition 3.12, this V-bundle is $\mathbf{L}_u$ for some $u \in SF(\Delta^*)$, and we have an equation

$$2u = -k + al + f, \quad \text{where } f \in M. \tag{7.1}$$

Evaluating (7.1) on σ_0 gives $2u(\sigma_0) = -1 + f(\sigma_0)$, and on $\sigma_2 = -\sigma_0$ gives $2u(\sigma_2) = -(1+a) - f(\sigma_0)$. The first equation implies $f(\sigma_0)$ is odd, the second that $f(\sigma_0)$ is even. Thus $w_2(M) \neq 0$. Theorem 6.2 then completes the proof. □

One can use arguments of M. Demazure (cf. [34] §3.4) as in the smooth case to show that $\mathrm{Aut}^o(X)$ is not reductive for these examples. The fan $\Delta_{k,p}^*$ has two roots α_1, α_2 and $\alpha_2 \neq -\alpha_1$ as the Hirzebruch surface F_1. The Lie algebra of Hamiltonian holomorphic vector fields of M is not reductive. By the proof of the Lichnerowicz theorem in the Sasakian case [10], the Sasakian structure can not be transversally deformed to constant scalar curvature.

Let $N_i, \bar{N}_i \subset \mathcal{S}, i = 1, \ldots, k+2$ be the Sasakian submanifolds of the 3-Sasakian manifold $\mathcal{S}$ as discussed after Theorem 6.4. Suppose $w_2(\mathcal{M}) \neq 0$. So $N_i, \bar{N}_i$ are principle S^1 subbundles of $\mathbf{L}^{-1}$ restricted to toric surfaces $D, \bar{D}$ with fans as in

Theorem 5.19. The augmented fan Δ^* of D has elements $\sigma_1 = \rho_1, \ldots, \sigma_i = \rho_i, \sigma_{i+1} = -\rho_i + \rho_{i+1}, \sigma_{i+2} = \rho_{n+i+1}, \ldots, \sigma_{k+3} = \rho_{2n}, n = k+2$, and $\mathbf{L}|_D$ is the line bundle associated to $l \in \mathrm{SF}(\Delta^*)$ with

$$l(\sigma_j) := \begin{cases} 0 & \text{if } j = i+1, \\ -1 & \text{otherwise.} \end{cases}$$

The same argument in the above proposition shows that N_i is diffeomorphic to $X_\infty \# (k-1) M_\infty$.

8 Higher Dimensional Examples

In this section, we employ the join construction of C. Boyer and K. Galicki [6, 9] to construct higher dimensional examples. Let $(M_i, g_i, \Phi_i, \xi_i, \eta_i), i = 1, 2$ be quasi-regular Sasakian manifolds of dimensions $2m_i + 1, i = 1, 2..$ Make homothetic deformations of the Sasakian structures so that the S^1-actions generated by $\xi_i, i = 1, 2$ have period 1. Then the transverse Kähler forms $\omega_i^T, i = 1, 2$ descend to forms $\omega_i, i = 1, 2$ on the leaf spaces $\mathcal{Z}_i, i = 1, 2$ with $[\omega_i] \in H_{orb}^2(\mathcal{Z}_i, \mathbb{Z}), i = 1, 2$. Then for a pair of positive integers (k_1, k_2), we have a Kähler form $k_1 \omega_1 + k_2 \omega_2$ on $\mathcal{Z}_1 \times \mathcal{Z}_2$ with $[k_1 \omega_1 + k_2 \omega_2] \in H_{orb}^2(\mathcal{Z}_1 \times \mathcal{Z}_2, \mathbb{Z})$. By Proposition 2.4, there is an S^1 V-bundle, denoted $M_1 *_{k_1, k_2} M_2$, with a homothetic family of Sasakian structures. We may assume that $\gcd(k_1, k_2) = 1$; for if $(k_1, k_2) = (lk_1', lk_2')$, $M_1 *_{k_1, k_2} M_2 = (M_1 *_{k_1', k_2'} M_2)/\mathbb{Z}_l$. Note that

$$M_1 *_{k_1, k_2} M_2 = (M_1 \times M_2)/S^1(k_1, k_2),$$

where S^1 acts by $(x, y) \to (e^{ik_2\theta}x, e^{-ik_1\theta}y)$.

In general, the join $M_1 *_{k_1, k_2} M_2$ is an orbifold of dimension $2(m_1 + m_2) + 1$. But a simple condition exists that implies smoothness. Let $v_i = \mathrm{Ord}(M_i), i = 1, 2$ denote the lcm of the orders of the leaf holonomy groups of the $\mathscr{F}_{\xi_i}$.

Proposition 8.1 ([6, 9]). *For each pair (k_1, k_2) of relatively prime positive integers, $M_1 *_{k_1, k_2} M_2$ is a smooth quasi-regular Sasakian manifold if, and only if, $\gcd(v_1 k_2, v_2 k_1) = 1$.*

Note that if M_1 and M_2 are positive, that is, $\mathrm{Ricci}_{g_i}^T, i = 1, 2$ are positive, then $M_1 *_{k_1, k_2} M_2$ is positive. Suppose M_1 and M_2 are Sasakian Einstein. We have $\mathrm{ind}(\mathcal{Z}_1 \times \mathcal{Z}_2) = \gcd(\mathrm{ind}(\mathcal{Z}_1), \mathrm{ind}(\mathcal{Z}_2))$. So we define the *relative indices* of M_1 and M_2 to be

$$l_i = \frac{\mathrm{ind}(M_i)}{\gcd(\mathrm{ind}(M_1), \mathrm{ind}(M_2))}, \quad \text{for } i = 1, 2. \tag{8.1}$$

Then the homothetic family of Sasakian structures on the join $M_1 *_{l_1, l_2} M_2$ has a Sasakian Einstein structure with transverse metric

$$g^T = \frac{(m_1 + 1)g_1^T + (m_2 + 1)g_2^T}{m_1 + m_2 + 1}.$$

Let $M_k \cong \#k(S^2 \times S^3)$ be one of the Sasaki–Einstein manifolds constructed in Section 6. Then $\mathrm{ind}(M_k) = 1$ or 2. Consider S^{2m+1} with its standard Sasakian structure. Then $\mathrm{ind}(S^{2m+1}) = m + 1$. Then the relative indices of S^{4j+3} and M_k are both 1. So Proposition 8.1 implies that $M_k * S^{4j+3}$, $j \geq 0$, is a Sasaki–Einstein $(4j+7)$-manifold for all examples M_k. One can iterate this procedure; for example

$$M_k * \overbrace{S^3 * \cdots * S^3}^{p \text{ times}},$$

is a $5 + 2p$ dimensional Sasaki–Einstein manifold. By making repeated joins to the examples in Corollary 6.6, we obtain the following.

Proposition 8.2. *For every possible dimension $n = 2m + 1 \geq 5$, there are infinitely many toric quasi-regular Sasaki–Einstein manifolds with arbitrarily high second Betti number.*

More precisely, in dimension 5, we have examples of every odd b_2, and infinitely many examples for each odd $b_2 \geq 3$.

In dimension $n = 2m + 1 \geq 7$ for m odd, we have examples of every even b_2, and infinitely many examples for each even $b_2 \geq 4$.

In dimension $n = 2m + 1 \geq 9$ for m even, we have examples of every odd b_2, and infinitely many examples for each odd $b_2 \geq 5$.

Example. Consider the 7-dimensional Sasaki–Einstein manifolds $S^3 * M_k$. The Gysin sequence of $\pi : S^3 * M_k \to \mathbb{C}P^1 \times X$ determines the cohomology in $\mathbb{Q}$-coefficients. The Leray spectral sequence of the fiber bundle $\varpi : S^3 * M_k \to \mathbb{C}P^1$ with fiber M_k can be used to show the following.

p	0	1	2	3	4	5	6	7
$H^p(S^3 * M_k, \mathbb{Z})$	$\mathbb{Z}$	0	$\mathbb{Z}^{k+1}$	0	T	$\mathbb{Z}^{k+1}$	0	$\mathbb{Z}$

Here T is a torsion group. This is quite similar to the cohomology of $\mathcal{S}_\Omega$ in Theorem 5.14. $\diamond$

Let $N = N_{k,p}$ denote the positive Ricci curvature Sasakian 5-manifold of Proposition 7.1 diffeomorphic to $X_\infty \#(k-1)M_\infty$. Then we see by Proposition 8.1 that for any regular Sasaki–Einstein manifold M, $N *_{1,l} M$ is smooth for any $l \geq 1$. Furthermore, we may transversally deform it to a positive Ricci curvature structure.

Let $\pi_i : \mathbf{L}_i \to Z_i, i = 1, 2$ be the holomorphic V-bundles whose S^1-principal bundles are N and M, respectively. Then $N *_{1,l} M$ is the principal S^1 V-bundle of $\pi : \mathbf{L}_1 \otimes \mathbf{L}_2^l$, and as in the proof of Proposition 7.1 we have

$$\mathbb{Z} \xrightarrow{\cup e} H^2_{orb}(Z_1 \times Z_2, \mathbb{Z}) \xrightarrow{\pi^*} H^2(N *_{1,l} M, \mathbb{Z}) \to 0,$$

where $e = c_1(\mathbf{L}_1) + lc_1(\mathbf{L}_2)$. We have $w_2(N *_{1,l} M) \equiv \pi^*(c_1(Z_1) + c_1(Z_2)) \mod 2$, and $w_2(N *_{1,l} M) = 0$ if, and only if, $\pi^*(c_1(Z_1) + c_1(Z_2))$ is divisible by 2 in $H^2(N *_{1,l} M, \mathbb{Z})$. If this is the case then there is an $u \in H^2_{orb}(Z_1 \times Z_2, \mathbb{Z})$ with

$$2u = c_1(Z_1) + c_1(Z_2) + s(c_1(\mathbf{L}_1) + lc_1(\mathbf{L}_2)), \quad \text{for } s \in \mathbb{Z}.$$

Let $\iota : Z_1 \to Z_1 \times Z_2$ be the inclusion $\iota(x) = (x, y)$ with $y \in Z_2$ a smooth point. Then $2\iota^*u = c_1(Z_1) + sc_1(\mathbf{L}_1)$, which contradicts Proposition 7.1. Thus $w_2(N *_{1,l} M) \neq 0$. By taking joins of the $N_{k,p}, k \geq 1, p \geq 0$ with spheres as above, we obtain the following.

Proposition 8.3. *For every dimension $n = 2m + 1 \geq 5$, there exist infinitely many toric quasi-regular positive Ricci curvature Sasakian manifolds of each $b_2 \geq 2$, and of each $b_2 \geq 1$ in dimension 5. These examples are simply connected and have $w_2 \neq 0$. Therefore, they do not admit a Sasaki–Einstein structure.*

References

1. M. F. Atiyah and N. J. Hitchin and I. M. Singer. Self-duality in four-dimensional Riemannian geometry. *Proc. Roy. Soc. London Ser. A*, 362(1711):425–461, 1978.
2. W. L. Baily. On the imbedding of V-manifolds in projective space. *Am. J. Math.*, 79:403–430, 1957.
3. D. Barden. Simply connected five-manifolds. *Ann. Math.*, 82(2):365–385, 1965.
4. R. Bielawski. Complete hyper-Kähler 4n-manifolds with a local tri-Hamiltonian $\mathbf{R}^n$-action. *Math. Ann.*, 314(3):505–528, 1999.
5. C. P. Boyer and K. Galicki. 3-Sasakian manifolds. In *Surveys in Differential Geometry: Einstein Manifolds*, Surv. Differ. Geom., VI, pages 123–184. Int. Press, Boston, MA, 1999.
6. C. P. Boyer and K. Galicki. On Sasakian-Einstein geometry. *Internat J. Math.*, 11(7):873–909, 2000.
7. C. P. Boyer, K. Galicki, and B. M. Mann. The geometry and topology of 3-Sasakian manifolds. *J. Reine Angew. Math.*, 455:184–220, 1994.
8. C. P. Boyer, K. Galicki, B. M. Mann, and E. G. Rees. Compact 3-Sasakian 7 manifolds with arbitrary second Betti number. *Invent. Math.*, 131(2):321–344, 1998.
9. C. P. Boyer, K. Galicki, and L. Ornea. Constructions in Sasakian geometry. *Math. Z.*, 257(4):907–924, 2007.
10. C.P. Boyer, K. Galicki, and S.R. Simanca. Canonical sasakian metrics. *Comm. Math. Phys.*, 279(3):705–733, 2008.
11. D. Calderbank, L. David, and P. Gauduchon. The Guillemin formula and Kähler metrics on toric symplectic manifolds. *J. Symplectic Geom.*, 1(4):767–784, 2003.
12. D. Calderbank and M. Singer. Toric selfdual Einstein metrics on compact orbifolds. *Duke Math. J.*, 133(2):237–258, 2006.
13. T. Delzant. Hamiltoniens périodiques et image convex de l'application moment. *Bull. Soc. Math. France*, 116:315–339, 1988.
14. J. H. Eschenburg. New examples of manifolds with strictly positive curvature. *Invent. Math.*, 66:469–480, 1982.
15. Th. Friedrich and H. Kurke. Compact four-dimensional self-dual Einstein manifolds with positive scalar curvature. *Math. Nachr.*, 106:271–299, 1982.
16. W. Fulton. *Introduction to Toric Varieties*. Number 131 in Annals of Mathematics Studies. Priceton University Press, Princeton, NJ, 1993.

17. A. Futaki. An obstruction to the existence of Kähler-Einstein metrics. *Invent. Math.*, 73:437–443, 1983.
18. A. Futaki, H. Ono, and G. Wang. Transverse Kähler geometry of Sasaki manifolds and toric Sasaki-Einstein manifolds. preprint arXiv:math.DG/0607586 v.5, January 2007.
19. K. Galicki and H. B. Lawson. Quaternionic reductio and quaternionic orbifolds. *Math. Ann.*, 282:1–21, 1988.
20. A. Grothendieck. Sur les faisceaux algébriques et faisceaux analytique cohérents, volume 16 of Séminaire Henri Cartan; 9e année:1956/57. Quelques questions de topology, Exposé no. 2. Secretariat mathématique, Paris, 1958.
21. A. Grothendieck. Géométrie algébriques et géométrie analytique (notes by Mme. M. Raymond), SGA 1960/61, Exp. XII. In *Revêtements étales et groupe fondamental*, volume 224 of *Lecture Notes in Math.* Springer-Verlag, Heidelberg, 1971.
22. V. Guillemin. Kähler structures on toric varieties. *J. Diff. Geo.*, 40:285–309, 1994.
23. V. Guillemin. *Moment Maps and Combinatorial Invariants of Hamiltonian T^n-spaces*, volume 122 of *Progress in Mathematics*. Birkhäuser, Boston, 1994.
24. A. Haefliger. Groupoïdes d'holonomie et classifiants. In *Transversal Structure of Foliations (Toulouse, 1982)*, number 116 in Astérisque, pages 70–97, 1984.
25. A. Haefliger and E. Salem. Actions of tori on orbifolds. *Ann. Global Anal. Geom.*, 9(1):37–59, 1991.
26. R. A. Hepworth. The topology of certain 3-Sasakian 7-manifolds. *Math. Ann.*, 339(4):733–755, 2007.
27. N. J. Hitchin. Kählerian twistor spaces. *Proc. London Math. Soc. (3)*, 43(1):133–150, 1981.
28. A. El Kacimi-Alaoui. Opérateurs transversalement elliptiques sur un feuilletage riemannien et applications. *Compositio Mathematica*, 79:57–106, 1990.
29. E. Lerman and S. Tolman. Hamiltonian torus actions on symplectic orbifolds and symplectic varieties. *Trans. Am. Math. Soc.*, 349:4201–4230, 1997.
30. T. Mabuchi. Einstein-Kähler forms, Futaki invariants and convex geometry on toric Fano varieties. *Osaka J. Math.*, 24:705–737, 1987.
31. D. Martelli, J. Sparks, and S. T. Yau. The geometric dual of a-maximization for toric Sasaki-Einstein manifolds. *Comm. Math. Phys.*, 268(1):39–65, 2006.
32. D. Martelli, J. Sparks, and S. T. Yau. Sasaki-Einstein manifolds and volume minimisation. *Comm. Math. Phys.*, 280(3):611–673, 2008.
33. T. Oda. *Lectures on Torus Embeddings and Applications (Based on Joint Work with Katsuya Miyake)*, volume 58 of *Tata Inst. Inst. of Fund. Research*. Springer, Berlin, Heidelberg, New York, 1978.
34. T. Oda. *Convex Bodies in Algebraic Geometry*, volume 15 of *Ergebnisse der Math. u. ihrer Grenz. Geb. 3 Folge*. Springer, New York, 1985.
35. H. S. Oh. Toral actions on 5-manifolds. *Trans. Am. Math.*, 278(1):233–252, 1983.
36. R. Penrose. Nonlinear gravitons and curved twistor theory. *Gen. Relativ. Gravitation*, 7(1):31–52, 1976.
37. S. Salamon. Quaternionic Kähler manifolds. *Invent. Math.*, 67:143–171, 1982.
38. S. Smale. On the structure of 5-manifolds. *Ann. Math.*, 75(1):38–46, 1962.
39. G. Tian and X. Zhu. A new holomorphic invariant and uniqueness of Kähler-Ricci solitons. *Comment. Math. Helv.*, 77:297–325, 2002.
40. C. van Coevering. Toric surfaces and Sasakian-Einstein 5-manifolds. preprint arXiv:math.DG/0607721 v1, July 2006.
41. X. J. Wang and X. Zhu. Kähler-Ricci solitons on toric manifolds with positive first Chern class. *Adv. Math.*, 188:87–103, 2004.

On the Geometry of Cohomogeneity
One Manifolds with Positive Curvature

Wolfgang Ziller

Abstract We discuss manifolds with positive sectial curvature on which a group acts isometrically with one dimensional quotient. A number of the known examples have this property, but some potential families for new examples in dimension 7 arise as well. We discuss the geometry of these known examples and the connection that the candidates have with self-dual Einstein metrics.

There are very few known examples of manifolds with positive sectional curvature. Apart from the compact rank one symmetric spaces, they exist only in dimensions 24 and below and are all obtained as quotients of a compact Lie group equipped with a biinvariant metric under an isometric group action. They consist of certain homogeneous spaces in dimensions $6, 7, 12, 13$, and 24 due to Berger [Be], Wallach [Wa], and Aloff–Wallach [AW], and of biquotients in dimensions $6, 7$, and 13 due to Eschenburg [E1],[E2],[E3] and Bazaikin [Ba].

When trying to find new examples, it is natural to search among manifolds with large isometry group, a program initiated by K. Grove in the 1990s, see [Wi] for a recent survey. Homogeneous spaces with positive curvature were classified in [Be],[Wa],[BB] in the 1970s. The next natural case to study is therefore manifolds on which a group acts isometrically with one dimensional quotient, so called cohomogeneity one manifolds. L. Verdiani [V1, V2] showed that in even dimensions, positively curved cohomogeneity one manifolds are equivariantly diffeomorphic to an isometric action on a rank one symmetric space. In odd dimensions, K. Grove and the author observed in 1998 that there are infinite families among the known nonsymmetric positively curved manifolds that admit isometric cohomogeneity one actions, and suggested a family of potential 7 dimensional candidates P_k. In [GWZ], a classification in odd dimensions was carried out, and another family Q_k and an isolated manifold R emerged in dimension 7. It is not yet known whether these manifolds admit a cohomogeneity one metric with positive curvature, although they all admit one with nonnegative curvature as a consequence of the main result in [GZ].

W. Ziller
University of Pennsylvania, Philadelphia, Pennsylvania, USA

K. Galicki and S.R. Simanca (eds.), *Riemannian Topology and Geometric Structures on Manifolds,* Progress in Mathematics 271, DOI 10.1007/978-0-8176-4743-8,

In [GWZ], the authors also discovered an intriguing connection that the manifolds P_k and Q_k have with a family of self-dual Einstein orbifold metrics constructed by Hitchin [Hi1] on $\mathbb{S}^4$. They naturally give rise to 3-Sasakian metrics on P_k and Q_k, which by definition have lots of positive curvature already.

The purpose of this survey is threefold. In Section 2, we study the positively curved cohomogeneity one metrics on known examples with positive curvature, including the explicit functions that define the metric. In Section 3, we describe the classification theorem in [GWZ]. It is remarkable that among 7-manifolds where $G = \mathbb{S}^3 \times \mathbb{S}^3$ acts by cohomogeneity one, one has the known positively curved Eschenburg spaces E_p, the Berger space B^7, the Aloff–Wallach space W^7, and the sphere $\mathbb{S}^7$, and that the candidates P_k, Q_k, and R all carry such an action as well. We thus carry out the proof in this most intriguing case where $G = \mathbb{S}^3 \times \mathbb{S}^3$ acts by cohomogeneity one on a compact 7-dimensional simply connected manifold. In Section 4, we describe the relationship to Hitchin's self-dual Einstein metrics. We also discuss some curvature properties of these Einstein metrics and the metrics they define on P_k and Q_k. The behavior of these metrics, as well as the known metrics with positive curvature, are illustrated in a series of pictures.

1 Preliminaries

In this section, we discuss the basic structure of cohomogeneity one actions and the significance of the Weyl group. For more details, we refer the reader to [AA, Br, GZ, Mo]. We assume from now on that the manifold M and the group G that acts on M are compact and will only consider the most interesting case, where $M/G = I = [0,L]$. If $\pi: M \to M/G$ is the orbit projection, the inverse images of the interior points are the regular orbits and $B_- = \pi^{-1}(0)$ and $B_+ = \pi^{-1}(L)$ are the two nonregular orbits. Choose a point $x_- \in B_-$ and let $\gamma: [0,L] \to M$ be a minimal geodesic from B_- to B_+, parameterized by arc length, which we can assume starts at x_-. The geodesic is orthogonal to B_- and hence to all orbits. Define $x_+ = \gamma(L) \in B_+$, $x_0 = \gamma(\frac{L}{2})$ and let $K^\pm = G_{x_\pm}$ be the isotropy groups at $x_\pm$ and $H = G_{x_0} = G_{\gamma(t)}, 0 < t < L$, the principal isotropy group. Thus $B_\pm = G \cdot x_\pm = G/K^\pm$ and $G \cdot \gamma(t) = G/H$ for $0 < t < L$. For simplicity, we denote the tangent space of $B_\pm$ at $x_\pm$ by $T_\pm$ and its normal space by $T_\pm^\perp$.

By the slice theorem, we have the following description of the tubular neighborhoods $D(B_-) = \pi^{-1}([0,\frac{L}{2}])$ and $D(B_+) = \pi^{-1}([\frac{L}{2},L])$ of the nonprincipal orbits:

$$D(B_\pm) = G \times_{K_\pm} D^{\ell_\pm},$$

where $D^{\ell_\pm}$ are disks of radius $\frac{L}{2}$ in $T_\pm^\perp$. Here the action of $K_\pm$ on $G \times D^{\ell_\pm}$ is given by $k * (g,p) = (gk^{-1}, kp)$ where k acts on $D^{\ell_\pm}$ via the slice representation, i.e., the restriction of the isotropy representation to $T_\pm^\perp$. Hence we have the decomposition

$$M = D(B_-) \cup_E D(B_+),$$

where $E = G \cdot x_0 = G/H$ is a principal orbit that is canonically identified with the boundaries $\partial D(B_\pm) = G \times_{K_\pm} \mathbb{S}^{\ell_\pm - 1}$, via the maps $g \cdot H \to [g, \dot{\gamma}(0)]$ respectively $g \cdot H \to [g, -\dot{\gamma}(L)]$. Note also that $\partial D^{\ell_\pm} = \mathbb{S}^{\ell_\pm - 1} = K^\pm/H$ since the boundary of the tubular neighborhoods must be a G orbit, and hence $\partial D^{\ell_\pm}$ is a $K^\pm$ orbit.

All in all we see that we can recover M from G and the subgroups H and $K^\pm$. We caution though that the isotropy types, i.e., the conjugacy classes of the isotropy groups $K^\pm$ and H do not determine M. The isotropy groups depend on the choice of a minimal geodesic between the two nonregular orbits, and thus on the metric as well. A different choice of a minimal geodesic corresponds to conjugating all isotropy groups by an element of G. A change of the metric corresponds to changing K^+ to nK^+n^{-1} for some $n \in N(H)_0$, the identity component of the normalizer (cf. [AA, GWZ]).

An important fact about cohomogeneity one actions is that there is a converse to the above construction. Suppose G is a compact Lie group with subgroups $H \subset K^\pm \subset G$ and assume furthermore that $K^\pm/H = \mathbb{S}^{\ell_\pm - 1}$ are spheres. We sometimes denote this situation by $H \subset \{K^-, K^+\} \subset G$ and call it a group diagram. It is well-known that a transitive action of a compact Lie group K on a sphere $\mathbb{S}^{\ell-1}$ is conjugate to a linear action. We can thus assume that $K^\pm$ acts linearly on $\mathbb{S}^{\ell_\pm}$ with isotropy group the chosen subgroup $H \subset K^\pm$ at some point $p_\pm \in \mathbb{S}^{\ell_\pm - 1}$. It hence extends to a linear action on the bounding disk $D^{\ell_\pm}$, and we can thus define a manifold

$$M = G \times_{K^-} D^{\ell_-} \cup_{G/H} G \times_{K^+} D^{\ell_+},$$

where we glue the two boundaries by sending $[g, p_-]$ to $[g, p_+]$. The group G acts on M via $g * [g', p] = [gg', p]$ on each half, and one easily checks that the gluing is G-equivariant, and that the action has isotropy groups $K^\pm$ at $[e, 0]$ and H at $[e, p_\pm]$. One may also choose an equivariant map $G/H \to G/H$ to glue the two boundaries together. But such equivariant maps are given by $gH \to gnH$ for some $n \in N(H)$, and the new manifold is alternatively obtained by replacing K^+ with nK^+n^{-1} in the group diagram. But we caution that this new manifold may not be equivariantly diffeomorphic to the old one if n does not lie in the identity component of $N(H)$.

Another important ingredient for understanding the geometry of a cohomogeneity one manifold is given by the Weyl group. The Weyl group W of the action is by definition the stabilizer of the geodesic γ modulo its kernel, i.e. the group elements which fix γ pointwise, which by construction is equal to H. It is easy to see (cf. [AA]) that W is a dihedral subgroup of $N(H)/H$ with $M/G = \text{Im}(\gamma)/W$ and is generated by involutions $w_\pm \in W$ with $w_-(\gamma(t)) = \gamma(-t)$ and $w_+(\gamma(-t+L)) = \gamma(t+L)$. Thus w_+w_- is a translation by $2L$, and has order $|W|/2$ when W is finite. The involutions $w_\pm$ can be represented by the unique element $a \in K_0^\pm$ mod H with $av = -v$, where $K_v^\pm = H$. If $\ell_\pm = 1$, they are also the unique element $a \in K_0^\pm$ mod H such that a^2 but not a lies in H. For simplicity, we denote such representatives $a \in K_0^\pm$ again by $w_\pm$.

Note that W is finite if and only if γ is a closed geodesic, and in that case the order $|W|$ is the number of minimal geodesic segments intersecting the regular part. In [GWZ] it was shown that a cohomogeneity one manifold with an invariant metric of positive curvature necessarily has finite Weyl group. Note also that any nonprincipal isotropy group along γ is of the form $wK^{\pm}w^{-1}$ for some $w \in N(H)$ representing an element of W, and that the isotropy types $K^{\pm}$ alternate along γ.

We now discuss how to describe cohomogeneity one metrics on M. For $0 < t < L$, $\gamma(t)$ is a regular point with constant isotropy group H, and the metric on the principal orbits $G\gamma(t) = G/H$ is a family of homogeneous metrics g_t. Thus on the regular part, the metric is determined by

$$g_{\gamma(t)} = dt^2 + g_t,$$

and since the regular points are dense, it also describes the metric on M. Using a fixed biinvariant inner product Q on $\mathfrak{g}$, we define the Q-orthogonal splitting $\mathfrak{g} = \mathfrak{h} \oplus \mathfrak{m}$, which thus satisfies $\mathrm{Ad}(H)(\mathfrak{m}) \subset \mathfrak{m}$. We identify the tangent space to G/H at $\gamma(t), t \in (0,L)$ with $\mathfrak{m}$ via action fields, $X \in \mathfrak{m} \to X^*(\gamma(t))$, which also identifies the isotropy representation with the action of $\mathrm{Ad}(H)_{|\mathfrak{m}}$. We can choose a Q-orthogonal decomposition $\mathfrak{m} = \mathfrak{m}_1 + \cdots + \mathfrak{m}_k$ of $\mathfrak{m}$ into $\mathrm{Ad}(H)$ invariant irreducible subspaces and thus $g_t|\mathfrak{m}_i = f_i(t)Q|\mathfrak{m}_i$ for some functions $f_1, \ldots, f_k$. If the modules $\mathfrak{m}_i$ are inequivalent to each other, they are automatically orthogonal, and the functions f_i describe the metric completely. In positive curvature, it typically happens, as we will see, that the modules are not orthogonal to each other, and further functions are necessary to describe their inner products. In order for the metric on M to be smooth, these functions must satisfy certain smoothness conditions at the endpoints $t = 0$ and $t = L$, which in general can be complicated. The action of $w_{\pm}$ on $T_{x_{\pm}}M$ (well determined only up to $\mathrm{Ad}(H)$) preserves $T_{\pm}$ and $T_{\pm}^{\perp}$, and the action on $T_{\pm}$, given by $\mathrm{Ad}(w_{\pm})$, relates the functions describing the metric: $(\mathrm{Ad}(w_-)(X))^*(\gamma(t)) = X^*(\gamma(-t))$. If, e.g., $\mathrm{Ad}(w_-)(\mathfrak{m}_i) \subset \mathfrak{m}_i$, the function f_i must be even, and if $\mathrm{Ad}(w_-)(\mathfrak{m}_i) \subset \mathfrak{m}_j$, then $f_i(t) = f_j(-t)$. In fact most, but not all, of the smoothness conditions at the endpoints can be explained in this fashion by the action of the Weyl group.

2 Known Examples of Cohomogeneity One Manifolds with Positive Curvature

In this section, we describe the cohomogeneity one actions on the known cohomogeneity one manifolds with positive curvature that were discovered by K. Grove and the author in 1998. Apart from a cohomogeneity one action by $\mathrm{SU}(4)$ on the infinite family of 13-dimensional Bazaikin spaces B_p^{13}, which we will not discuss in this survey, they are all cohomogeneity one under an action of $S^3 \times S^3$ or one of its finite quotients. We start with the well-known action of $\mathrm{SO}(3)$ on $\mathbb{S}^4$ and $\mathrm{SO}(3)$ on $\mathbb{CP}^2$ since they are important in understanding the remaining examples and determine much of their geometry. We then study the action of $\mathrm{SO}(4)$ on $\mathbb{S}^7$, and of $\mathrm{SO}(4)$ on the Berger space $B^7 = \mathrm{SO}(5)/\mathrm{SO}(3)$. This latter action was also discovered by Podesta–Verdiani in [PV2]. Of a different nature is the action of $\mathrm{SU}(2) \times \mathrm{SO}(3)$ on the infinite family of Eschenburg biquotients E_p, which contains as a special case a homogeneous Aloff–Wallach space. We finish with a second cohomogeneity one action on the same Aloff–Wallach space, which shares some features of both actions. To distinguish them, we denote the first one by $W^7_{(1)}$ and the second by $W^7_{(2)}$. For a survey of the known examples of positive curvature, see [Zi2].

All actions described here are by groups locally isomorphic to S^3 or $S^3 \times S^3$. For comparison, we will describe them ineffectively so that $G = S^3$ or $G = S^3 \times S^3$ acts on the manifold. The effective version of the action, which we denote by $\bar{G}$, is obtained by dividing G by the ineffective kernel, which is the intersection of the center of G with the principal isotropy group H.

$$M = S^4 \text{ or } \mathbb{CP}^2 \text{ with } \bar{G} = SO(3).$$

We begin by describing the well-known cohomogeneity one action by $SO(3)$ on S^4. Let V be the 5-dimensional vector space of real 3×3 matrices with $A = A^t, \operatorname{tr}(A) = 0$ and with inner product $\langle A, B \rangle = \operatorname{tr} AB$. The group $SO(3)$ acts on V via conjugation $g \cdot A = g A g^{-1}$ preserving the inner product and hence acts on $S^4(1) \subset V$. Every point in $S^4(1)$ is conjugate to a matrix in the great circle $F = \{\operatorname{diag}(\lambda_1, \lambda_2, \lambda_3) \mid \sum \lambda_i = 0, \sum \lambda_i^2 = 1\}$ and hence the quotient space is one-dimensional. For the purpose of computations, we choose an orthonormal basis

$$e_1 = \frac{\operatorname{diag}(1,1,-2)}{\sqrt{6}} \ , \ e_2 = \frac{\operatorname{diag}(1,-1,0)}{\sqrt{2}} \ , \ e_3 = \frac{S_{12}}{\sqrt{2}} \ , \ e_4 = \frac{S_{13}}{\sqrt{2}} \ , \ e_5 = \frac{S_{23}}{\sqrt{2}}, \quad (2.1)$$

where S_{ij} is a symmetric matrix with a one in entries ij and ji and 0 everywhere else. If we choose $x_- = e_1$, then clearly $K^- = S(O(2)O(1))$ and the orbit G/K^- is the set of all symmetric matrices with 2 equal positive eigenvalues. Furthermore, $T_- = \operatorname{span}(e_4, e_5)$ with $T_-^\perp = \operatorname{span}(e_2, e_3)$ and thus $\gamma(t) = \cos(t)e_1 + \sin(t)e_2$ is a geodesic orthogonal to B_- and hence to all orbits. Clearly $x_+ = \gamma(\pi/3) = \operatorname{diag}(2,-1,-1)/\sqrt{6}$ is the first point along the geodesic γ that lies on the second singular orbit, consisting of the set of all symmetric matrices with 2 equal negative eigenvalues. Thus $L = \pi/3$ and $K^+ = S(O(1)O(2))$. For $\gamma(t)$ with $0 < t < \frac{\pi}{3}$, all eigenvalues λ_i are distinct, and hence the principal isotropy group is $H = S(O(1)O(1)O(1)) = \mathbb{Z}_2 \times \mathbb{Z}_2$.

If we denote by E_{ij}, the usual basis of the set of skew symmetric matrices, the above action of $SO(3)$ on $S^4(1)$ induces 3 action fields E_{ij}^*. A computation shows that:

$$E_{12}^* = 2\sin(t)e_3 \ , \ E_{23}^* = (\sqrt{3}\cos(t) - \sin(t))e_5 \ , \ E_{13}^* = (\sqrt{3}\cos(t) + \sin(t))e_4.$$

For the functions $f_1 = |E_{12}^*|^2$, $f_2 = |E_{23}^*|^2$, $f_3 = |E_{13}^*|^2$, which describe the metric, we thus obtain:

$$f_1 = 4\sin^2(t), \ f_2 = (\sqrt{3}\cos(t) - \sin(t))^2, \ f_3 = (\sqrt{3}\cos(t) + \sin(t))^2,$$

and all other inner products are 0.

For later purposes, it will be convenient to lift the isotropy groups into S^3 under the twofold cover $S^3 = \operatorname{Sp}(1) \to SO(3)$, given by conjugation on $\mathbb{H}$, which sends

$q \in Sp(1)$ into a rotation in the 2-plane $Im(q)^\perp \subset Im(\mathbb{H})$ with angle 2θ, where θ is the angle between q and 1 in S^3. After renumbering the coordinates, the group K^- lifts to $Pin(2) = \{e^{i\theta} \mid \theta \in \mathbb{R}\} \cup \{je^{i\theta} \mid \theta \in \mathbb{R}\}$, which we abbreviate to $e^{i\theta} \cup je^{i\theta}$. Similarly, K^+ lifts to $Pin(2) = e^{j\theta} \cup ie^{j\theta}$, and $H = S(O(1)O(1)O(1)) \subset SO(3)$ lifts to the quaternion group $Q = \{\pm 1, \pm i, \pm j, \pm k\}$. Thus the cohomogeneity one manifold S^4 can also be represented by the group diagram

$$Q \subset \{e^{i\theta} \cup je^{i\theta}, e^{j\theta} \cup ie^{j\theta}\} \subset S^3.$$

We will now discuss the Weyl symmetry using this group picture. Clearly $w_- = e^{i\frac{\pi}{4}}$ since $e^{i\frac{\pi}{2}} = i$ lies in H and similarly $w_+ = e^{j\frac{\pi}{4}}$. Thus $a = w_- w_+ = \frac{1}{2}(1+i+j+k)$ represents a translation by $-2L$ along the geodesic and since $a^3 = -1 \in H$, the Weyl group is $W = D_3$. This is consistent with the fact that the angle between x_- and x_+ is $\pi/3$ and hence $\gamma(t)$ intersects the regular part in $6 = |W|$ components. Notice now that $aia^{-1} = j, aja^{-1} = k$, and $aka^{-1} = i$. Thus $f_1(t+2L) = f_2(t)$ and $f_1(t+4L) = f_3(t)$. By applying only w_+ at $t = L$, we also obtain $f_3(t) = f_1(-t+2L)$. Thus the function $f_1(t)$ on the interval $[0, 3L]$ determines the full geometry of the cohomogeneity one manifold:

$$f_2(t) = f_1(t+2L), \ f_3(t) = f_1(t+4L) = f_1(-t+2L), \quad 0 < t < L \qquad (2.2)$$

There is a related cohomogeneity one action by $SO(3) \subset SU(3)$ on $\mathbb{CP}^2$, which has singular orbits the real points $B_- = \mathbb{RP}^2 \subset \mathbb{CP}^2$ and the quadric $B_+ = S^2 = \{[z_0, z_1, z_2] \mid \sum z_i^2 = 0\}$. Here we use the metric on $\mathbb{CP}^2$ induced by the biinvariant metric $\langle A, B \rangle = -\frac{1}{2} Re\, tr AB$ on $SU(3)$, which has curvature $1 \le sec \le 4$. One easily shows that the unit speed geodesic $\gamma(t)$, given in homogeneous coordinates by $[(\cos(t), i\sin(t), 0)]$, is orthogonal to all orbits and that the isotropy group at $x_- = \gamma(0)$ is $K^- = S(O(1)O(2))$, at $x_+ = \gamma(\pi/4)$ is $K^+ = SO(2)$, embedded in the first two coordinates, and that $H = G_{\gamma(t)} = \mathbb{Z}_2 = \langle diag(-1,-1,1) \rangle$ for $0 < t < \pi/4$. Hence $L = \pi/4$ and the group diagram, lifted to S^3, is given by:

$$\{\pm 1, \pm j\} \subset \{e^{i\theta} \cup je^{i\theta}, e^{j\theta}\} \subset S^3.$$

Thus a projection along the orbits gives rise to a twofold branched cover $\mathbb{CP}^2 \to S^4$ with branching locus the singular orbit $\mathbb{RP}^2 = S^3/(e^{i\theta} \cup je^{i\theta})$. One easily shows that the functions describing the metric are given by

$$f_1 = \sin^2(t), \ f_2 = \cos^2(2t), \ f_3 = \cos^2(t).$$

The Weyl group symmetry changes as $w_- = i$ but $w_+ = e^{j\frac{\pi}{4}}$ and hence $W = D_2$. The functions are thus related by

$$f_3(t) = f_1(t+2L) = f_1(-t+2L), \ f_2(t) = f_2(-t) = f_2(-t+2L). \qquad (2.3)$$

Hence in this case, the functions f_1 and f_2 on $[0, 2L]$ determine the full geometry of the cohomogeneity one manifold.

We finally mention the cohomogeneity one action by $\mathrm{SU}(2)$ on $\mathbb{CP}^2$, which fixes a point p_0. The second singular orbit is then the cut locus of p_0, and hence $L = \pi/2$ in this case.

$$M = \mathbb{S}^7 \text{ with } \bar{G} = \mathrm{SO}(4).$$

The action of $\mathrm{SO}(4)$ on $\mathbb{R}^8$ that induces a cohomogeneity one action on $\mathbb{S}^7$ is given by the isotropy representation of the rank 2 symmetric space $G_2 / \mathrm{SO}(4)$ on its tangent space. As a complex representation, it is the representation of $\mathrm{SU}(2) \times \mathrm{SU}(2)$ obtained by taking the tensor product of the unique 2-dimensional irreducible representation of $\mathrm{SU}(2)$ with the 4-dimensional one on the second factor. Thus the first $\mathrm{SU}(2)$ factor acts as the Hopf action on $\mathbb{S}^7$, and the second factor induces an action by $\mathrm{SO}(4)/\mathrm{SU}(2) \simeq \mathrm{SO}(3)$ on $\mathbb{S}^7(1)/\mathrm{SU}(2) = \mathbb{S}^4(\frac{1}{2})$. This action of $\mathrm{SO}(3)$ on $\mathbb{R}^5$ is irreducible as the representation of $\mathrm{SU}(2) \times \mathrm{SU}(2)$ on $\mathbb{R}^8$ is irreducible. Thus it agrees with the cohomogeneity one action of $\mathrm{SO}(3)$ in the previous example, and hence the $\bar{G}$ action on $\mathbb{S}^7$ is cohomogeneity one as well. Since $\mathrm{SU}(2)$ acts freely, this implies in particular that both actions have isomorphic isotropy groups, i.e., $K^{\pm} \simeq \mathrm{O}(2)$ and $H \simeq \mathbb{Z}_2 \oplus \mathbb{Z}_2$ considered as subgroups of $\mathrm{SO}(4)$. We now need to determine their explicit embeddings into $\mathrm{SO}(4)$ respectively $\mathbb{S}^3 \times \mathbb{S}^3$.

If we let V_k be the vector space of homogeneous polynomials of degree k in two complex variables z, w, then $\mathrm{SU}(2)$ acting on vectors (z, w) via matrix multiplication induces an irreducible representation on V_k of (complex) dimension $k + 1$ and preserves the inner product, which makes $z^m w^n$ into an orthogonal basis with $|z^m w^n|^2 = m! n!$. The isotropy representation (complexified) is thus $V_1 \otimes V_3$. The map $(z, w) \rightarrow (w, -z)$, extended to be a complex antilinear map $J_k \colon V_k \rightarrow V_k$, satisfies $J_k^2 = (-1)^k \mathrm{Id}$ and hence $(J_1 \otimes J_3)^2 = \mathrm{Id}$. Thus $J_1 \otimes J_3$ induces a real structure on $\mathbb{C}^8$ and hence its $+1$ eigenspace W is invariant under the action of $G = \mathrm{SU}(2) \times \mathrm{SU}(2)$, and is spanned by:

$$xz^3 + yw^3 \,,\ i(xz^3 - yw^3) \,,\ xzw^2 + ywz^2 \,,\ i(xzw^2 - ywz^2),$$
$$yz^3 - xw^3 \,,\ i(yz^3 + xw^3) \,,\ xz^2 w - yw^2 z \,,\ i(xz^2 w + yw^2 z),$$

which is our desired representation of G on $\mathbb{R}^8$.

Now let ΔQ be the diagonal embedding of the quaternion group into $\mathrm{SU}(2) \times \mathrm{SU}(2) = \mathbb{S}^3 \times \mathbb{S}^3$, i.e., $\Delta Q = \{\pm(1,1), \pm(i,i), \pm(j,j), \pm(k,k)\}$. Here we identify $a + bj \in \mathbb{S}^3$ with $\begin{pmatrix} a & b \\ -\bar{b} & \bar{a} \end{pmatrix} \in \mathrm{SU}(2)$. One easily checks that ΔQ fixes the two plane spanned by the orthonormal vectors $a = (xz^3 + yw^3)/2\sqrt{3}$, $b = (xzw^2 + ywz^2)/2$. By the above, this is then also the principal isotropy group H since the image of ΔQ in $\mathrm{SO}(4)$ is $\mathbb{Z}_2 \oplus \mathbb{Z}_2$. The great circle $\gamma(t) = \cos(t)a + \sin(t)b$ in this 2-plane meets all orbits orthogonally since it agrees with the fixed point set of H on $\mathbb{S}^7$. Now one easily checks that $x_- = \gamma(0)$ is fixed by the circle $(e^{-3it}, e^{it}) \subset \mathbb{S}^3 \times \mathbb{S}^3$ and hence $K^- = (e^{-3it}, e^{it}) \cdot H = (e^{-3it}, e^{it}) \cup (j, j) \cdot (e^{-3it}, e^{it})$. The first singular point along γ occurs at $x_+ = \gamma(\pi/6)$ since the projection of γ is a normal geodesic in the cohomogeneity one manifold $\mathbb{S}^4(\frac{1}{2})$. Thus $L = \pi/6$, and a computation shows that $\gamma(\pi/6)$ is fixed by the circle (e^{jt}, e^{jt}) and hence $K^+ = (e^{jt}, e^{jt}) \cup (i, i) \cdot (e^{jt}, e^{jt})$. Thus the group picture is given by

$$H = \Delta Q \subset \{(e^{-3it}, e^{it}) \cdot H \ , \ (e^{jt}, e^{jt}) \cdot H\} \subset S^3 \times S^3.$$

We identify the Lie algebra $\mathfrak{g}$ with $\mathrm{Im}\,\mathbb{H} \oplus \mathrm{Im}\,\mathbb{H}$ and will use the basis $X_1 = (i,0)$, $X_2 = (j,0)$, $X_3 = (k,0)$ and $Y_1 = (0,i)$, $Y_2 = (0,j)$, $Y_3 = (0,k)$ of $\mathfrak{g}$ and define $f_i(t) = |X_i^*(\gamma(t))|^2$, $g_i(t) = |Y_i^*(\gamma(t))|^2$, $h_i(t) = \langle X_i^*(\gamma(t)), Y_i^*(\gamma(t))\rangle$. Using the above action of $S^3 \times S^3$ applied to $\gamma(t)$, we obtain the action fields

$$X_1^*(\gamma(t)) = i\cos(t)\frac{xz^3 - yw^3}{2\sqrt{3}} + i\sin(t)\frac{xzw^2 - ywz^2}{2},$$

$$Y_1^*(\gamma(t)) = i\cos(t)\frac{3xz^3 - 3yw^3}{2\sqrt{3}} + i\sin(t)\frac{-xzw^2 + ywz^2}{2},$$

and thus

$$f_1 = 1 \ , \ g_1 = 8\cos^2(t) + 1 \ , \ h_1 = 4\cos^2(t) - 1.$$

As in the case of $\mathbb{S}^4$, the remaining functions are now determined via Weyl symmetry. We have $w_- = e^{i\frac{\pi}{4}}(-1,1)$ and $w_+ = e^{j\frac{\pi}{4}}(1,1)$ and thus $a = w_- w_+ = \frac{1}{2}(1 + i + j + k)(-1,1)$ with $a^3 = (-1,1)$. Hence $W = D_6$ corresponding to the fact that γ meets B_+ at $t = \pi/6$ and hence intersects the regular part in 12 components. Conjugation with a behaves as in the case of $\mathbb{S}^4$ on each component and hence f_i, as well g_i and h_i, satisfy the symmetry relations (2.2). Finally, we observe that all remaining inner products are necessarily 0 since the actions of the isotropy group ΔQ on the 3 subspaces $\mathrm{span}\{X_i, Y_i\}, i = 1,2,3$ are inequivalent to each other.

$$M = B^7 \text{ with } \bar{G} = \mathrm{SO}(4).$$

As we saw in our first example, $\mathrm{SO}(3)$ acts orthogonally on the vector space V, consisting of the set of traceless symmetric 3×3 matrices, via conjugation and hence isometrically on $\mathbb{S}^4$. This gives rise to an embedding $\phi \colon \mathrm{SO}(3) \to \mathrm{SO}(5)$ and defines a homogeneous space $B^7 = \mathrm{SO}(5)/\mathrm{SO}(3)$, also known as the Berger space. Berger showed in [Be] that a biinvariant metric on $\mathrm{SO}(5)$ induces a metric on B^7 with positive sectional curvature.

The subgroup $\mathrm{SO}(4) \subset \mathrm{SO}(5)$ acts on B^7 via left multiplication, and we claim it is cohomogeneity one. Using the basis (2.1) from Example 1, we let $\mathrm{SO}(4) = \mathrm{SO}(5)_{e_1}$ be the subgroup fixing e_1. The isotropy groups are then given by $\mathrm{SO}(4)_{g\,\mathrm{SO}(3)} = \mathrm{SO}(4) \cap g\,\mathrm{SO}(3)g^{-1} = g(\mathrm{SO}(3)_{g^{-1}e_1})g^{-1}$. Hence it follows from our first example that $K^\pm \simeq O(2)$ and $H \simeq \mathbb{Z}_2 \oplus \mathbb{Z}_2$ and thus the action is cohomogeneity one. We now need to compute the explicit embeddings of $K^\pm$ in $\mathrm{SO}(4)$ respectively $S^3 \times S^3$.

To avoid confusion, we let E_{ij} be the basis of skew symmetric matrices in $\mathrm{SO}(3)$ and F_{ij} the one in $\mathrm{SO}(5)$. If we set $\phi_*(E_{12}) = H_1$, $\phi_*(E_{23}) = H_2$, $\phi_*(E_{13}) = H_3$, one easily shows, using the explicit description of the action of $\mathrm{SO}(3)$ on V, that

$$H_1 = 2F_{23} + F_{45} \ , \ H_2 = F_{34} + \sqrt{3}F_{15} - F_{25} \ , \ H_3 = F_{35} + \sqrt{3}F_{14} + F_{24}.$$

Thus H_i defines an orthogonal basis of the Lie algebra of $\phi(\mathrm{SO}(3))$ with $|H_i|^2 = 5$.

For the point x_-, we choose $x_- = e \cdot SO(3)$, the identity coset in $SO(5)/SO(3)$. The group $K_0^- \simeq SO(2)$ then acts by rotation with angle 2θ in the e_2, e_3 plane and angle θ in the e_4, e_5 plane as $\phi_*(E_{12}) = 2F_{23} + F_{45}$, which determines its embedding into $SO(4)$. For K^+, we need to follow a normal geodesic. Clearly, F_{12} and F_{13} are orthogonal to H_i and the orbit of $SO(4)$ and hence lie in the normal space of B_-. In B^7, being normal homogeneous, a geodesic is the image of a one parameter group with initial vector orthogonal to H_i. Thus we can let $\gamma(t) = \exp(tF_{12}) \cdot SO(3) = (\cos(t)e_1 + \sin(t)e_2) \cdot SO(3)$ be the geodesic orthogonal to all orbits. From Example 1, it follows that the isotropy at $\gamma(t)$ is isomorphic to $\mathbb{Z}_2 \oplus \mathbb{Z}_2$ for $0 < t < \pi/3$ and to $O(2)$ at $\gamma(\pi/3)$ and thus $L = \pi/3$. If $g = \exp(\frac{\pi}{3}F_{12})$, we have $g^{-1}e_1 = \frac{1}{2}e_1 + \frac{\sqrt{3}}{2}e_2 = \mathrm{diag}(2, -1, -1)/\sqrt{6}$ and hence $K^+ = g(S(O(1)O(2)))g^{-1}$. From the embedding ϕ, it is clear that $SO(2) \subset S(O(1)O(2)) \subset SO(3)$ fixes $g^{-1}e_1$ and rotates by θ in the e_3, e_4 plane and by 2θ in the plane spanned by $\frac{\sqrt{3}}{2}e_1 - \frac{1}{2}e_2$ and e_5. Conjugating with g gives a rotation that fixes e_1, rotates by θ in the e_3, e_4 plane and by -2θ in the e_2, e_5 plane.

We now lift these groups into $G = S^3 \times S^3$ using the identification $e_2 \leftrightarrow 1$, $e_3 \leftrightarrow i$, $e_4 \leftrightarrow j$, $e_5 \leftrightarrow k$ and the twofold cover $S^3 \times S^3 \to SO(4)$ given by left and right multiplication of unit quaternions. It sends $X_1 = (i, 0) \to F_{23} + F_{45}$, $Y_1 = (0, i) \to -F_{23} + F_{45}$ and similarly for $X_i, Y_i, i = 2, 3$. Thus, after renumbering the coordinates, it follows that $K_0^- = (e^{-3it}, e^{it})$ and $K_0^+ = (e^{jt}, e^{-3jt})$. For the group picture to be consistent, we are left with:

$$\Delta Q \subset \{(e^{-3it}, e^{it}) \cdot H ,\ (e^{jt}, e^{-3jt}) \cdot H\} \subset S^3 \times S^3 .$$

In order to determine the functions describing the metric, let $\bar{X}_i, \bar{Y}_i$ be the action fields on $SO(5)$ and X_i^*, Y_i^* those on $SO(5)/SO(3)$. To compute their length at $\gamma(t)$, we translate them back to the identity with the isometric left translation by $\exp(tF_{12})^{-1}$. We thus obtain:

$$\bar{X}_1(t) = \mathrm{Ad}(\exp(-tF_{12}))(F_{23} + F_{45}) = -\sin(t)F_{13} + \cos(t)F_{23} + F_{45},$$
$$\bar{Y}_1(t) = \mathrm{Ad}(\exp(-tF_{12}))(-F_{23} + F_{45}) = \sin(t)F_{13} - \cos(t)F_{23} + F_{45}.$$

Since $X_1^* = \bar{X}_1 - \frac{1}{5}\langle \bar{X}_1, H_1 \rangle H_1 = \bar{X}_1 - \frac{1}{5}(2\cos(t) + 1)H_1$ and $Y_1^* = \bar{Y}_1 - \frac{1}{5}(-2\cos(t) + 1)H_1$ we have:

$$f_1 = \frac{1}{5}(5 + 4\sin^2(t) - 4\cos(t)) ,\ g_1 = \frac{1}{5}(5 + 4\sin^2(t) + 4\cos(t)),$$
$$h_1 = -\frac{1}{5}(1 - 4\cos^2(t)).$$

The remaining functions are determined by Weyl group symmetry. Similarly to the example of S^7, we see that $w_- = e^{i\frac{\pi}{4}}(1, 1)$ and $w_+ = e^{j\frac{\pi}{4}}(1, 1)$ and hence $(w_- w_+)^3 = (-1, -1) \in H$. Thus in this case $W = D_3$. But conjugation by $w_- w_+$ behaves as before, and hence all functions satisfy the same Weyl symmetry as in (2.2).

It is now interesting to compare the metrics in these two examples, which we do in a sequence of pictures. Figure 1 shows all 9 functions between two singular orbits, clearly not very instructive. Figure 2 illustrates the effects of Weyl symmetry in these pictures in a typical case of the 3 g functions for the Berger space. Thus Figure 3, which shows $f = f_1$, $g = g_1$, $h = h_1$ on $[0, 3L]$, encodes all the geometry of the space. As was discovered by K. Grove, B. Wilking and the author, the positivity of the sectional curvatures $\sec(\gamma'(t), X^*)$, $X \in \mathfrak{g}$ implies that the inverse of the metric tensor is a convex matrix. Figure 4 shows the functions F_1, G_1, H_1 in the inverse of $\begin{pmatrix} f_1 & h_1 \\ h_1 & g_1 \end{pmatrix}$ on the interval $[0, 3L]$. Smoothness conditions are now encoded in the growth behavior of the functions as $t \to 0$ and $t \to 3L$.

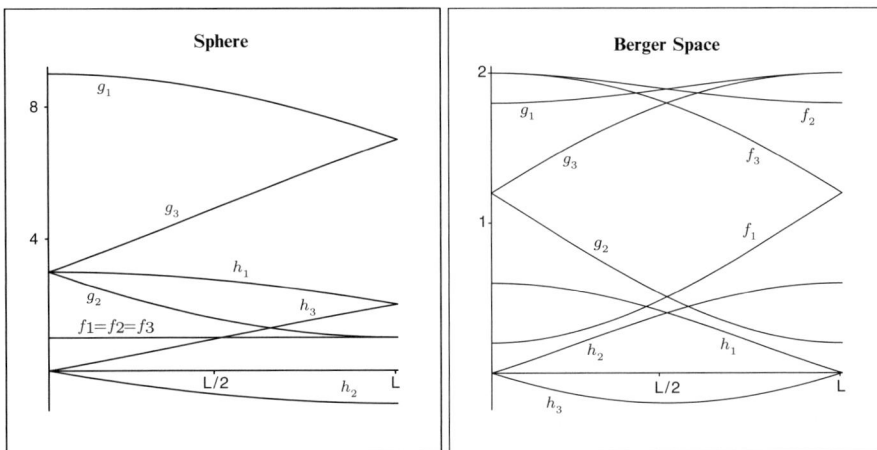

Fig. 1 All 9 functions on $[0, L]$.

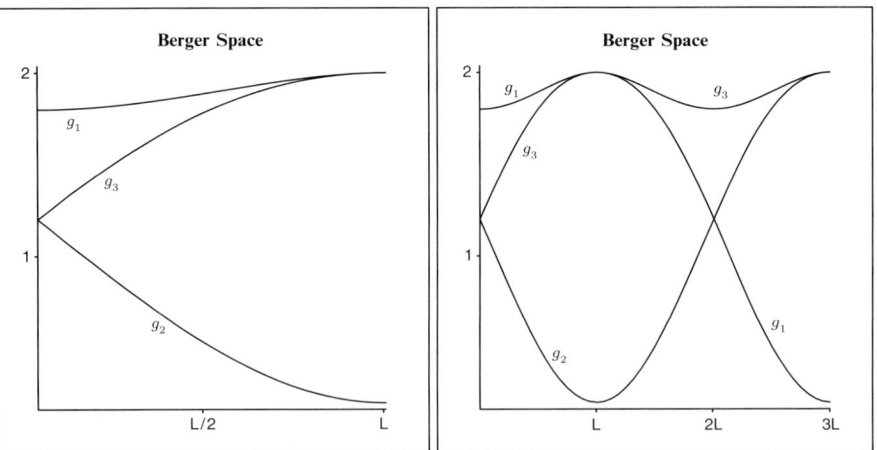

Fig. 2 The g functions on $[0, L]$ and $[0, 3L]$.

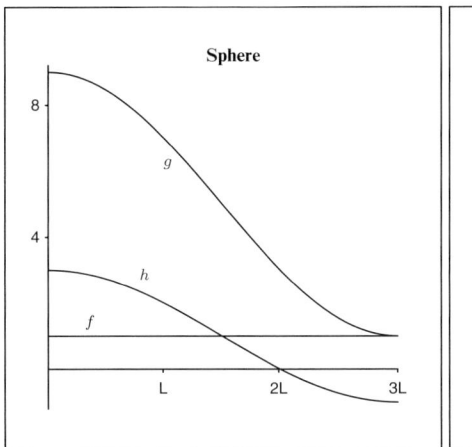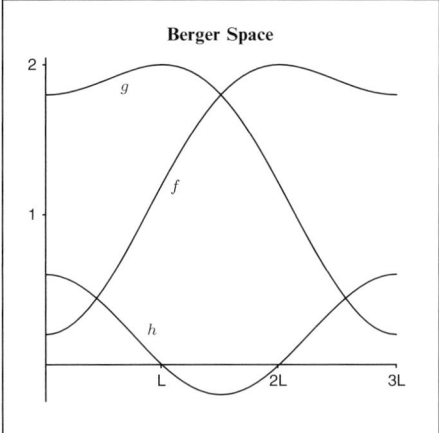

Fig. 3 All functions on $[0, 3L]$.

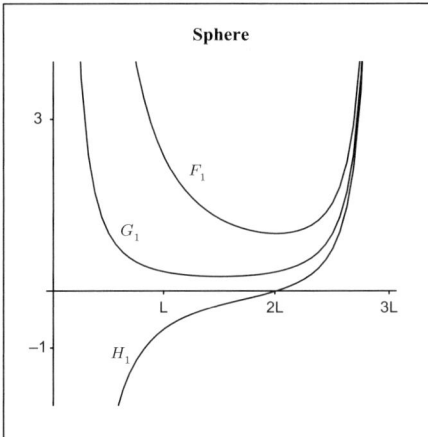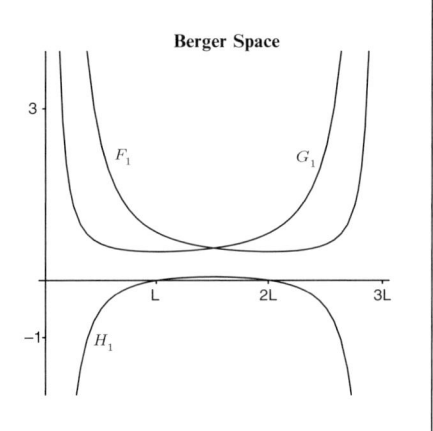

Fig. 4 The inverse functions on $[0, 3L]$.

$$M = E_p^7 \text{ with } \bar{G} = \mathrm{SO}(3) \times \mathrm{S}^3.$$

Next, we examine a family of biquotients among the 7-dimensional Eschenburg spaces. Define

$$E_p := \mathrm{SU}(3) /\!/ \, \mathrm{S}_p^1 = \mathrm{diag}(z, z, z^p) \backslash \mathrm{SU}(3) / \mathrm{diag}(1, 1, z^{p+2})^{-1},$$

where it is understood that $\mathrm{S}^1 = \{z \mid |z| = 1\}$ acts on $\mathrm{SU}(3)$ simultaneously on the left and on the right. Up to equivalence, we can assume that $p \geq 0$, and Eschenburg showed that it admits a metric with positive sectional curvature if $p \geq 1$. The positively curved metric is obtained by scaling the biinvariant metric

$\langle A, B \rangle = -\frac{1}{2} \operatorname{Re} \operatorname{tr} AB$ on $\mathrm{SU}(3)$ in direction of the subgroup $\operatorname{diag}(A, \det \bar{A})$, $A \in \mathrm{U}(2)$ by an amount $\varepsilon < 1$. The group $G = \mathrm{SU}(2) \times \mathrm{SU}(2)$ acts on E_p isometrically by multiplying on the left and on the right in the first two coordinates since it clearly commutes with the circle action, and we claim this action is cohomogeneity one. Indeed, we can first divide by the second $\mathrm{SU}(2)$ and the action of the first $\mathrm{SU}(2)$, since $(\{e\} \times \mathrm{SU}(2)) \cdot \mathrm{S}_p^1 = \mathrm{U}(2)$ is then an action on $\mathrm{SU}(3)/\mathrm{U}(2) = \mathbb{CP}^2$, which fixes the identity coset $e \cdot \mathrm{U}(2)$ and acts transitively on the normal sphere. Thus this action on $\mathbb{CP}^2$, and hence also the action of G on E_p, is cohomogeneity one. Notice that if p is even, the action becomes effectively an action by $\mathrm{SO}(3) \times \mathrm{SU}(2)$, whereas if p is odd, by $\mathrm{SU}(2) \times \mathrm{SO}(3)$.

One singular point is clearly the image of the identity matrix, $x_- = e \cdot \mathrm{S}_p^1$, with $K^- = \{(g, \pm g) \mid g \in \mathrm{SU}(2)\}$. One easily checks that in the modified metric on $\mathrm{SU}(3)$, the one parameter group $\exp(t E_{13})$, being orthogonal to $\mathrm{U}(2)$, is still a (unit speed) geodesic in $\mathrm{SU}(3)$ (see, e.g., [DZ, p. 28]). Since it is also orthogonal to the orbit of G at x_-, its projection $\gamma(t)$ into $\mathrm{SU}(3)/\!/\mathrm{S}_p^1$ is a geodesic orthogonal to all orbits. Its projection to $\mathbb{CP}^2$ is also a normal geodesic, and the induced $\mathrm{SU}(2)$ action has isotropy $\mathrm{SU}(2)$ at $t = 0$ and isotropy S^1 at $t = \pi/2$ and the principal isotropy is trivial. Thus the same holds for the action of $\bar{G}$ on E_p, in particular $L = \pi/2$. For the explicit embeddings, one easily checks ([Zi1],[GSZ]) that the isotropy group of G at $x_+ = \gamma(\pi/2)$ is equal to $K^+ = (e^{i(p+1)t}, e^{ipt})$. Hence we obtain the group diagram:

$$\mathbb{Z}_2 = ((-1)^{p+1}, (-1)^p) \subset \{\Delta \mathrm{S}^3 \cdot H, \ (e^{i(p+1)t}, e^{ipt})\} \subset \mathrm{S}^3 \times \mathrm{S}^3.$$

To compute the functions describing the metric, we identify $\mathrm{S}^3 \times \mathrm{S}^3$ as before with $\mathrm{SU}(2) \times \mathrm{SU}(2)$ and translate the action fields at $\gamma(t)$ back to the identity. We then obtain:

$$\bar{X}_1^* = \operatorname{Ad}(\exp(-t E_{13})) \operatorname{diag}(i, -i, 0) = \operatorname{diag}(i \cos^2(t), -i, i \sin^2(t)) + \cos(t) \sin(t) I_{13},$$
$$\bar{X}_2^* = \operatorname{Ad}(\exp(-t E_{13})) E_{12} = \cos(t) E_{12} - \sin(t) E_{23},$$
$$\bar{X}_3^* = \operatorname{Ad}(\exp(-t E_{13})) I_{12} = \cos(t) I_{12} + \sin(t) I_{23},$$
$$\bar{Y}_1^* = -\operatorname{diag}(i, -i, 0), \ \bar{Y}_2^* = -E_{12}, \ \bar{Y}_3^* = -I_{12},$$

where we used the notation I_{kl} for a matrix in $\mathfrak{su}(3)$ with i in entry kl and lk and 0 elsewhere. The vertical space of the Riemannian submersion $\pi \colon \mathrm{SU}(3) \to \mathrm{SU}(3)/\!/\mathrm{S}_p^1$ (translated back to the identity) is spanned by

$$v = \operatorname{Ad}(\exp(-t E_{13})) i \operatorname{diag}(1, 1, p) - i \operatorname{diag}(0, 0, p+2)$$
$$= i \operatorname{diag}(\cos^2(t) + p \sin^2(t), 1, \sin^2(t) + p \cos^2(t) - p - 2) + (1-p) \cos(t) \sin(t) I_{13},$$

whose length in the Eschenburg metric is

$$|v|^2 = 3\varepsilon + (1-p)^2(1-\varepsilon)\cos^2(t)\sin^2(t) + \varepsilon(p+2)(p-1)\sin^2(t).$$

Notice that $\bar{X}_2, \bar{X}_3, \bar{Y}_2, \bar{Y}_3$ are already horizontal with respect to the Riemannian submersion π but that we need to subtract the vertical component from $\bar{X}_1$ and $\bar{Y}_1$.

A computation now shows that:

$$f_1 = \frac{\varepsilon}{4\alpha} \left[(3\varepsilon(p-2)^2 - 4p^2 + 8p - 16) \cos^4(t) + \right.$$
$$\left. + (4p^2 - 8p + 16 - 6\varepsilon p(p-2)) \cos^2(t) + 3\varepsilon p^2 \right],$$

$$g_1 = \frac{\varepsilon}{4\alpha} \left[(p-1)^2(3\varepsilon-4) \cos^4(t) + (2p-2)(2p-2-3\varepsilon(p+1)) \cos^2(t) + 3\varepsilon(p+1)^2 \right],$$

$$h_1 = -\frac{\varepsilon}{4\alpha} \left[(p-1)(3\varepsilon(p-2) - 4p + 4) \cos^4(t) + \right.$$
$$\left. + (4(p-1)^2 - 6\varepsilon(p^2 - p - 1)) \cos^2(t) + 3\varepsilon p(p+1) \right],$$

$$f_2 = f_3 = 1 + (\varepsilon - 1) \cos^2(t) , \; g_2 = g_3 = \varepsilon , \; h_2 = h_3 = -\varepsilon \cos(t).$$

where $\alpha = (p-1)^2(\varepsilon-1)\cos^4(t) + (p-1)(p-1-\varepsilon(2p+1))\cos^2(t) + \varepsilon(p^2 + p + 1)$ and all other inner products are 0.

Notice that $p = 1$ is a special case since the Eschenburg space is now simply the homogeneous Aloff–Wallach space $W^7 = SU(3)/\operatorname{diag}(z,z,\bar{z}^2)$, which, with the above cohomogeneity one action, we denote by $W_{(1)}^7$. The functions are given by:

$$f_1 = \frac{1}{4} \left[(\varepsilon-4)\cos^4(t) + 2(\varepsilon+2)\cos^2(t) + \varepsilon \right] , \; f_2 = f_3 = 1 + (\varepsilon - 1)\cos^2(t),$$

$$g_1 = g_2 = g_3 = \varepsilon , \; h_1 = -\frac{\varepsilon}{2}(\cos^2(t) + 1) , \; h_2 = h_3 = -\varepsilon \cos(t).$$

A major difference with the previous two cases lies in the Weyl group. Clearly $w_- = (-1,-1)$ and $w_+ = (i^{p+1}, i^p)$. Thus $(w_- w_+)^2 \in H$ and hence $W = D_2$. The Weyl group elements multiply each of the natural basis vectors in $T_{p_\pm} B_\pm$ with ± 1 and hence Weyl group symmetry simply says that all (noncollapsing) functions must be even at $t = 0$ and at $t = L$, in particular their first derivatives must vanish. Thus any of the relationships between different functions that was so useful in the previous cases is lost. Also, notice that, unlike in the previous two examples, the vanishing of the remaining inner products is not forced anymore by the action of the isotropy group since it acts trivially on G/H. The basic behavior of the functions is illustrated in Figure 5 for the Aloff–Wallach space and the Eschenburg space with a typical value of $p = 10$ and $\varepsilon = \frac{1}{2}$. Here we have drawn the graphs on $[0, 4L]$, i.e., once around the closed geodesic, for better comparison. As $p \to \infty$, the functions converge, but the limiting metric is not smooth at the singular orbits.

$$W_{(2)}^7 \text{ with } \bar{G} = SO(3) SO(3).$$

The Aloff–Wallach space $W^7 = SU(3)/\operatorname{diag}(z,z,\bar{z}^2)$ has a second cohomogeneity one action by combining right multiplication by $SU(2)$ as in the previous example with left multiplication by $SO(3) \subset SU(3)$. Observe that the right action by $SU(2)$ is effectively an action of $SO(3) = U(2)/Z(U(2))$ and that the action is free with quotient $SU(3)/U(2) = \mathbb{CP}^2$. The left action by $SO(3)$ then induces an action on the quotient, which has to be the cohomogeneity one action by $SO(3)$ on $\mathbb{CP}^2$

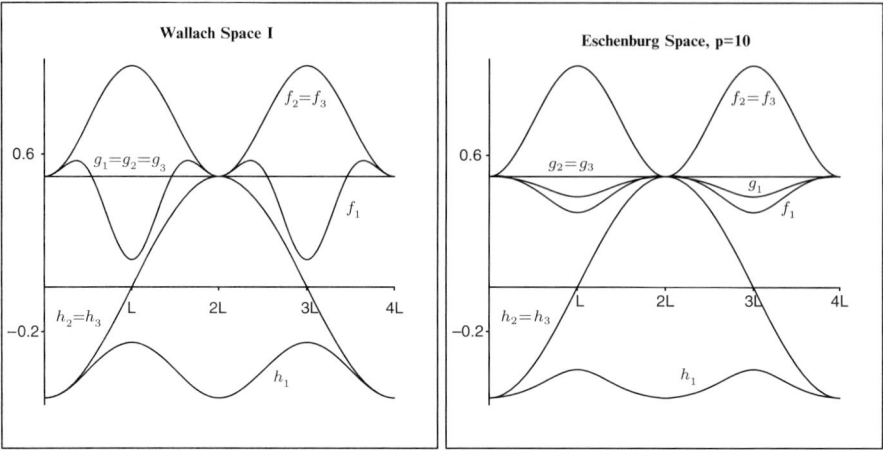

Fig. 5 Wallach space $W^7_{(1)}$ and Eschenburg space E_{10} on $[0, 4L]$.

mentioned earlier, since there exists only one SO(3) in SU(3). In particular, the $\bar{G} = $ SO(3) SO(3) action on W^7 is cohomogeneity one, which also determines the isomorphism type of the isotropy groups: $K^- = $ O(2), $K^+ = $ SO(2) and $H = \mathbb{Z}_2$.

We can choose x_- again to be the identity coset since the isotropy is SO(3) $\cap$ SU(2) $\cdot$ diag$(z, z, \bar{z}^2) = $ SO(3) $\cap$ U(2) $= $ O(2). The tangent space to B_- is spanned by E_{ij}, I_{12} and diag$(i, -i, 0)$ and hence $\gamma(t) = \exp(t I_{13}) \cdot $ diag$(z, z, \bar{z}^2)$ is a unit speed geodesic in SU(3)/ diag$(z, z, \bar{z}^2)$ orthogonal to all orbits. Since the projection to $\mathbb{CP}^2$ is a normal geodesic orthogonal to the orbits of the cohomogeneity one action on $\mathbb{CP}^2$, it follows that the singular isotropy groups occur at $t = 0$ and $t = \pi/4$ and thus $L = \pi/4$. Instead of trying to compute the embedding of these isotropy groups directly, we will use the functions describing the metric instead.

We first compute the action fields X_i^*. For comparison, we again consider the action as an (ineffective) action by $S^3 \times S^3$, and since the twofold cover SU(2) $\to$ SO(3) multiplies the natural basis vectors by 2, we find:

$$\bar{X}_1(t) = \text{Ad}(-\exp(t I_{13}))(2E_{12}) = -2\cos(t)E_{12} - 2\sin(t)I_{23},$$
$$\bar{X}_2(t) = \text{Ad}(-\exp(t I_{13}))(2E_{13}) = -2\cos(2t)E_{13} + 2\sin(2t)\,\text{diag}(i, 0, -i),$$
$$\bar{X}_3(t) = \text{Ad}(-\exp(t I_{13}))(2E_{23}) = 2\sin(t)I_{12} - 2\cos(t)E_{23}.$$

Notice that the component of diag$(i, 0, -i)$ orthogonal to diag$(z, z, \bar{z}^2)$ is $\frac{1}{2}$ diag $(i, -i, 0)$ and that $\bar{X}_1(t)$ and $\bar{X}_3(t)$ are orthogonal already.

For the right action fields, we have $\bar{Y}_1^* = -E_{12}$, $\bar{Y}_2^* = -$ diag$(i, -i, 0)$, $\bar{Y}_3^* = -I_{12}$ and thus:

$$f_1 = 4\sin^2(t) + 4\varepsilon\cos^2(t)\ ,\ g_1 = \varepsilon\ ,\ h_1 = 2\varepsilon\cos(t),$$
$$f_2 = 4\cos^2(2t) + \varepsilon\sin^2(2t)\ ,\ g_2 = \varepsilon\ ,\ h_2 = -\varepsilon\sin(2t),$$
$$f_3 = 4\cos^2(t) + 4\varepsilon\sin^2(t)\ ,\ g_3 = \varepsilon\ ,\ h_3 = -2\varepsilon\sin(t),$$

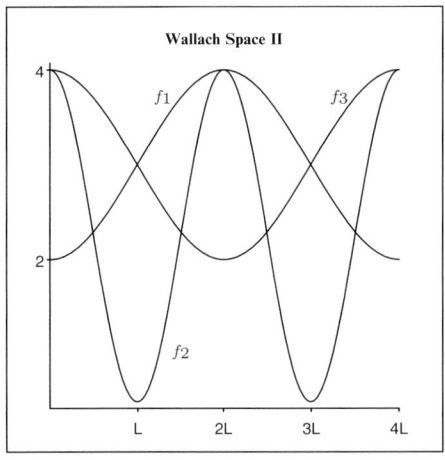 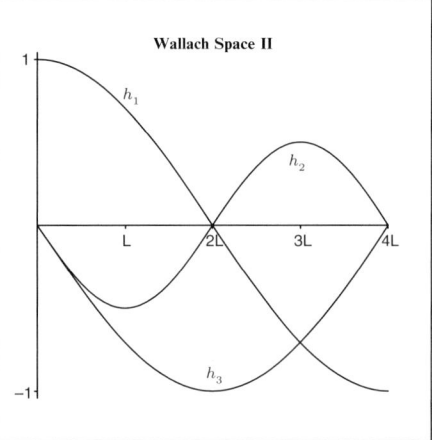

Fig. 6 Wallach space $W_{(2)}^7$ on $[0, 4L]$.

with all other inner products being 0. Since $X_1^*(0) - 2Y_1^*(0) = 0$ and $X_2^*(\pi/4) + Y_2^*(\pi/4) = 0$ it follows that $K_0^- = (e^{it}, e^{-2it}) \cdot H$ and $K_0^+ = (e^{jt}, e^{jt}) \cdot H$. Since the isomorphism type of the isotropy groups of the action is already determined, this leaves only the following possibility for its group diagram:

$$\mathbb{Z}_4 \oplus \mathbb{Z}_2 = \{(\pm 1, \pm 1), (\pm j, \pm j)\} \subset \{(e^{it}, e^{-2it}) \cdot H, (e^{jt}, e^{jt}) \cdot H\} \subset S^3 \times S^3.$$

For the Weyl group, we have $w_- = (i, -1)$, $w_+ = (e^{j\frac{\pi}{4}}, e^{j\frac{\pi}{4}})$ and $(w_- w_+)^2 = (-1, j)$ and hence $W = D_4$. The functions f_i and g_i satisfy the same Weyl symmetry as in (2.3), but for h_i we have:

$$h_3(t) = -h_1(-t + 2L) = h_1(t + 2L), \; h_2(t) = -h_2(-t) = h_2(-t + 2L).$$

It is interesting to observe this modified Weyl symmetry behavior in Figure 6 (for a typical value of $\varepsilon = \frac{1}{2}$).

The above metric with $\varepsilon = 1$ is the one induced by the biinvariant metric, which has nonnegative curvature, but not positive. For $\varepsilon = 2$, we obtain a 3-Sasakian metric (see Section 4) after dividing the metric by 2, i.e., multiplying all functions by $\frac{1}{2}$ and replacing the parameter t by $\sqrt{2}\, t$. The second SU(2) factor is then the 3-Sasakian action.

For later purposes we note that, up to conjugation, this group diagram can also be written as:

$$\mathbb{Z}_4 \oplus \mathbb{Z}_2 = \{(\pm 1, \pm 1), (\pm i, \pm i)\} \subset \{(e^{it}, e^{it}) \cdot H, (e^{jt}, e^{2jt}) \cdot H\} \subset S^3 \times S^3.$$

In Figure 7, we show the graphs for the functions in the inverse matrix on $W_{(1)}^7$ and $W_{(2)}^7$. We only include the most interesting case of the one's with index 1, although these do not determine the remaining ones as was the case for the sphere and the Berger space.

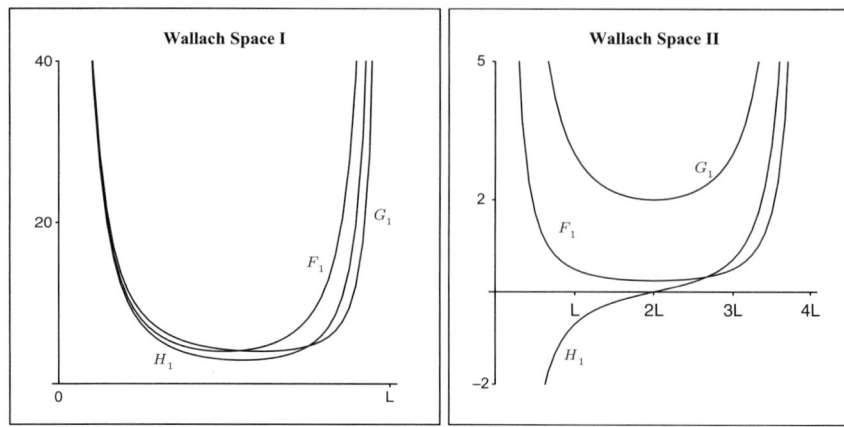

Fig. 7 Inverse functions for $W^7_{(1)}$ and $W^7_{(2)}$.

3 Classification of Cohomogeneity One Manifolds with Positive Curvature

In this section, we describe the classification result in [GWZ]. In even dimensions, positively curved cohomogeneity one manifolds were classified by L. Verdiani [V1, V2]. Here only rank one symmetric spaces occur. The actions for these spaces though are numerous and have been classified in [HL, Iw1, Iw2, Uc]. In odd dimensions, cohomogeneity one actions on spheres are even more numerous. The classification of course must also contain all of the examples described in the previous section, as well as the cohomogeneity one action by SU(4) on the Bazaikin spaces $B^{13}_p = \mathrm{diag}(z,z,z,z,z^p) \backslash \mathrm{SU}(5) / \mathrm{diag}(z^{p+4},A)^{-1}$, where $A \in \mathrm{Sp}(2) \subset \mathrm{SU}(4) \subset \mathrm{SU}(5)$. Here SU(4) acts by multiplication on the left in the first 4 coordinates. Encouragingly, a series of "candidates" P_k, Q_k and R emerges in dimension 7 for which it is not yet known whether they can carry a cohomogeneity one metric with positive curvature. They are cohomogeneity one under an action of $S^3 \times S^3$, and are defined as cohomogeneity one manifolds in terms of their isotropy groups. For P_k, the group diagram is

$$\Delta Q \subset \{(e^{it}, e^{it}) \cdot H \,,\, (e^{j(1+2k)t}, e^{j(1-2k)t}) \cdot H\} \subset S^3 \times S^3,$$

and for Q_k it is

$$\{(\pm 1, \pm 1), (\pm i, \pm i)\} \subset \{(e^{it}, e^{it}) \cdot H \,,\, (e^{jkt}, e^{j(k+1)t}) \cdot H\} \subset S^3 \times S^3,$$

whereas for R we have

$$\{(\pm 1, \pm 1), (\pm i, \pm i)\} \subset \{(e^{it}, e^{2it}) \cdot H \,,\, (e^{3jt}, e^{jt}) \cdot H\} \subset S^3 \times S^3.$$

Notice that the action of $S^3 \times S^3$ on P_k is effectively an action by $SO(4)$, and the one on Q_k and R by $SO(3) \times SO(3)$.

Theorem 3.1 (L. Verdiani, K. Grove–B. Wilking–W. Ziller). *A simply connected compact cohomogeneity one manifold with an invariant metric of positive sectional curvature is equivariantly diffeomorphic to one of the following:*

- *A compact rank one symmetric space with an isometric action,*
- *One of E_p^7, B_p^{13} or B^7,*
- *One of the 7-manifolds P_k, Q_k, or R,*

with one of the actions described above.

The first in each sequence P_k, Q_k admit an invariant metric with positive curvature since from the group diagrams in Section 2 it follows that $P_1 = \mathbb{S}^7$ and $Q_1 = W_{(2)}^7$. By the main result in [GZ], the manifolds P_k, Q_k, R all carry an invariant metric with nonnegative curvature since the cohomogeneity one actions have singular orbits of codimensions 2. Recall that the cohomogeneity one action on B^7 looks like those for P_k with slopes $(-3, 1)$ and $(1, -3)$. In some tantalizing sense then, the exceptional Berger manifold B^7 is associated with the P_k family in an analogues way as the exceptional candidate R is associated with the Q_k family.

The candidates also have interesting topological properties. In [GWZ] it was shown that the manifolds P_k are two-connected with $\pi_3(P_k) = \mathbb{Z}_k$ and that Q_k has the same cohomology ring as E_k. The fact that the manifolds P_k are 2-connected is particularly significant since by the finiteness theorem of Petrunin–Tuschmann [PT] and Fang–Rong [FR], there exist only finitely many diffeomorphism types of 2-connected positively curved manifolds, if one specifies the dimension and the pinching constant, i.e., δ with $\delta \leq sec \leq 1$. Thus, if P_k admit positive curvature metrics, the pinching constants δ_k necessarily go to 0 as $k \to \infty$, and P_k would be the first example of this type.

It is remarkable that all nonlinear actions in Theorem 3.1, apart from the Bazaikin spaces B_p^{13}, occur in dimension 7 and are cohomogeneity one under a group locally isomorphic to $S^3 \times S^3$. It is also remarkable that in positive curvature, only the above slopes are allowed, whereas for arbitrary slopes one has an invariant metric with nonnegative curvature by [GZ]. We will give a proof of Theorem 3.1 in this special case of $G = S^3 \times S^3$ since this case is clearly of particular interest.

Three important ingredients in the proof of the classification Theorem 3.1 are given by:

- The normal geodesic is closed, or equivalently, the Weyl group is finite.
- The action is linearly primitive, i.e., the Lie algebras of all singular isotropy groups along a fixed normal geodesic generate $\mathfrak{g}$ as vector spaces.
- The action is group primitive, i.e., the groups K^- and K^+ generate G as subgroups and so do K^- and nK^+n^{-1} for any $n \in N(H)_0$.

For $G = S^3 \times S^3$, the classification is based on the following lemma.

Lemma 3.2. *Let* $G = S^3 \times S^3$ *act by cohomogeneity one on the positively curved manifold* M. *If* G/K *is a singular orbit,* $G/K_0 = S^3 \times S^3 / (e^{ipt}, e^{iqt})$ *with* $p, q \neq 0, (p, q) = 1$, *and* $H \cap K_0 = \mathbb{Z}_k$, *we have:*

(a) $k \geq 2$ *and if* $k = 2$, *then* $|p + q| = 1$ *or* $|p - q| = 1$,
(b) *If* $k > 2$, *then* $(p, q) = (\pm 1, \pm 1)$ *or* $|2p + 2q| = k$ *or* $|2p - 2q| = k$.
(c) *If furthermore* $G/K_0^- = S^3 \times S^3 / (e^{ip_- t}, e^{iq_- t})$, $G/K_0^+ = S^3 \times S^3 / (e^{ip_+ t}, e^{iq_+ t})$ *and, up to conjugacy,* $H = \Delta Q$ *or* $H = \{(\pm 1, \pm 1), (\pm i, \pm i)\}$, *then* $\min\{|p_+|, |p_-|\} = \min\{|q_+|, |q_-|\} = 1$.

Proof. The main ingredient in the proof of (a) and (b) is the equivariance of the second fundamental form of G/K regarded as a K equivariant linear map $B \colon S^2 T \to T^\perp$. The nontrivial irreducible representations of $S^1 = \{e^{i\theta} \mid \theta \in \mathbb{R}\}$ consist of two-dimensional representations given by multiplication by $e^{in\theta}$ on $\mathbb{C}$, called a weight n representation. The action of K_0 on $T^\perp = \mathbb{R}^2$ will have weight k if $H \cap K_0 = \mathbb{Z}_k$, since $\mathbb{Z}_k$ is necessarily the ineffective kernel. The action of K_0 on T on the other hand has weights 0 on $W_0 = \mathrm{span}\{(-qi, pi)\}$, weight $2p$ on the two plane $W_1 = \mathrm{span}\{(j, 0), (k, 0)\}$, and weight $2q$ on $W_2 = \mathrm{span}\{(0, j), (0, k)\}$. The action on $S^2(W_1 \oplus W_2)$ has therefore weights 0 and $4p$ on $S^2 W_1$, 0 and $4q$ on $S^2 W_2$, and $2p + 2q$ and $2p - 2q$ on $W_1 \otimes W_2$.

Let us first assume that $(p, q) \neq (\pm 1, \pm 1)$. Then for any homogeneous metric on G/K_0, there exists a vector $w_1 \in W_1$ and $w_2 \in W_2$ such that the 2-plane spanned by w_1 and w_2 tangent to G/K has curvature 0 intrinsically. Indeed, since $p \neq \pm q$, $\mathrm{Ad}(K_0)$ invariance of the metric on G/K_0 implies that the 2-planes $\mathrm{span}\{(j, 0), (0, j)\}$ and $\mathrm{span}\{(k, 0), (0, k)\}$ and the line W_0 are orthogonal to each other. Hence $\mathrm{Ad}((j, j))$ induces an isometry on G/K_0, which implies that the two plane spanned by $w_1 = (j, 0) \in W_1$ and $w_2 = (0, j) \in W_2$ is the tangent space of the fixed point set of $\mathrm{Ad}((j, j))$ and thus has curvature 0 since the fixed point set in G/K_0 is a 2-torus with a left invariant metric. Since $(p, q) \neq (\pm 1, \pm 1)$ at least one of the numbers $4p$ or $4q$ is not equal to the normal weight $k > 0$. The equivariance of the second fundamental form then implies that $B_{S^2 W_i}$ vanishes for at least one i and hence by the Gauss equations $B(w_1, w_2) \neq 0$ for the above vectors w_1 and w_2. Using the equivariance of the second fundamental form once more, we see that $W_1 \otimes W_2$ contains a subrepresentation whose weight is equal to the normal weight k. Hence, $|2p + 2q| = k$ or $|2p - 2q| = k$. In particular, $k \geq 2$.

It remains to show that if $(p, q) = (\pm 1, \pm 1)$, then $k > 2$. We first show that we still have a 2-plane as above with 0-curvature. Indeed, $\mathrm{Ad}(K_0)$ invariance implies that the inner products between W_1 and W_2 are given by $\langle (X, 0), (0, Y) \rangle = \langle \phi(X), Y \rangle$ where $\phi \colon W_1 \to W_2$ is an $\mathrm{Ad}(K_0)$ equivariant map. Hence, if we choose $j' = \phi(j)$ and $k' = \phi(k)$, the two planes $\mathrm{span}\{(j, 0), (0, j')\}$ and $\mathrm{span}\{(k, 0), (0, k')\}$ are orthogonal to each other, so that by the same argument $w_1 = (j, 0) \in W_1$ and $w_2 = (0, j') \in W_2$ span a 2-plane with curvature 0. We thus obtain again that $B_{S^2 W_i} = 0$, hence $B(w_1, w_2) \neq 0$, which gives a contradiction unless $k = 4$.

To prove part (c), we use the following general fact about an isometric G action on M. The strata, i.e., components in M/G of orbits of the same type (K), are (locally) totally geodesic (cf. [Gr]). Indeed, by the slice theorem, such a component near the image of $p \in M$ with $G_p = K$ is given by the fixed point set $D^K \subset D/K \subset M/G$ where D is a slice at p. In the case of the $S^3 \times 1$ action on M, the isotropy groups are given by the intersections of $S^3 \times 1$ with $K^\pm$ and H since $S^3 \times 1$ is normal in G. Using the special form of the principal isotropy group H, it follows that on the regular part the isotropy groups are effectively trivial. On the other hand, if $\min\{|q_+|, |q_-|\} > 1$, they are nontrivial along $B_\pm$. This implies that the image of both $B_\pm$ in $M/S^3 \times 1$ are totally geodesic. Since these strata are two-dimensional and M/S^3 is four-dimensional, both strata cannot be totally geodesic according to Petrunin's analogue [Pe] of Frankel's theorem for Alexandrov spaces. This finishes our claim.   □

We now use the classification of 7-dimensional compact simply connected (group) primitive cohomogeneity one manifolds in [Ho]. Although the use of this classification is not necessary, it simplifies the argument and brings out its main points. Surprisingly there are, in addition to numerous linear actions on $\mathbb{S}^7$, only 6 primitive families in the classification, 5 of them with $G = S^3 \times S^3$, and we apply Lemma 3.2 to exclude from them the ones that do not admit an invariant metric with positive curvature.

Example 1. The simplest primitive group diagram on a 7-manifold with $G = S^3 \times S^3$ is given by

$$\{e\} \subset \{\Delta S^3, (e^{ip\theta}, e^{iq\theta})\} \subset S^3 \times S^3,$$

for any p, q. A modification of this example is

$$\mathbb{Z}_2 = \{(1, -1)\} \subset \{\Delta S^3 \cdot H, (e^{ip\theta}, e^{iq\theta})\} \subset S^3 \times S^3,$$

with p even and q odd. Notice that the second family is a twofold branched cover of the first.

In the first case, the normal weight is $k = 1$, and thus Lemma 3.2 (a) implies that there are no positively curved invariant metrics. In the second case $k = 2$, and it follows that $|p - q| = 1$ or $|p + q| = 1$, which, up to automorphisms of $S^3 \times S^3$, is the Eschenburg space E_p. Notice though that in both cases, we also need to exclude the possibility that one of p or q is 0, which is not covered by Lemma 3.2. For this special case, one uses the product lemma [GWZ, Lemma 2.6], which we will not discuss here.

Example 2. The second family has $H = \mathbb{Z}_4$:

$$\langle(i, i)\rangle \subset \{(e^{ip_-\theta}, e^{iq_-\theta}) \cdot H, (e^{jp_+\theta}, e^{jq_+\theta}) \cdot H\} \subset S^3 \times S^3,$$

where $p_-, q_- \equiv 1 \mod 4$, and p_+, q_+ are arbitrary. This family is excluded altogether. Indeed, if p_+ and q_+ are both odd, the normal weight at B_+ is $k = 2$, which implies that $|p_+ \pm q_+| = 1$, which is clearly impossible. If one is even, the other odd, $k = 1$, which is excluded by Lemma 3.2. If $p_+ = 0$ or $q_+ = 0$, the action is not group primitive.

Example 3. The third family has $H = \mathbb{Z}_2 \oplus \mathbb{Z}_4$:

$$\{(\pm 1, \pm 1), (\pm i, \pm i)\} \subset \{(e^{ip_- \cdot \theta}, e^{iq_- \cdot \theta}) \cdot H, (e^{jp_+ \cdot \theta}, e^{jq_+ \cdot \theta}) \cdot H\} \subset S^3 \times S^3,$$

where p_-, q_- is odd, p_+ even and q_+ odd. On the left, B_- has normal weight $k = 4$ and thus $(p_-, q_-) = (\pm 1, \pm 1)$ or $|p_- + q_-| = 2$ or $|p_- - q_-| = 2$. On the right, B_+ has normal weight $k = 2$ and thus $|p_+ + q_+| = 1$ or $|p_+ - q_+| = 1$. Notice also that we can assume that all integers are positive by conjugating all groups by i or j in one of the components, and observing that $p_+ = 0$ is not group primitive. Together with Lemma 3.2 (c), this leaves only the possibilities $\{(p_-, q_-), (p_+, q_+)\} = \{(1, 3), (2, 1)\}$ or $\{(p_-, q_-), (p_+, q_+)\} = \{(1, 1), (p_+, q_+)\}$ with $|q_+ - p_+| = 1$. The first case is the exceptional manifold R. In the second case, we can also assume that $q_+ > p_+$ by interchanging the two S^3 factors if necessary, and hence $(p_+, q_+) = (p, p + 1)$ with $p > 0$. This gives us the family Q_k.

Example 4. The last family has $H = \Delta Q$:

$$\Delta Q \subset \{(e^{ip_- \cdot \theta}, e^{iq_- \cdot \theta}) \cdot H, (e^{jp_+ \cdot \theta}, e^{jq_+ \cdot \theta}) \cdot H\} \subset S^3 \times S^3,$$

where $p_\pm, q_\pm \equiv 1 \mod 4$. Now the weights on both normal spaces are 4 and hence $|q_\pm + p_\pm| = 2$ or $(p_\pm, q_\pm) = (1, 1)$. Combining with Lemma 3.2 (c) yields only two possibilities. Either $\{(p_-, q_-), (p_+, q_+)\} = \{(1, -3), (-3, 1)\}$ or $\{(1, 1), (p_+, q_+)\}$ with $q_+ + p_+ = 2$. The first case is the Berger space B^7, and in the second case we can arrange that $\{(p_-, q_-), (p_+, q_+)\} = \{(1, 1), (1 + 2k, 1 - 2k)\}$ with $k \geq 0$. The case $k = 0$ is excluded since it would not be group primitive. Thus we obtain the family P_k.

Remark. There is only one further family of compact simply connected 7-dimensional primitive cohomogeneity one manifolds, given by the action of $S^1 \times S^3 \times S^3$ on the Kervaire sphere. For this action, it was shown in [BH] (see also [Se] for the 5-dimensional case) that it cannot admit an invariant metric with positive curvature unless it is a linear action on a sphere. In [GVWZ], it was shown that in most cases it does not even admit an invariant metric with nonnegative curvature. If we also allow nonprimitive 7-dimensional cohomogeneity one manifolds, one finds 9 further families in [Ho]. He also shows that the only cohomogeneity one manifold in dimension 7 or below (primitive or not), where it is not yet known if it admits an invariant metric with nonnegative curvature, are the two families in Example 1.

We also mention that in dimension 7, one finds a classification of positively curved cohomogeneity one manifold in [PV1, PV2] in the case where the group is not locally isomorphic to $S^3 \times S^3$, and that a classification in dimensions 6 and below was obtained in [Se].

4 Candidates and Hitchin Metrics

The two families of cohomogeneity one manifolds P_k and Q_k have another remarkable and unexpected property. They carry a natural metric on them that is 3-Sasakian and can be regarded as an orbifold principal bundle over $\mathbb{S}^4$ or $\mathbb{CP}^2$, equipped with a self-dual Einstein metric.

Before we discuss this metric, we give a description of our candidates using the language of self-duality. If we consider an oriented four manifold M^4 equipped with a metric, we can use the Hodge star operator $*: \Lambda^2 T^* \to \Lambda^2 T^*$ to define self-dual and anti–self-dual 2-forms, i.e., forms ω with $*\omega = \pm\omega$. They each form a 3-dimensional orientable vector bundle $\Lambda_{\pm}^2 T^*(M)$ over M with a fiber metric induced by the metric on M. Thus their oriented orthonormal frame bundles are two natural $SO(3)$ principal bundles associated with the tangent bundle of M. The star operator depends on the conformal class of the metric, but the isomorphism type of the vector bundles only depend on the given orientation. The same construction can be carried out for oriented orbifolds equipped with an orbifold metric as such a metric is locally the quotient of a smooth Riemannian metric under a finite group of isometries. The principal bundles are then orbifold bundles. In particular, $SO(3)$ will in general only act almost freely (i.e., all isotropy groups are finite) and the total space may only be an orbifold.

We use this construction now for the following orbifold structure O_k on $\mathbb{S}^4$. Consider the cohomogeneity one action of $SO(3)$ on $\mathbb{S}^4$ described in Section 2. The two singular orbits are Veronese embeddings of $\mathbb{RP}^2$ into $\mathbb{S}^4$. We define an orbifold O_k by requiring that it be smooth along the regular orbits and one singular orbit, but along the second singular orbit it is smooth in the orbit direction and has an angle $2\pi/k$ normal to it. Since the normal space is two-dimensional, this orbifold is still homeomorphic to $\mathbb{S}^4$. When k is even, we have the $SO(3)$ equivariant twofold branched cover $\mathbb{CP}^2 \to \mathbb{S}^4$ mentioned in Section 2, and O_{2k} pulls back to an orbifold structure on $\mathbb{CP}^2$ with angle $2\pi/k$ normal to the real points $\mathbb{RP}^2 \subset \mathbb{CP}^2$. If we equip the orbifold with an orientation and with an orbifold metric invariant under the $SO(3)$ action, we can define the $SO(3)$ principal bundle H_k of the vector bundle of self-dual 2-forms on O_k. The orientation we choose is adapted to the cohomogeneity one action as follows. Recall that we have a natural basis in the orbit direction corresponding to the action fields $X_1^* = E_{12}^*, X_2^* = E_{13}^*, X_3^* = E_{23}^*$ and we choose the orientation defined by $\gamma', X_1^*, X_2^*, X_3^*$ where γ is the normal geodesic chosen in Section 2. Along the singular orbits X_1^* respectively X_2^* vanishes and should be replaced by the derivative of the Jacobi field induced by their action fields. The isometric cohomogeneity one action of $SO(3)$ on O_k clearly lifts to an action on the bundle of self-dual 2-forms and thus onto the principal bundle H_k. It commutes with the principal bundle action of $SO(3)$ and together they form an $SO(3) \times SO(3)$ cohomogeneity one action on H_k. We now show:

Theorem 4.1. *The total space H_k of the $SO(3)$ principal orbifold bundle of self-dual 2-forms on O_k is smooth, and the cohomogeneity one manifolds P_k and Q_k are equivariantly diffeomorphic to the (twofold) universal covers of H_{2k-1} and H_{2k}, respectively.*

Proof. Recall that for the SO(3) cohomogeneity one action on $\mathbb{S}^4$ in Section 2, the isotropy groups are given by $H = \mathbb{Z}_2 \oplus \mathbb{Z}_2 \subset \{O(2), O'(2)\} \subset SO(3)$ where the two singular isotropy groups are embedded in two different blocks. Since the metric on O_k is smooth near B_-, it follows that we still have $K^- \cong O(2)$, which we can assume is embedded in the upper block, and hence $H \cong \mathbb{Z}_2 \oplus \mathbb{Z}_2$ embedded as the set of diagonal matrices in SO(3). The normal angle along B_+ is $2\pi/k$ and a neighborhood of B_+ can be described as follows. The homomorphism $\phi_k \colon SO(2) \to SO(2)$, $\phi_k(A) = A^k$ gives rise to a homomorphism $j_k \colon K_0^+ \simeq SO(2) \to SO(3)$, where $A \in K_0^+$ goes to $\phi_k(A)$ followed by an embedding into the lower block of SO(3). We can now define $K^+ = O(2)$ for k odd, $K^+ = O(2) \times \mathbb{Z}_2$ for k even and extend the homomorphism to $j_k \colon K^+ \to SO(3)$ such that $\text{diag}(1,-1) \in O(2)$ goes to $\text{diag}(-1,1,-1)$ and the nontrivial element in $\mathbb{Z}_2$ goes to $\text{diag}(1,-1,-1)$ when k is even. A neighborhood of the singular orbit on the right is then given by $D(B_+) = SO(3) \times_{K^+} D_+^2$ where K^+ acts on SO(3) via j_k, K_0^+ acts on D_+^2 via ϕ_2, $\text{diag}(1,-1) \in O(2)$ acts as a reflection, and $\mathbb{Z}_2$ acts trivially. Indeed, we then have $SO(3) \times_{K^+} D_+^2 = SO(3) \times_{(K^+/\ker j_k)} (D_+^2/\ker j_k)$ with singular orbit $SO(3)/O(2)$ and normal disk $D_+^2/\ker j_k = D_+^2/\mathbb{Z}_k$. Furthermore, $\partial D(B_+) = SO(3) \times_{K^+} \mathbb{S}_+^1 = SO(3)/H$. Notice that the element $-\text{Id} \in K^+$ acts trivially on $SO(3) \times D_+^2$ when k is even. Thus the extra element we added in this case guarantees that for the effective version of the action, we still have $\mathbb{S}_+^1 = K^+/H$.

The vector bundle of self-dual 2-forms can be viewed as follows: Let P be the SO(4) principal bundle of oriented orthonormal frames in the orbifold tangent bundle of O_k. This frame bundle is a smooth manifold since the finite isometric orbifold groups act freely on frames. SO(4) has two normal subgroups $SU(2)_-$ and $SU(2)_+$, given by left and right multiplication of unit quaternions, with $SO(4)/SU(2)_\pm \simeq SO(3)$. The SO(3) principal bundles $P/SU(2)_+$ and $P/SU(2)_-$ are then the principal bundles for the vector bundle of self-dual and the vector bundle of anti–self-dual 2-forms. This is due to the fact that the splitting $\Lambda^2 V \cong \Lambda_-^2 V \oplus \Lambda_+^2 V$ for an oriented four-dimensional vector space corresponds to the splitting of Lie algebra ideals $\mathfrak{so}(4) \cong \mathfrak{so}(3) \oplus \mathfrak{so}(3)$ under the isomorphism $\Lambda^2 V \cong \mathfrak{so}(4)$. Alternatively, we can first project under the twofold cover $SO(4) \to SO(3)\,SO(3)$ and then divide by one of the SO(3) factors. The action of SO(4) on P is only almost free since $P/SO(4) = O_k$, but we will show that both $SU(2)_+$ and $SU(2)_-$ act freely on P, or equivalently, each SO(3) factor in $SO(3)\,SO(3)$ acts freely on $P/\{-\text{Id}\}$. This then implies that $P/SU(2)_+ = H_k$ is indeed a smooth manifold.

The description of the disc bundle $D(B_+)$ gives rise to a description of the corresponding SO(4) frame bundle $SO(3) \times_{K^+} SO(4)$ where the action of K^+ on SO(3) is given by j_k as above, and the action of K_0^+ on SO(4) is given via $SO(2) \subset SO(4) \colon A \in SO(2) \to (\phi_k(A), \phi_2(A))$ acting on the splitting $T_+ \oplus T_+^\perp$ into tangent space and normal space of the singular orbit. Similarly for the left-hand side where $k = 1$. On the left-hand side the X_1^* direction collapses, T_- is oriented by X_2^*, X_3^* and $T_-^\perp$ by $\gamma'(0), X_1^*$. On the right-hand side the X_2^* direction collapses, T_+ is oriented by X_3^*, X_1^* and $T_+^\perp$ by $\gamma'(L), X_2^*$. Furthermore, $SO(2) \subset O(2)$ has negative weights on $T_\pm$, where we have endowed the isotropy groups on the left and on the right with orientations induced by X_1 and X_2, respectively. Indeed, $[E_{12}, E_{13}] = -E_{23}$

on the left and $[E_{13}, E_{23}] = -E_{12}$ on the right. On $T_{-}^{\perp}$, the weight is positive, and on $T_{+}^{\perp}$ negative. Hence $K_0^{\pm} \subset SO(3) SO(4)$ sits inside the natural maximal torus in $SO(3) SO(4)$ with slopes $(1, -1, 2)$ on the left, and $(k, -k, -2)$ on the right. To make this precise metrically, we can choose as a metric on H_k, the natural connection metric induced by the Levi–Cevita connection on O_k. A parallel frame is then a geodesic in this metric. By equivariance under H, the unit vectors $X_i^*/|X_i^*|$ form such a parallel frame.

Under the homomorphism $SO(4) \to SO(3) SO(3)$ and the natural maximal tori in $SO(4)$ and in $SO(3) SO(3)$, a slope (p, q) circle goes into one with slope $(p + q, -p + q)$. Hence the slopes of $K_0^{\pm}$ in $SO(3) SO(3) SO(3)$ are $(1, 1, 3)$ on the left, and $(k, -(k+2), k-2)$ on the right. This also implies that the second and third $SO(3)$ factor each act freely on $P/\{-\text{Id}\}$. Here we have used the fact that we already know that $SO(4)$, and thus each $SO(3)$, acts freely on the regular part and hence freeness only needs to be checked in K_0^+. Notice also that for k even, all slopes in K_0^+ should be divided by 2 to make the circle description effective. If we divide by the third $SO(3)$ to obtain H_k, the slopes are $(1, 1)$ on the left and $(k, -(k+2))$ on the right. Finally, notice that the principle isotropy group of the $SO(3) SO(3)$ action on H_k is again $\mathbb{Z}_2 \oplus \mathbb{Z}_2$ as this is true for the $SO(3)$ action on O_k and $SO(4)$ acts freely on the regular points in P. This determines the group diagram. For $k = 2m - 1$, it is the group diagram of the twofold subcover of P_m obtained by dividing $G = S^3 \times S^3$ by its center. Here we have used the fact that conjugation of all groups by (i, i) does not change K^- or H, but changes the signs of the slopes in K^+. For $k = 2m$, this is the group picture of the twofold subcover of Q_m obtained by adding a component to all 3 isotropy groups, generated, e.g., by (j, j). In this case, we can change the slope from $(2m, -2m - 2)$ to $(m, m + 1)$ by conjugating all groups with $(1, i)$. This finishes our proof. $\square$

Remarks. (a) The proof also shows that the $SO(3)$ principal bundles $P/SU(2)_-$ corresponding to the vector bundle of anti–self-dual two forms is smooth and has slopes $(1, 3)$ on the left and $(k, k-2)$ on the right. Note that in the case of $k = 3$, one obtains the slopes for the exceptional manifold B^7 and in the case of $k = 4$ the ones for R (up to twofold covers).

(b) In the case of $k = 2\ell$, we can regard O_k as an orbifold metric O_ℓ on $\mathbb{CP}^2$. In this case, it follows that the $SO(3)$ principal bundle of the bundle of self-dual 2-forms is Q_ℓ itself.

We now explain the relationship to the Hitchin metrics. Recall that a metric on M is called 3-Sasakian if $G = SU(2)$ or $G = SO(3)$ acts isometrically and almost freely with totally geodesic orbits of curvature 1. Moreover, for U tangent to the $SU(2)$ orbits and X perpendicular, $X \wedge U$ is required to be an eigenvector of the curvature operator $\hat{R}$ with eigenvalue 1, in particular the sectional curvatures $\sec(X, U)$ are equal to 1. In the case we are interested in, where the dimension of M is 7, the quotient $B = M^7/G$ is 4-dimensional and its induced metric is self-dual Einstein with positive scalar curvature, although it is in general only an orbifold metric. Recall that a metric is called self-dual if the curvature operator satisfies $\hat{R} \circ * = * \circ \hat{R}$. Conversely, given a self-dual Einstein orbifold metric on B^4 with positive scalar curvature, the SO(3) principal orbifold bundle of self-dual 2-forms on B^4 has a 3-Sasakian orbifold metric given by the naturally defined Levi–Cevita connection metric. See [BG] for a survey on this subject.

Recall that $\mathbb{S}^4$ and $\mathbb{CP}^2$, according to Hitchin, are the only smooth self-dual Einstein 4-manifolds. The 3-Sasakian metrics they give rise to are the metric on $\mathbb{S}^7(1)$ in the first case, and in the second case the metric on the Wallach space $W_{(2)}^7$ described in Section 2. However, in the more general context of orbifolds, Hitchin constructed in [Hi1] a sequence of self-dual Einstein orbifolds O_k homeomorphic to $\mathbb{S}^4$, one for each integer $k > 0$. The metric is invariant under the cohomogeneity one action by SO(3) from Section 2 and has an orbifold singularity as in the orbifold O_k discussed earlier. The cases of $k = 1, 2$ correspond to the smooth standard metrics on $\mathbb{S}^4$ and on $\mathbb{CP}^2$, respectively. Hence the Hitchin metrics give rise to 3-Sasakian orbifold metrics on the seven-dimensional orbifold H_k^7. Here one needs to check that the orientation we chose above agrees with the orientation in [Hi1]. As we saw in Theorem 4.1, H_k is actually smooth and the 3-Sasakian metric, as a quotient of the smooth connection metric on the principal frame bundle, is also smooth. Thus our candidates P_k and Q_k all admit a smooth 3-Sasakian metric. In the context of 3-Sasakian geometry, the examples P_k are particularly interesting since they are two connected, and so far, the only known 2-connected example in dimension 7 was $\mathbb{S}^7$.

It was shown by O. Dearricott in [De] (see also [CDR]) that a 3-Sasakian metric, scaled down in direction of the principal $G = SO(3)$ or $SU(2)$ orbits, has positive sectional curvature if and only if the self-dual Einstein orbifold base has positive curvature. It is therefore interesting to examine the curvature properties of the Hitchin metrics, which we will now discuss shortly. The metric is described by the 3 functions $T_i(t) = |X_i^*(\gamma(t))|^2$ along a normal geodesic γ, since invariance under the isotropy group implies that these vectors are orthogonal. It turns out that in order to solve the ODE along γ given by the condition that the metric is self-dual Einstein, it is convenient to change the arc length parameter from t to r. The metric is thus described by

$$g_{\gamma(r)} = f(r)dr^2 + T_1(r)d\theta_1^2 + T_2(r)d\theta_2^2 + T_3(r)d\theta_3^2,$$

where $d\theta_i$ is dual to X_i. In order to solve the ODE, Hitchin uses complex algebraic geometry on the twistor space of O_k. For general k, the solutions are explicit only in principal and it is thus a tour de force to prove the required smoothness properties of the metric. For small values of k though, one finds explicit solutions in [Hi1] and [Hi2]:

Example 1. The first nonsmooth example is the Hitchin metric with normal angle $2\pi/3$. Here the functions are algebraic:

$$T_1 = \frac{80r^2(r^6 - 2r^5 - 5r^4 - 15r^3 - 20r^2 + 13r + 4)}{(3r^3 + 7r^2 + r + 1)^2(3r^3 - 13r^2 + r + 1)},$$

$$T_2 = \frac{5r(3r - 1)(r - \beta)(r + \beta + 2)(2 - 3r - r^2 + r^3 + 5\beta r)^2}{(3r^3 + 7r^2 + r + 1)^2(r^2 + r - 1)(r^2 + r + 4)},$$

$$T_3 = \frac{5r(3r-1)(r+\beta)(r-\beta+2)(2-3r-r^2+r^3-5\beta r)^2}{(3r^3+7r^2+r+1)^2(r^2+r-1)(r^2+r+4)},$$

$$f = \frac{5(3r-1)(r^2+r+4)(r+1)^2}{(r+r^2-1)/(3r^3+7r^2+r+1)^2},$$

with $\beta = \sqrt{\frac{r+r^2-1}{r}}$ and $\frac{\sqrt{5}-1}{2} \le r \le 1$.

Example 2. The simplest example of a nonsmooth Hitchin metric has normal angle $2\pi/4$, where the functions are given by:

$$T_1 = \frac{(1-r^2)^2}{(1+r+r^2)(r+2)(2r+1)} \; , \; T_2 = \frac{1+r+r^2}{(r+2)(2r+1)^2} \; , \; T_3 = \frac{r(1+r+r^2)}{(r+2)^2(2r+1)},$$

$$f = \frac{1+r+r^2}{r(r+2)^2(2r+1)^2},$$

with $1 \le r < \infty$.

Example 3. Finally, we have the Hitchin metric with normal angle $2\pi/6$:

$$T_1 = \frac{(3r^2+2r+1)(r^2+2r-1)^2(r^2-2r+3)(r^2+1)}{(3r^2-2r+1)(r^2-2r-1)^2(r^2+2r+3)^2},$$

$$T_2 = \frac{(3r^2-2r+1)(r^2-2r+3)(r+1)^3(r-1)}{(3r^2+2r+1)(r^2+2r+3)^2(r^2-2r-1)},$$

$$T_3 = \frac{-4(3r^2-2r+1)(3r^2+2r+1)r}{(r^2+2r+3)^2(r^2-2r-1)(r^2-2r+3)},$$

$$f = \frac{(r+1)(r^2-2r+3)(3r^2+2r+1)(3r^2-2r+1)}{r(1-r)(r^2-2r-1)^2(r^2+1)(r^2+2r+3)^2},$$

with $\sqrt{2}-1 \le r \le 1$.

In the above formulas, we have corrected and simplified the functions given in [Hi1] in the case of $k=6$. In the case of $k=3$, one uses the solution of the Painlevé equation in [Hi2, p. 28] (with the corrected value $w^2 = 1+u-u^{-1}$) and applies the general recipe described in [Hi1, p. 194–195] to obtain the form of the metric.

Although one can in principle use the methods in [Hi1] to determine the functions for larger values of k, they quickly become even more complicated. The above 3 cases are sufficient though to understand the behavior in general. If the functions are given in arc length parameter, one has $\sec(\gamma, X_i^*) = -f_i''/f_i$, where $f_i(t) = |X_i^*(\gamma(t))|$, and hence positive curvature is equivalent to the concavity of f_i. In Figures 8–10, we therefore have drawn a graph of the length functions f_i in arc length parameter, together with a graph of $\sec(\gamma, X_i^*)$. The pictures are similar, but notice the difference in scale. One sees that the nonsmooth singular orbit must occur

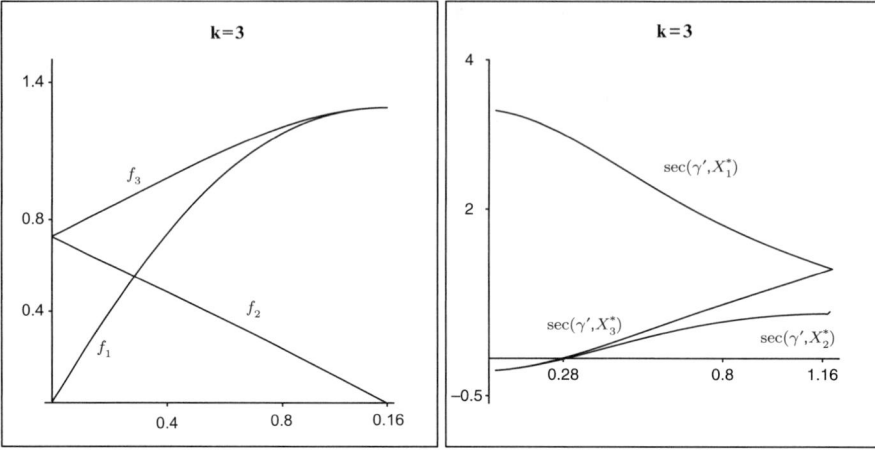

Fig. 8 Hitchin metric with normal angle $2\pi/3$.

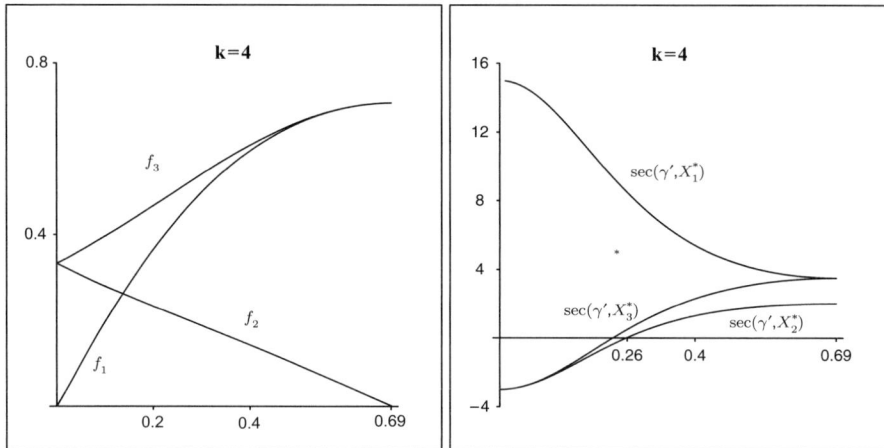

Fig. 9 Hitchin metric with normal angle $2\pi/4$.

at $t = L$ since it is necessarily totally geodesic, which implies that the noncollapsing functions have 0 derivative. The pictures show that the function f_1, which vanishes at the smooth singular orbit, is concave, whereas the other two are slightly convex near the smooth singular orbit.

For each of the 3 elements $g \in H$, the fixed point set of g is a 2-sphere, since this is clearly true for the linear action on $\mathbb{S}^4$ corresponding to $k = 1$. They are isometric to each other via an element of the Weyl group. Since the circle that commutes with g acts by isometries on the 2-sphere, it is rotationally symmetric with an orbifold point at one of the poles. It has positive curvature, except in a small region two thirds toward this pole. Figure 11 shows the length of the action field induced by the circle action on this 2-sphere (which is equal to $f_i/2$) in the case of $k = 3$ and $k = 6$. Notice

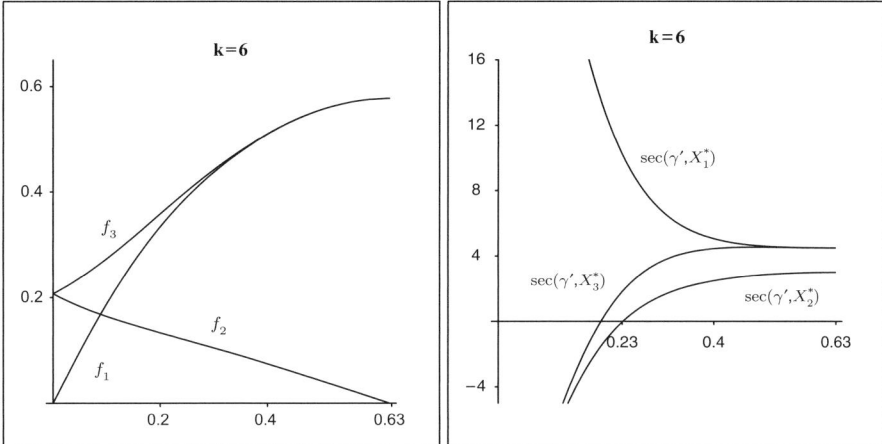

Fig. 10 Hitchin metric with normal angle $2\pi/6$.

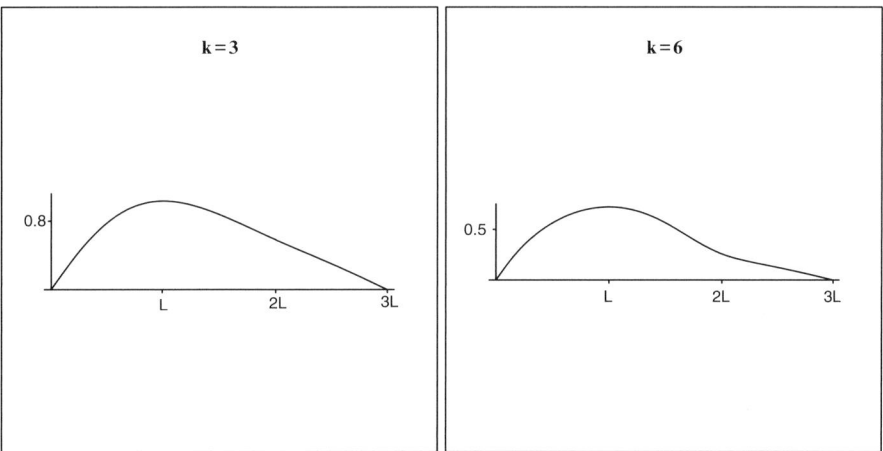

Fig. 11 Hitchin metrics on $[0, 3L]$.

that it extends from 0 to $3L$. These 2-spheres can also be isometrically embedded as surfaces of revolution in 3-space, which we exhibit in Figure 12.

One can show that, as a consequence of being self-dual Einstein, $\sec(X_i^*, X_j^*) = \sec(\gamma', X_k^*)$, when i, j, k are distinct, and that if all 3 are positive, the curvature of any 2-plane is indeed positive also. Thus the Hitchin metric has positive curvature wherever the above orbifold 2-sphere has positive curvature. By Dearricott's theorem, this implies that the induced 3-Sasakian metric on our candidates, scaled down in direction of the principal $SO(3)$ orbits, has positive sectional curvature on half of the manifold.

Thus this metric does not yet give the desired metrics of positive curvature on P_k and Q_k. It is also tempting to think that, as in the case of the known actions in

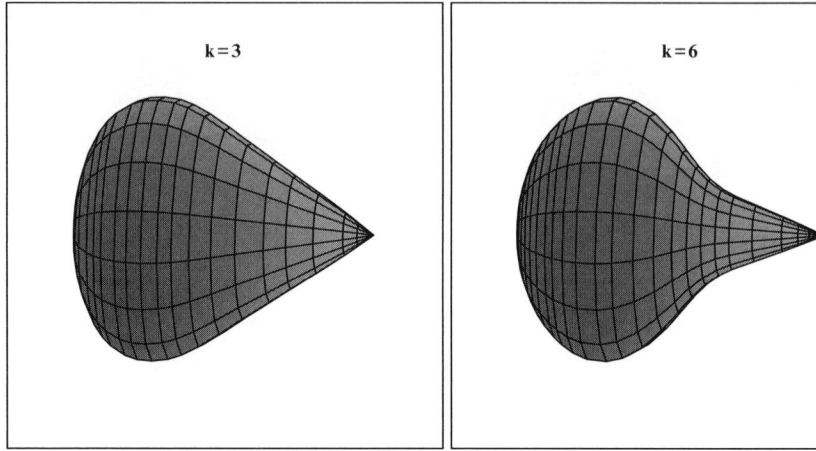

Fig. 12 Fixed point orbifold 2-spheres in Hitchin metrics.

Section 2, simple trigonometric expressions for the 9 functions describing a metric on P_k, Q_k, or R might yield a metric with positive curvature on our candidates. But this does not seem to be the case either, as already the smoothness conditions and simple necessary convexity properties require trigonometric functions that are quite complicated.

It is intriguing that the (noncompact) space of 2-monopoles studied by Atiyah and Hitchin in [AH] has surprisingly similar properties to the above metric. It carries a self-dual Einstein orbifold metric, which in this case is Ricci flat, i.e., is hyperkähler. It is invariant under $SO(3)$ with principal orbits $SO(3)/\mathbb{Z}_2 \oplus \mathbb{Z}_2$ and a singular orbit $\mathbb{RP}^2$ with normal angle $2\pi/k$. Of the 3 functions describing the metric, one is concave as well, and the other two are not. Thus one of the fixed point sets of elements in H is a (noncompact) surface of revolution with positive curvature.

Acknowledgments The author was supported by a grant from the National Science Foundation, by IMPA in Rio de Janeiro, and by the Max Planck Institute in Bonn and would like to thank the institutes for their hospitality.

References

[AA]   A.V. Alekseevsy and D.V. Alekseevsy, *G-manifolds with one dimensional orbit space,* Adv. Sov. Math. **8** (1992), 1–31.

[AW]   S. Aloff and N. Wallach, *An infinite family of 7–manifolds admitting positively curved Riemannian structures,* Bull. Am. Math. Soc. **81**(1975), 93–97.

[AH]   M. Atiyah and N. Hitchin, *T*he geometry and dynamics of magnetic monopoles, Princeton University Press, 1988.

[BH]   A. Back and W.Y. Hsiang, *Equivariant geometry and Kervaire spheres,* Trans. Am. Math. Soc. **304** (1987), no. 1, 207–227.

[Ba]   Y. Bazaikin, *On a family of* 13-*dimensional closed Riemannian manifolds of positive curvature*, Siberian Math. J., **37** (1996), 1068–1085.

[BB]   L. Bérard Bergery, *Les variétés riemanniennes homogènes simplement connexes de dimension impaire à courbure strictement positive*, J. Math. pure et appl. **55** (1976), 47–68.

[Be]   M. Berger, *Les variétés riemanniennes homogènes normales simplement connexes à courbure strictement positive*, Ann. Scuola Norm. Sup. Pisa **15** (1961), 179–246.

[BG]   C. P. Boyer and K. Galicki 3-*Sasakian manifolds*, Surveys in differential geometry: essays on Einstein manifolds, Surv. Differ. Geom., **VI**, (1999), 123–184.

[Br]   G. E. Bredon, *Introduction to compact transformation groups*, Academic Press, New York, 1972, Pure and Applied Mathematics, Vol. 46.

[CDR]   L. Chaves, A.Derdzinski and A. Rigas *A condition for positivity of curvature*, Bol. Soc. Brasil. Mat. **23** (1992), 153–165.

[DZ]   D'Atri and W. Ziller, *Naturally reductive metrics and Einstein metrics on compact Lie groups*, Memoir of Am. Math. Soc. **215** (1979).

[De]   O. Dearricott, *Positive sectional curvature on 3-Sasakian manifolds*, Ann. Global Anal. Geom. **25** (2004), 59–72.

[E1]   J. H. Eschenburg, *New examples of manifolds with strictly positive curvature*, Invent. Math. **66** (1982), 469–480.

[E2]   J. H. Eschenburg, *Freie isometrische Aktionen auf kompakten Lie-Gruppen mit positiv gekrümmten Orbiträumen,* Schriftenr. Math. Inst. Univ. Münster **32** (1984).

[E3]   J.-H. Eschenburg, *Inhomogeneous spaces of positive curvature*, Diff. Geom. Appl. **2** (1992), 123–132.

[FR]   F. Fang and X. Rong, *Positive pinching, volume and second Betti number*, Geom. Funct. Anal. **9** (1999), 641–674.

[Gr]   K. Grove, *Geometry of, and via, Symmetries*, Conformal, Riemannian and Lagrangian geometry (Knoxville, TN, 2000), Am. Math. Soc. Univ. Lecture Series **27** (2002).

[GSZ]   K. Grove, K. Shankar and W. Ziller, *Symmetries of Eschenburg spaces and the Chern Problem*, Special Issue in honor of S. S. Chern, Asian J. Math. **10** (2006), 647–662.

[GZ]   K. Grove and W. Ziller, *Curvature and symmetry of Milnor spheres*, Ann. Math. **152** (2000), 331–367.

[GVWZ] K. Grove, L. Verdiani, B. Wilking and W. Ziller, *Non-negative curvature obstruction in cohomogeneity one and the Kervaire spheres*, Ann. del. Scuola Norm. Sup. 5 (2006), 159–170.

[GWZ]   K. Grove, B. Wilking and W. Ziller, *Positively curved cohomogeneity one manifolds and 3-Sasakian geometry*, J. Diff. Geom., **78** (2008), 33–111.

[Hi1]   N. Hitchin, *A new family of Einstein metrics*, Manifolds and geometry (Pisa, 1993), 190–222, Sympos. Math., XXXVI, Cambridge Univ. Press, Cambridge, 1996.

[Hi2]   N. Hitchin, *Poncelet Polygons and the Painlevé equations*, Geometry and analysis (Bombay, 1992), Tata Inst. Fundam. Res. **13** (1995), 151–185.

[Ho]   C. Hoelscher, *Cohomogeneity one manifolds in low dimensions*, Ph.D. thesis, University of Pennsylvania, 2007.

[HL]   W.Y. Hsiang and B. Lawson, *Minimal submanifolds of low cohomogeneity*, J. Diff. Geom. **5** (1971), 1–38.

[Iw1]   K. Iwata, *Classification of compact transformation groups on cohomology quaternion projective spaces with codimension one orbits*, Osaka J. Math. **15** (1978), 475–508.

[Iw2]   K. Iwata, *Compact transformation groups on rational cohomology Cayley projective planes*, Tohoku Math. J. **33** (1981), 429–442.

[Mo]   P. Mostert, *On a compact Lie group acting on a manifold*, Ann. Math. (2) **65** (1957), 447–455; Errata, Ann. Math. (2) **66** (1957), 589.

[Pe]   A. Petrunin, *Parallel transportation for Alexandrov spaces with curvature bounded below,* Geom. Funct. Anal. **8** (1998), 123–148

[PT]   A. Petrunin and W. Tuschmann, *Diffeomorphism finiteness, positive pinching, and second homotopy*, Geom. Funct. Anal. **9** (1999), 736–774.

[PV1]   F. Podesta and L. Verdiani, *Totally geodesic orbits of isometries*, Ann. Glob. Anal. Geom. **16** (1998), 399–412. erratum ibid. **19** (2001), 207–208.

[PV2]   F. Podesta and L. Verdiani, *Positively curved 7-dimensional manifolds*, Quat. J. Math. Oxford **50** (1999), 497–504.

[Se]   C. Searle, *Cohomogeneity and positive curvature in low dimensions,* Math. Z. **214** (1993), 491–498: Err. ibid. **226** (1997), 165–167.

[Uc]   F. Uchida, *Classification of compact transformation groups on cohomology complex projective spaces with codimension one orbits*, Japan J. Math. **3** (1977), 141–189.

[V1]   L. Verdiani, *Cohomogeneity one Riemannian manifolds of even dimension with strictly positive sectional curvature, I,* Math. Z. **241** (2002), 329–339.

[V2]   L. Verdiani, *Cohomogeneity one manifolds of even dimension with strictly positive sectional curvature,* J. Diff. Geom. **68** (2004), 31–72.

[Wa]   N. Wallach, *Compact homogeneous Riemannian manifolds with strictly positive curvature*, Ann. Math., **96** (1972), 277–295.

[Wi]   B. Wilking, *Nonnegatively and Positively Curved Manifolds*, in: Metric and Comparison Geometry, Surv. Differ. Geom. 11, ed. K. Grove and J. Cheeger, International Press, 2007.

[Zi1]   W. Ziller, *Homogeneous spaces, biquotients, and manifolds with positive curvature*, Lecture Notes 1998, unpublished.

[Zi2]   W. Ziller, *Examples of Riemannian Manifolds with non-negative sectional curvature*, in: Metric and Comparison Geometry, Surv. Differ. Geom. 11, ed. K. Grove and J. Cheeger, International Press, 2007.

The Sasaki Cone and Extremal Sasakian Metrics

Charles P. Boyer, Krzysztof Galicki, and Santiago R. Simanca

Abstract We study the Sasaki cone of a CR structure of Sasaki type on a given closed manifold. We introduce an energy functional over the cone and use its critical points to single out the strongly extremal Reeb vector fields. Should one such vector field be a member of the extremal set, the scalar curvature of a Sasaki extremal metric representing it would have the smallest L^2-norm among all Sasakian metrics of fixed volume that can represent vector fields in the cone. We use links of isolated hypersurface singularities to produce examples of manifolds of Sasaki type, many of these in dimension five, whose Sasaki cone coincides with the extremal set, and examples where the extremal set is empty. We end up by proving that a conjecture of Orlik concerning the torsion of the homology groups of these links holds in the five-dimensional case.

1 Introduction

The study of a Sasakian structure goes along with the one-dimensional foliation on the manifold associated to the Reeb vector field. In the case where the orbits are all closed, the manifold has the structure of an orbifold circle bundle over a compact Kähler orbifold, which must be algebraic, and which has at most cyclic quotient singularities. For a long time, techniques from Riemannian submersions, suitably extended, were used to find interesting canonical Sasakian metrics in this orbibundle setting. This led to the common belief that the only interesting such metrics occurred precisely in this setting. Compact Sasakian manifolds with non-closed leaves—irregular Sasakian structures—were known to exist, but there was no evidence to suspect they could be had with Einstein metrics as well. It was thought

C.P. Boyer, K. Galicki, and S.R. Simanca
Department of Mathematics and Statistics, University of New Mexico, Albuquerque,
New Mexico, USA

K. Galicki and S.R. Simanca (eds.), *Riemannian Topology and Geometric Structures on Manifolds,* Progress in Mathematics 271, DOI 10.1007/978-0-8176-4743-8,

reasonable that all Sasaki–Einstein metrics could be understood well by simply studying the existence of Kähler–Einstein metrics on compact cyclic orbifolds.

The existence of Sasaki–Einstein metrics is of great interest in the physics of the famous CFT/AdS Duality Conjecture, and the discovery of irregular Sasaki–Einstein structures [18] was a rather remarkable feat, in particular if we take into account their rather explicit depiction in coordinates. An attempt to study them using a treatment inspired in the use of classical Lagrangians, replaced in this context by a Riemannian functional whose critical points define the *canonical metrics* we seek, was started in [13] using the squared L^2-norm of the scalar curvature as the functional in question, with its domain restricted to the space of metrics adapted to the geometry of the underlying Sasakian structure under consideration. This choice of Lagrangian is well-known in Kähler geometry, which was and remains a guiding principle to us, and is quite natural since Sasakian geometry sits naturally in between two Kähler geometries, that of the transversals to the one-dimensional foliations associated to the Reeb vector field, and that of the metric cones inside which the Sasaki manifolds sits as a base. Not surprisingly, the Sasakian metrics that are critical points of the Lagrangian we use are those that are transversally extremal in the sense of Calabi, and several important results known in Kähler geometry have now a counterpart in the Sasakian context [13, 17] also.

In the theory we developed [13], a given CR structure $(\mathcal{D}, J)$ of Sasaki type is the analogue in the Sasaki context of a complex structure on a manifold of Kähler type, and the Reeb vector field ξ in a Sasakian structure with $(\mathcal{D}, J)$ as underlying CR structure is the analogue of a choice of a class in the Kähler cone. The Sasaki cone $\kappa(\mathcal{D}, J)$ of the CR structure arises naturally, and $\xi \in \kappa(\mathcal{D}, J)$ is said to be canonical or extremal if there exists a Sasakian structure (ξ, η, Φ, g), with underlying CR structure $(\mathcal{D}, J)$, for which the metric g in the structure is Sasaki extremal. The set of vector fields in the cone for which this happens is denoted by $\mathfrak{e}(\mathcal{D}, J)$. In this chapter, we present a number of examples where $\kappa(\mathcal{D}, J) = \mathfrak{e}(\mathcal{D}, J)$, as well as a number of examples where $\mathfrak{e}(\mathcal{D}, J)$ is the empty set.

In the cases where $\kappa(\mathcal{D}, J) = \mathfrak{e}(\mathcal{D}, J)$, it is natural to ask which is the optimal choice we could make of a Reeb vector field, be as it may that each one admits a Sasaki extremal representative. In effect, this question turns out to be natural in general, even in the cases where $\mathfrak{e}(\mathcal{D}, J)$ is the empty set. We briefly discuss this here also, introducing the notion of strongly extremal Reeb vector field. A detailed elaboration of this idea will appear elsewhere.

The specific examples we give are links of isolated hypersurface singularities. For these manifolds, there is a long-standing conjecture of Orlik whose validity would describe the torsion of their homology groups. We present a proof of this conjecture in dimension five, as obtained recently by the second author.

This article developed from partial results obtained by the three authors while working on an ongoing project. It is presented here by the first and last author in memory of their friend and collaborator Kris Galicki.

2 Preliminaries on Sasakian Geometry

A contact metric structure (ξ, η, Φ, g) on a manifold M is said to be a *Sasakian structure* if (ξ, η, Φ) is normal. A smooth manifold provided with one such structure is said to be a *Sasakian manifold*, or a *manifold of Sasaki type*. We briefly spell out the meaning of this definition in order to recall the relevant geometric set-up it brings about. A detailed discussion of Sasakian and contact geometry can be found in [7].

The triple (ξ, η, Φ) defines an almost contact structure on M. That is to say, ξ is a nowhere zero vector field, η is a one form, and Φ is a tensor field of type $(1,1)$, such that

$$\eta(\xi) = 1, \quad \Phi^2 = -\mathbb{1} + \xi \otimes \eta.$$

The vector field ξ defines the *characteristic foliation* $\mathcal{F}_\xi$ with one-dimensional leaves, and the kernel of η defines the codimension one sub-bundle $\mathcal{D}$. We have the canonical splitting

$$TM = \mathcal{D} \oplus L_\xi, \tag{1}$$

where L_ξ is the trivial line bundle generated by ξ.

The sub-bundle $\mathcal{D}$ inherits an almost complex structure J by restriction of Φ, and the dimension of M must be an odd integer, which we set be $2n+1$ here. When forgetting the tensor Φ and characteristic foliation, the sub-bundle $\mathcal{D}$ by itself defines what is called a *contact structure* on M.

The Riemannian metric g is compatible with the almost contact structure (ξ, η, Φ), and so we have that

$$g(\Phi(X), \Phi(Y)) = g(X, Y) - \eta(X)\eta(Y).$$

Thus, g induces an almost Hermitian metric on $\mathcal{D}$. This fact makes of (ξ, η, Φ, g) an almost contact metric structure, and in that case, the canonical decomposition (1) is orthogonal. The orbits of the field ξ are geodesics, a condition that can be re-expressed as $\xi \lrcorner d\eta = 0$, and so $d\eta$ is a basic 2-form. We have the relation

$$g(\Phi X, Y) = d\eta(X, Y), \tag{2}$$

which says that (ξ, η, Φ, g) is a contact metric structure. Notice that the volume element defined by g is given by

$$d\mu_g = \frac{1}{n!} \eta \wedge (d\eta)^n. \tag{3}$$

On the cone $C(M) = M \times \mathbb{R}^+$ we have the metric

$$g_C = dr^2 + r^2 g,$$

and the radial vector field $r\partial_r$ satisfies the relation

$$\pounds_{r\partial_r} g_C = 2g_C.$$

We have also an almost complex structure I on $C(M)$ given by

$$I(Y) = \Phi(Y) - \eta(Y)r\partial_r, \quad I(r\partial_r) = \xi.$$

The contact metric structure (ξ, η, Φ) is said to be *normal* if the pair $(C(M), I)$ is a complex manifold. In that case, the induced almost complex structure J on $\mathcal{D}$ is integrable. Further, g_C is a Kähler metric on $(C(M), I)$.

2.1 Transverse Kähler Structure

For a Sasakian structure (ξ, η, Φ, g), the integrability of the almost complex structure I on the cone $C(M)$ implies that the Reeb vector field ξ leaves both, η and Φ, invariant. We obtain a codimension one integrable strictly pseudo-convex CR structure $(\mathcal{D}, J)$, where $\mathcal{D} = \ker \eta$ is the contact bundle and $J = \Phi|_{\mathcal{D}}$, and the restriction of g to $\mathcal{D}$ defines a symmetric form on $(\mathcal{D}, J)$ that we shall refer to as the transverse Kähler metric g^T. By (2), the Kähler form of the transverse Kähler metric is given by the form $d\eta$. Therefore, the Sasakian metric g is determined fully in terms of (ξ, η, Φ) by the expression

$$g = d\eta \circ (\mathbb{1} \otimes \Phi) + \eta \otimes \eta. \tag{4}$$

The Killing field ξ leaves invariant η and Φ, and the decomposition (1) is orthogonal. Despite its dependence on the other elements of the structure, we insist on explicitly referring to g as part of the Sasakian structure (ξ, η, Φ, g).

We consider the set

$$\mathcal{S}(\xi) = \{\text{Sasakian structure } (\tilde{\xi}, \tilde{\eta}, \tilde{\Phi}, \tilde{g}) \mid \tilde{\xi} = \xi\}, \tag{5}$$

and provide it with the C^∞ compact-open topology as sections of vector bundles. For any element $(\tilde{\xi}, \tilde{\eta}, \tilde{\Phi}, \tilde{g})$ in this set, the 1-form $\zeta = \tilde{\eta} - \eta$ is basic, and so $[d\tilde{\eta}]_B = [d\eta]_B$. Here, $[\cdot]_B$ stands for a cohomology class in the basic cohomology ring, a ring that is defined by the restriction d_B of the exterior derivative d to the subcomplex of basic forms in the de Rham complex of M. Thus, all of the Sasakian structures in $\mathcal{S}(\xi)$ correspond to the same basic cohomology class. We call $\mathcal{S}(\xi)$ the *space of Sasakian structures compatible with* ξ, and say that the Reeb vector field ξ *polarizes* the Sasakian manifold M [13].

Given a Reeb vector field ξ, we have its characteristic foliation $\mathcal{F}_\xi$, so we let $\nu(\mathcal{F}_\xi)$ be the vector bundle whose fiber at a point $p \in M$ is the quotient space T_pM/L_ξ, and let $\pi_\nu : TM \rightarrow \nu(\mathcal{F}_\xi)$ be the natural projection. The background structure $\mathcal{S} = (\xi, \eta, \Phi, g)$ induces a complex structure $\bar{J}$ on $\nu(\mathcal{F}_\xi)$. This is defined by $\bar{J}\bar{X} := \overline{\Phi(X)}$, where X is any vector field in M such that $\pi(X) = \bar{X}$. Furthermore, the underlying CR structure $(\mathcal{D}, J)$ of $\mathcal{S}$ is isomorphic to $(\nu(\mathcal{F}_\xi), \bar{J})$ as a complex vector bundle. For this reason, we refer to $(\nu(\mathcal{F}_\xi), \bar{J})$ as the complex normal bundle of the Reeb vector field ξ, although its identification with $(\mathcal{D},J)$ is not canonical. In this sense, M is polarized by $(\xi,\bar{J})$.

We define $\mathcal{S}(\xi,\bar{J})$ to be the subset of all structures $(\tilde{\xi},\tilde{\eta},\tilde{\Phi},\tilde{g})$ in $\mathcal{S}(\xi)$ such that the diagram

$$
\begin{array}{ccc}
TM & \xrightarrow{\tilde{\Phi}} & TM \\
\downarrow{\pi_v} & & \downarrow{\pi_v} \\
\nu(\mathcal{F}_\xi) & \xrightarrow{\bar{J}} & \nu(\mathcal{F}_\xi),
\end{array}
\tag{6}
$$

commutes. This set consists of elements of $\mathcal{S}(\xi)$ with the same transverse holomorphic structure $\bar{J}$, or with more precision, the same complex normal bundle $(\nu(\mathcal{F}_\xi),\bar{J})$.

2.2 The Sasaki Cone

If we look at the Sasakian structure (ξ,η,Φ,g) from the point of view of CR geometry, its underlying strictly pseudo-convex CR structure $(\mathcal{D},J)$, with associated contact bundle $\mathcal{D}$, has Levi form $d\eta$.

If $(\mathcal{D},J)$ is a strictly pseudo-convex CR structure on M of codimension one, we say that $(\mathcal{D},J)$ is of *Sasaki type* if there exists a Sasakian structure $\mathcal{S} = (\xi,\eta,\Phi,g)$ such that $\mathcal{D} = \ker\eta$ and $\Phi|_{\mathcal{D}} = J$. We consider the set

$$
\mathcal{S}(\mathcal{D},J) = \left\{
\begin{array}{c}
\mathcal{S} = (\xi,\eta,\Phi,g) : \mathcal{S} \text{ a Sasakian structure} \\
(\ker\eta,\Phi|_{\ker\eta}) = (\mathcal{D},J)
\end{array}
\right\}
\tag{7}
$$

of Sasakian structures *with underlying* CR *structure* $(\mathcal{D},J)$.

We denote by $\mathfrak{con}(\mathcal{D})$ the Lie algebra of infinitesimal contact transformations, and by $\mathfrak{cr}(\mathcal{D},J)$ the Lie algebra of the group $\mathfrak{CR}(\mathcal{D},J)$ of CR automorphisms of $(\mathcal{D},J)$. If $\mathcal{S} = (\xi,\eta,\Phi,g)$ is a contact metric structure whose underlying CR structure is $(\mathcal{D},J)$, then $\mathcal{S}$ is a Sasakian structure if, and only if, $\xi \in \mathfrak{cr}(\mathcal{D},J)$.

The vector field X is said to be positive for $(\mathcal{D},J)$ if $\eta(X) > 0$ for any $(\xi,\eta,\Phi,g) \in \mathcal{S}(\mathcal{D},J)$. We denote by $\mathfrak{cr}^+(\mathcal{D},J)$ the subset of all of these positive fields, and consider the projection

$$
\begin{array}{ccc}
\mathcal{S}(\mathcal{D},J) & \xrightarrow{\iota} & \mathfrak{cr}^+(\mathcal{D},J) \\
(\xi,\eta,\Phi,g) & \mapsto & \xi
\end{array}.
\tag{8}
$$

This mapping identifies naturally $\mathfrak{cr}^+(\mathcal{D},J)$ with $\mathcal{S}(\mathcal{D},J)$. Furthermore, $\mathfrak{cr}^+(\mathcal{D},J)$ is an open convex cone in $\mathfrak{cr}(\mathcal{D},J)$ that is invariant under the adjoint action of the Lie group $\mathfrak{CR}(\mathcal{D},J)$. The identification (8) gives the moduli space $\mathcal{S}(\mathcal{D},J)/\mathfrak{CR}(\mathcal{D},J)$ of all Sasakian structures whose underlying CR structure is $(\mathcal{D},J)$. The conical structure of $\mathfrak{cr}^+(\mathcal{D},J)$ justifies the following definition:

Definition 2.1. *Let $(\mathcal{D}, J)$ be a CR structure of Sasaki type on M. The Sasaki cone $\kappa(\mathcal{D}, J)$ is the moduli space of Sasakian structures compatible with $(\mathcal{D}, J)$,*

$$\kappa(\mathcal{D}, J) = \mathcal{S}(\mathcal{D}, J) / \mathfrak{CR}(\mathcal{D}, J).$$

The isotropy subgroup of an element $\mathcal{S} \in \mathcal{S}(\mathcal{D}, J)$ is, by definition, $\mathfrak{Aut}(\mathcal{S})$. It contains a maximal torus T_k. In fact, we have the following result [13]:

Theorem 2.2. *Let M be a closed manifold of dimension $2n+1$, and let $(\mathcal{D}, J)$ be a CR structure of Sasaki type on it. Then the Lie algebra $\mathfrak{cr}(\mathcal{D}, J)$ decomposes as $\mathfrak{cr}(\mathcal{D}, J) = \mathfrak{t}_k + \mathfrak{p}$, where $\mathfrak{t}_k$ is the Lie algebra of a maximal torus T_k of dimension k, $1 \leq k \leq n+1$, and $\mathfrak{p}$ is a completely reducible T_k-module. Furthermore, every $X \in \mathfrak{cr}^+(\mathcal{D}, J)$ is conjugate to a positive element in the Lie algebra $\mathfrak{t}_k$.*

Theorem 2.2 and (8) imply that each orbit can be represented by choosing a positive element in the Lie algebra $\mathfrak{t}_k$ of a maximal torus T_k. So let us fix a maximal torus T_k of a maximal compact subgroup G of $\mathfrak{CR}(\mathcal{D}, J)$, and let $\mathcal{W}$ denote the Weyl group of G. If $\mathfrak{t}_k^+ = \mathfrak{t}_k \cap \mathfrak{cr}^+(\mathcal{D}, J)$ denotes the subset of positive elements in $\mathfrak{t}_k$, we have the identification $\kappa(\mathcal{D}, J) = \mathfrak{t}_k^+ / \mathcal{W}$. Each Reeb vector field in $\mathfrak{t}_k^+ / \mathcal{W}$ corresponds to a unique Sasakian structure, so we can view $\mathfrak{t}_k^+ / \mathcal{W}$ as a subset of $\mathcal{S}$, and we have

Lemma 2.3. *Let $(\mathcal{D}, J)$ be a CR structure of Sasaki type on M, and fix a maximal torus $T_k \in \mathfrak{CR}(\mathcal{D}, J)$. Then we have that*

$$\bigcap_{\mathcal{S} \in \mathfrak{t}_k^+ / \mathcal{W}} \mathfrak{Aut}_0(\mathcal{S}) = T_k,$$

where $\mathfrak{Aut}_0(\mathcal{S})$ denotes the identity component of the isotropy group $\mathfrak{Aut}(\mathcal{S})$. In particular, T_k is contained in the isotropy subgroup of every $\mathcal{S} \in \mathfrak{t}_k^+ / \mathcal{W}$.

Now the basic Chern class of a Sasakian structure $\mathcal{S} = (\xi, \eta, \Phi, g)$ is represented by the Ricci form $\rho^T / 2\pi$ of the transverse metric g_T. Although the notion of basic changes with the Reeb vector field, the complex vector bundle $\mathcal{D}$ remains fixed. Hence, for any Sasakian structure $\mathcal{S} \in \mathcal{S}(\mathcal{D}, J)$, the transverse 2-form $\rho^T / 2\pi$ associated to $\mathcal{S}$ represents the first Chern class $c_1(\mathcal{D})$ of the complex vector bundle $\mathcal{D}$.

When $k = 1$, the maximal torus of $\mathfrak{CR}(\mathcal{D}, J)$ is one-dimensional. Since the Reeb vector field is central, this implies that $\dim \mathfrak{aut}(\mathcal{S}) = \dim \mathfrak{cr}(\mathcal{D}, J) = 1$. Hence, we have that $\mathcal{S}(\mathcal{D}, J) = \mathfrak{cr}^+(\mathcal{D}, J) = \mathfrak{t}_1^+ = \mathbb{R}^+$, and $\mathcal{S}(\mathcal{D}, J)$ consists of the 1-parameter family of Sasakian structures given by $\mathcal{S}_a = (\xi_a, \eta_a, \Phi_a, g_a)$, where

$$\xi_a = a^{-1}\xi, \quad \eta_a = a\eta, \quad \Phi_a = \Phi, \quad g_a = ag + (a^2 - a)\eta \otimes \eta, \tag{9}$$

and $a \in \mathbb{R}^+$, the 1-parameter family of transverse homotheties.

In effect, the *transverse homotheties* (9) are the only deformations $(\xi_t, \eta_t, \Phi_t, g_t)$ of a given structure $\mathcal{S} = (\xi, \eta, \Phi, g)$ in the Sasakian cone $\kappa(\mathcal{D}, J)$ where the Reeb vector field varies in the form $\xi_t = f_t \xi$, f_t a scalar function. For we then have that the family of tensors Φ_t is constant, and since $\pounds_{\xi_t}\Phi_t = 0$, we see that f_t must be annihilated by any section of the sub-bundle $\mathcal{D}$. But then (2) implies that $df_t = (\xi f_t)\eta$, and we conclude that the function f_t is constant. Thus, in describing fully the tangent space of $\mathcal{S}(\mathcal{D},J)$ at $\mathcal{S}$, it suffices to describe only those deformations $(\dot{\xi}_t, \eta_t, \Phi_t, g_t)$ where $\dot{\xi} = \partial_t \xi_t|_{t=0}$ is g-orthogonal to ξ. These correspond to deformations where the volume of M in the metric g_t remains constant in t and are parametrized by elements of $\kappa(\mathcal{D},J)$ that are g-orthogonal to ξ.

The terminology we use here is chosen to emphasize the analogy that the Sasaki cone is to a CR structure of Sasaki type what the Kähler cone is to a complex manifold of Kähler type. Indeed, for any point $\mathcal{S} = (\xi,\eta,\Phi,g)$ in $\kappa(\mathcal{D},J)$, the complex normal bundle $(v(\mathfrak{F}_\xi), \bar{J})$ is isomorphic to $(\mathcal{D},J)$. In this sense, the complex structure $\bar{J}$ is fixed with the fixing of $(\mathcal{D},J)$, the Reeb vector field ξ polarizes the manifold, and the Sasaki cone $\kappa(\mathcal{D},J)$ represents the set of all possible polarizations. We observe though that when we fix $\xi \in \kappa(\mathcal{D},J)$, the underlying CR structure of elements in $\mathcal{S}(\xi,\bar{J})$ might change, even though their normal bundles are all isomorphic to $(\mathcal{D},J)$.

Let $\xi \in \kappa(\mathcal{D},J)$. We may ask if there are canonical representatives of $\mathcal{S}(\xi,\bar{J})$. We can answer this question using a variational principle, as done in [13]. Let $\mathfrak{M}(\xi,\bar{J})$ be the set of all compatible Sasakian metrics arising from structures in $\mathcal{S}(\xi,\bar{J})$, and consider the functional

$$\mathfrak{M}(\xi,\bar{J}) \to \mathbb{R},$$
$$g \mapsto \int_M s_g^2 d\mu_g. \tag{10}$$

Its critical points define the canonical representatives that we seek. Furthermore, by the identification of $\kappa(\mathcal{D},J)$ with $\mathfrak{t}_k^+$, we may then single out the set of Reeb vector fields ξ in $\mathfrak{t}_k^+$ for which the functional (10) admits critical points at all.

Definition 2.4. *We say that $(\xi,\eta,\Phi,g) \in \mathcal{S}(\xi,\bar{J})$ is an extremal Sasakian structure if g is a critical point of* (10). *A vector field $\xi \in \kappa(\mathcal{D},J)$ is said to be extremal if there exists an extremal Sasakian structure in $\mathcal{S}(\xi,\bar{J})$. We denote by $\mathfrak{e}(\mathcal{D},J)$ the set of all extremal elements of the Sasaki cone and refer to it as the extremal Sasaki set of the* CR *structure $(\mathcal{D},J)$.*

Notice that $(\xi,\eta,\Phi,g) \in \mathcal{S}(\xi,\bar{J})$ is extremal if, and only if, the transversal metric g^T is extremal in the Kähler sense [13]. We also have that $\mathfrak{e}(\mathcal{D},J)$ is an open subset of $\kappa(\mathcal{D},J)$ [13], a Sasakian version of the the openness theorem in Kähler geometry [27, 38].

3 The Energy Functional in the Sasaki Cone

By Lemma 2.3, given any Sasakian structure (ξ,η,Φ,g) in $\mathcal{S}(\mathcal{D},J)$, the space of g^T-Killing potentials associated to a maximal torus has dimension $k-1$. We denote by $\mathcal{H}_\xi = \mathcal{H}_{\xi,g}(\mathcal{D},J)$ the vector subspace of $C^\infty(M)$ that they and the constant functions span. This is the space of functions whose transverse gradient is holomorphic.

Given a Sasakian metric g such that $(\xi, \eta, \Phi, g) \in \mathcal{S}(\mathcal{D}, J)$, we define the map

$$C^\infty(M) \to \mathcal{H}_\xi$$
$$f \mapsto \pi_g f \tag{11}$$

to be the L^2-orthogonal projection with $\mathcal{H}_\xi$ as its range. Then the metric g in $\mathfrak{M}(\xi, \bar{J})$ is extremal if, and only if, the scalar curvature $s_g = s_g^T - 2n$ is equal to its projection onto $\mathcal{H}_\xi$ [39]. Or said differently, if, and only if, s_g is an affine function of the space of Killing potentials. The functional (10) then admits the lower bound

$$\mathfrak{M}(\xi, \bar{J}) \ni g \mapsto \int s_g^2 d\mu_g \geq \int (\pi_g s_g)^2 d\mu_g,$$

where the right side only depends upon the polarization $(\xi, \mathcal{D}, J)$, and it is only reached if the metric g is extremal. This reproduces in this context a situation analyzed in the Kähler case already [40, 41].

We define the energy of a Reeb vector field in the Sasakian cone to be the functional given by this optimal lower bound:

$$\kappa(\mathcal{D}, J) \xrightarrow{E} \mathbb{R}$$
$$\xi \mapsto \int (\pi_g s_g)^2 d\mu_g, \tag{12}$$

Definition 3.1. *A Reeb vector field $\xi \in \kappa(\mathcal{D}, J)$ is said to be strongly extremal if it is a critical point of the functional (12) over the space of Sasakian structures in $\kappa(\mathcal{D}, J)$ that fix the volume of M. The Sasakian structure (ξ, η, Φ, g) is said to be a strongly extremal representative of $\kappa(\mathcal{D}, J)$ if ξ is strongly extremal, and if $g \in \mathcal{S}(\xi, \bar{J})$ is an extremal representative of the polarized manifold (M, ξ, J).*

4 The Euler–Lagrange Equation for Strongly Extremal Reeb Vector Fields

We now discuss some of the variational results that play a role in the derivation of the Euler–Lagrange equation of a strongly extremal Reeb vector field. Complete details of this derivation will appear elsewhere.

Let us consider a path $\xi_t \in \kappa(\mathcal{D}, J)$, and a corresponding path of Sasakian structures $(\xi_t, \eta_t, \Phi_t, g_t)$. The path of contact forms must be of the type

$$\eta_t = \frac{1}{\eta(\xi_t)} \eta + \frac{1}{2} d^c \varphi_t, \tag{13}$$

where φ_t is a ξ_t-basic function. For convenience, we drop the subscript when we consider the value of quantities at $t = 0$. Then, we have that $\eta(\xi) = 1$, φ is constant, and $\dot{\varphi}$ a ξ-basic function.

Proposition 4.1. *Let* $(\xi_t, \eta_t, \Phi_t, g_t)$ *be a path of Sasakian structures that starts at* (ξ, η, Φ, g), *such that* $\xi_t \in \kappa(\mathcal{D}, J)$ *and* η_t *is of the form* (13). *Then we have that*

$$d\mu_t = 1 - t\left((n+1)\eta(\dot{\xi}) + \frac{1}{2}\Delta_B\dot{\varphi}\right)d\mu + O(t^2),$$

and

$$\frac{d}{dt}\mu_{g_t}(M) = -(n+1)\int(\eta(\dot{\xi}))\,d\mu.$$

Volume preserving Sasakian deformations are given by variations ξ_t *of the Reeb vector field whose infinitesimal deformation is globally orthogonal to it, that is to say, such that* $\eta(\dot{\xi})$ *is orthogonal to the constants. Finally,*

$$D^2\mu(\dot{\xi}, \dot{\beta}) = (n+1)(n+2)\int(\eta(\dot{\xi}))(\eta(\dot{\beta}))\,d\mu.$$

Proof. By differentiation of (3), we obtain the variational expression for the volume form. The expression for the infinitesimal variation of the volume follows by Stokes' theorem and the one for the Hessian by an iteration of this argument. $\square$

Proposition 4.2. *Let* $(\xi_t, \eta_t, \Phi_t, g_t)$ *be a path of Sasakian structures that starts at* (ξ, η, Φ, g), *such that* $\xi_t \in \kappa(\mathcal{D}, J)$ *and* η_t *is of the form* (13). *Then we have the expansion*

$$s_t = s^T - 2n - t\left(n\Delta_B(\eta(\dot{\xi})) - s^T\eta(\dot{\xi}) + \frac{1}{2}\Delta_B^2\dot{\varphi} + 2(\rho^T, i\partial\overline{\partial}\dot{\varphi})\right) + O(t^2),$$

for the scalar curvature of g_t.

Proof. We have the relation $s_t = s_t^T - 2n$, where s_t^T is the scalar curvature of the transverse Kähler metric $d\eta_t$. Since the latter is

$$s_t^T = -2(d\eta_t, i\partial\overline{\partial}\log\det(d\eta_t))_{d\eta_t},$$

taking the t-derivative, and evaluating at $t = 0$, we obtain the desired result. Just notice that the $d\eta$-trace of $d\dot{\eta}$ is equal to the $d\eta$-trace of $-\eta(\dot{\xi})d\eta + i\partial\overline{\partial}\dot{\varphi}$, as the trace of $d(\eta(\dot{\xi})) \wedge \eta$ is zero. $\square$

We may now sketch the computation of the variation of the functional (12) along the curve ξ_t. If the subscript t denotes quantities associated to the Sasakian structure $(\xi_t, \eta_t, \Phi_t, g_t)$, we have that

$$\frac{d}{dt}E(\xi_t) = 2\int(\pi_t s_t)\left(\frac{d}{dt}(\pi_t s_t)\right)d\mu_t + \int(\pi_t s_t)^2\frac{d}{dt}d\mu_t,$$

and by the results of the previous propositions, the value at $t = 0$ is given by

$$\frac{d}{dt}E(\xi_t)\,|_{t=0} = 2\int(\pi s)\left[(-n\Delta_B + s^T)\eta(\dot{\xi}) + \dot{\pi}s\right]d\mu - (n+1)\int(\pi s)^2\eta(\dot{\xi})d\mu.$$

$$(14)$$

We may compute the variation of π, but need less than that for our purposes. Indeed, if G_g is the Green's operator of the metric g, we have that

$$\int (\pi s)\dot{\pi}s\,d\mu = \int (s^T - \pi s^T)2iG_g\overline{\partial}_g^*(\partial^\# \pi s\lrcorner d\dot{\eta})\,,$$

where given a basic function f, $\partial^\# f$ is the operator obtained by raising the indices of the $(0,1)$ form $\overline{\partial}f$ using the transversal Kähler metric $d\eta$, and since

$$i\partial_g^\# \pi s\lrcorner d\dot{\eta} = -\overline{\partial}(\pi s\,\eta(\dot{\xi}))\,,$$

we obtain that

$$\int (\pi s)\dot{\pi}s\,d\mu = \int (s^T - \pi s^T)2iG_g\overline{\partial}_g^*(\partial^\# \pi s\lrcorner d\dot{\eta}) = - \int (s^T - \pi s^T)\pi s\,\eta(\dot{\xi})d\mu\,,\tag{15}$$

where the last equality follows because $s^T - \pi s^T$ is orthogonal to the constants.

Combining these results, we obtain the following:

Theorem 4.3. *Let $(\xi_t, \eta_t, \Phi_t, g_t)$ be a path of volume preserving Sasakian structures that starts at (ξ, η, Φ, g), such that $\dot{\xi}_t \in \kappa(\mathcal{D}, J)$, and η_t is of the form (13) with φ_t a constant. Then we have that*

$$\frac{d}{dt}E(\xi_t)\,|_{t=0} = 2\int (\pi s)\left[(-n\Delta_B + \pi s^T)\eta(\dot{\xi})\right]d\mu - (n+1)\int (\pi s)^2\eta(\dot{\xi})d\mu\,.\tag{16}$$

The Reeb vector field $\xi \in \kappa(\mathcal{D}, J)$ is a critical point of (12) over the space of deformations with g_t of fixed volume if, and only if,

$$\pi_g\left[-2n\Delta_B\pi_g s_g - (n-1)(\pi_g s_g)^2\right] + 4n\pi_g s_g = \lambda\,,\tag{17}$$

for λ a constant. □

4.1 Some Examples of Critical Reeb Vector Fields

We may define a character that obstructs the existence of extremal Sasakian structures $(\xi, \eta, \Phi, g) \in \mathcal{S}(\xi, \bar{J})$ where the scalar curvature s_g of g is constant [13]. This is analogous to the Futaki invariant [14, 16] in the Kähler category.

Indeed, for any transversally holomorphic vector field $\partial_g^\# f$ with Killing potential f, the Sasaki–Futaki invariant is given by

$$\mathfrak{F}_\xi(\partial_g^\# f) = - \int f(s_g - s_0)d\mu_g\,,\tag{18}$$

where g is any Sasakian metric in $\mathfrak{M}(\xi, \bar{J})$, s_g is its scalar curvature, s_0 is the projection of s_g onto the constants. This expression defines uniquely a character on the entire algebra of transversally holomorphic vector fields, and if (ξ, η, Φ, g) is an extremal Sasakian structure in $\mathcal{S}(\xi, \bar{J})$, the scalar curvature s_g is constant if, and only if, $\mathfrak{F}_\xi \equiv 0$.

Proposition 4.4. *Suppose that* $(\xi, \eta, \Phi, g) \in \mathcal{S}(\xi, \bar{J})$ *is such that* $\pi_g s_g$ *is a constant. Then, for any other Sasakian structure* $(\xi, \tilde{\eta}, \tilde{\Phi}, \tilde{g})$ *in* $\mathcal{S}(\xi, \bar{J})$, *we have that* $\pi_{\tilde{g}} s_{\tilde{g}}$ *is constant also, and equal to* $\pi_g s_g$.

Proof. The proof follows by the same reasoning used to prove an analogous statement in the Kähler context (cf. [42], Proposition 5). □

Theorem 4.5. *Let* $\xi \in \kappa(\mathcal{D}, J)$ *be such that the Sasaki–Futaki invariant is identically zero,* $\mathfrak{F}_\xi(\cdot) = 0$. *Then* ξ *is a critical point of* (12).

Proof. If $\mathfrak{F}_\xi(\cdot) = 0$, then $\pi_g s_g$ is constant, and ξ satisfies (17). □

Example 4.6. Under the assumption that the cone $C(M) = M \times \mathbb{R}^+$ is a Calabi–Yau manifold, the Hilbert action over Sasakian metrics has been extensively analyzed [29]. When considered as a function of Reeb vector fields of charge $n + 1$, this functional is a multiple of the volume,[1] and its variation over such a domain is a constant multiple of the Sasaki–Futaki invariant. Thus, a critical Reeb vector field ξ has vanishing Sasaki–Futaki invariant, and if (ξ, η, Φ, g) is a Sasakian structure corresponding to it, the function $\pi_g s_g$ is necessarily constant. By the convexity of the volume functional (see Proposition 4.1), the critical Reeb vector field of this functional is unique. We thus see that the critical Reeb field singled out by the Hilbert action, on this type of special manifolds, must be a critical point of our functional (12) also, as was to be expected. □

5 Sasakian Geometry of Links of Isolated Hypersurface Singularities

Many of the results we discuss from here on involve the natural Sasakian geometry occurring on links of isolated hypersurface singularities. In the spirit of self-containment, we provide a brief review of the geometry of links in arbitrary dimension, which we shall use mostly in dimension 5 later on.

Let us recall (cf. Chapters 4 and 9 of [7]) that a polynomial $f \in \mathbb{C}[z_0, \ldots, z_n]$ is said to be a *weighted homogeneous polynomial* of *degree* d and *weight* $\mathbf{w} = (w_0, \ldots, w_n)$ if for any $\lambda \in \mathbb{C}^\times$

$$f(\lambda^{w_0} z_0, \ldots, \lambda^{w_n} z_n) = \lambda^d f(z_0, \ldots, z_n).$$

---

[1] In earnest, this is done in [29] assuming that M is quasi-regular, but the result holds in further generality. Notice also that in the statement for the *charge* of the Reeb vector field, we have made adjustments between the convention we use for the dimension of M and the one in the said reference.

We are interested in those weighted homogeneous polynomials f whose zero locus in $\mathbb{C}^{n+1}$ has only an isolated singularity at the origin. We define the *link* $L_f(\mathbf{w}, d)$ as $f^{-1}(0) \cap \mathbb{S}^{2n+1}$, where $\mathbb{S}^{2n+1}$ is the $(2n+1)$-sphere in Euclidean space. By the Milnor fibration theorem [30], $L_f(\mathbf{w}, d)$ is a closed $(n-2)$-connected manifold that bounds a parallelizable manifold with the homotopy type of a bouquet of n-spheres. Furthermore, $L_f(\mathbf{w}, d)$ admits a Sasaki–Seifert structure $\mathcal{S} = (\xi_{\mathbf{w}}, \eta_{\mathbf{w}}, \Phi_{\mathbf{w}}, g_{\mathbf{w}})$ in a natural way [5, 44]. This structure is quasi-regular, and its bundle $(\mathcal{D}, J)$ has $c_1(\mathcal{D}) = 0$. This latter property implies that

$$c_1(\mathcal{F}_{\xi_{\mathbf{w}}}) = a[d\eta_{\mathbf{w}}]_B \tag{19}$$

for some constant a, where $\mathcal{F}_{\xi_{\mathbf{w}}}$ is the characteristic foliation. The sign of a determines the negative, null, and positive cases that we shall refer to below.

The Sasaki–Futaki invariant (18) is the obstruction for the scalar curvature of the metric g in an extremal Sasakian structure $\mathcal{S} = (\xi, \eta, \Phi, g)$ to be constant [13]. One particular case where this obstruction vanishes is when there are no transversally holomorphic vector fields other than those generated by ξ itself. In that case, the Sasaki cone of the corresponding CR structure turns out to be of dimension one. If the Sasakian structure is quasi-regular, the transversally holomorphic vector fields are the lifts of holomorphic vector fields on the projective algebraic variety $\mathcal{Z} = M/\mathbb{S}^1_\xi$, where $\mathbb{S}^1_\xi$ is the circle generated by the Reeb vector field.

Let us recall that a Sasakian metric g is said to be η-Einstein if

$$\text{Ric}_g = \lambda g + \nu \eta \otimes \eta$$

for some constants λ and ν that satisfy the relation $\lambda + \nu = 2n$. They yield particular examples of Sasakian metrics of constant scalar curvature, a function that in these cases is given in terms of λ by the expression $s_g = 2n(1 + \lambda)$.

The only transversally holomorphic vector fields of a negative or a null link of an isolated hypersurface singularity are those generated by the Reeb vector field. These links have a one-dimensional Sasaki cone, and by the transverse Aubin–Yau theorem [11], any point in this cone can be represented by a Sasakian structure whose metric is η-Einstein (see examples of such links in [7,8]). We thus obtain the following

Theorem 5.1. *Let $L_f(\mathbf{w}, d)$ be either a negative or null link of an isolated hypersurface singularity with underlying CR structure $(\mathcal{D}, J)$. Then its Sasaki cone $\kappa(\mathcal{D}, J)$ is one-dimensional, and it coincides with the extremal Sasaki set $\mathfrak{e}(\mathcal{D}, J)$.*

The case of positive links is more complicated. This can be deduced already from the fact that if we were to have a Sasakian structure $\mathcal{S} = (\xi, \eta, \Phi, g)$ such that $c_1(\mathcal{F}_\xi) = a[d\eta]_B$ with $a > 0$ and where g is η-Einstein, then there would exist a transverse homothety (9) that would yield a Sasaki–Einstein metric of positive scalar curvature, and generally speaking, these types of metrics are difficult to be had.

In order to understand this case better, we begin by stating the following result that applies to links in general, and which follows by Corollary 5.3 of [13]:

Proposition 5.2. *Let $L_f(\mathbf{w}, d)$ be the link of an isolated hypersurface singularity with its natural Sasakian structure $\mathcal{S} = (\xi_{\mathbf{w}}, \eta_{\mathbf{w}}, \Phi_{\mathbf{w}}, g_{\mathbf{w}})$. If $g \in \mathcal{S}(\xi_{\mathbf{w}}, \bar{J})$ is an extremal Sasakian metric with constant scalar curvature, then g is η-Einstein.*

By Lemma 5.5.3 of [7], if the weight vector $\mathbf{w} = (w_0, \ldots, w_n)$ of a link $L_f(\mathbf{w}, d)$ is such that $2w_i < d$ for all but at most one of the indices, then the Lie algebra $\mathfrak{aut}(\mathcal{S})$ is generated by the Reeb vector field $\xi_{\mathbf{w}}$. We have the following

Theorem 5.3. *Let $L_f(\mathbf{w}, d)$ be the link of an isolated hypersurface singularity with $2w_i < d$ for all but at most one of the indices, and let $(\mathcal{D}, J)$ be its underlying CR structure. Then the dimension of the Sasaki cone $\kappa(\mathcal{D}, J)$ is one.*

Even if we assume that for a given positive link $L_f(\mathbf{w}, d)$ the hypothesis of Theorem 5.3 applies, and so its Sasaki–Futaki invariant vanishes identically, there could be other obstructions preventing the existence of extremal Sasakian metrics of constant scalar curvature on it. Indeed, as was observed in [19], classical estimates of Bishop and Lichnerowicz may obstruct the existence of Sasaki–Einstein metrics, an observation that has produced a very effective tool to rule out the existence of these metrics in various cases. Employing the Lichnerowicz estimates [19] in combination with the discussion above and some results in [13], we obtain the following:

Proposition 5.4. *Let $L_f(\mathbf{w}, d)$ be a link of an isolated hypersurface singularity with its natural Sasakian structure, and let $I := |\mathbf{w}| - d = (\sum_j w_j) - d$ be its index. If*

$$I > n \min_i \{w_i\},$$

then $L_f(\mathbf{w}; d)$ cannot admit any Sasaki–Einstein metric. In that case, $\mathcal{S}(\xi_{\mathbf{w}}, \bar{J})$ does not admit extremal representatives.

Combining Theorem 5.3 and Proposition 5.4, we may obtain examples of positive links whose Sasaki cones are one-dimensional and whose extremal sets are empty.

Theorem 5.5. *Let $L_f(\mathbf{w}, d)$ be a positive link of an isolated hypersurface singularity whose index I satisfies the relation $I > n \min_i \{w_i\}$, and such that $2w_i < d$ for all but at most one of the indices. If $(\mathcal{D}, J)$ is its underlying CR structure, we have that the dimension of the Sasaki cone $\kappa(\mathcal{D}, J)$ is one, and that the extremal set $\mathfrak{e}(\mathcal{D}, J)$ is empty.*

Besides the conditions given above on the weight vector of a link, there are other interesting algebraic conditions that suffice to prove the existence of extremal Sasakian metrics on it. These arise when we view the link as a Seifert fibration over a base X with orbifold singularities. If the Kähler class of the projective algebraic variety X is a primitive integral class in the second orbifold cohomology group, and the constant a in (19) is positive, they serve to measure the singularity of the pair $(X, K^{-1} + \Delta)$, where K^{-1} is an anticanonical divisor, and Δ is a branch divisor, and are known as *Kawamata log terminal* (Klt) conditions (cf. Chapter 5 of [7]).

A weighted homogeneous polynomial of the form

$$f = z_0^{a_0} + \cdots + z_n^{a_n}, \quad a_i \geq 2, \tag{20}$$

is called a *Brieskorn–Pham* (BP) polynomial. In this case the exponents a_i, the weights w_i, and degree are related by $d = a_i w_i$ for each $i = 0, \ldots, n$. We change slightly the notation in this case, and denote by $L_f(\mathbf{a})$, $\mathbf{a} = (a_0, \ldots, a_n)$, the link that a BP polynomial f defines. These are special but quite important examples of links. Their Klt conditions were described in [9]. The base of a BP link $L_f(\mathbf{a})$ admits a positive Kähler–Einstein orbifold metric if

$$1 < \sum_{i=0}^{n} \frac{1}{a_i} < 1 + \frac{n}{n-1} \min_{i,j} \left\{ \frac{1}{a_i}, \frac{1}{b_i b_j} \right\}. \tag{21}$$

where $b_j = \gcd(a_j, C^j)$ and $C^j = \mathrm{lcm}\{a_i : i \neq j\}$. This condition leads to the finding of a rather large number of examples of Sasaki–Einstein metrics on homotopy spheres [9, 10] and rational homology spheres [4, 25].

In the special case when the integers $(a_0, \ldots, a_n)$ are pairwise relatively prime, Ghigi and Kollár [20] obtained a sharp estimate. In this case, the BP link is always a homotopy sphere, and if we combine the now sharp Klt estimate with Proposition 5.4 above, we see that, when the a_is are pairwise relatively prime, a BP link $L_f(\mathbf{a})$ admits an extremal Sasakian metric if, and only if,

$$\sum_{i=0}^{n} \frac{1}{a_i} < 1 + n \min_i \left\{ \frac{1}{a_i} \right\}.$$

Other applications of the Klt estimate (21) exist [7, 8]. If f is neither a BP polynomial nor a perturbation thereof, alternative estimates must be developed. Further details can be found in [7].

Notice that in all the link examples we have discussed above, information about the topology of the underlying manifold is absent. We end this section addressing that issue in part.

Topological information about links was first given by Milnor and Orlik in [31] through the study of the Alexander polynomial. They computed the Betti numbers of the manifold underlying the link of any weighted homogeneous polynomial of an isolated hypersurface singularity, and in the case of a rational homology sphere, the order of the relevant homology groups. A bit later, Orlik [32] postulated a combinatorial conjecture for computing the torsion, an algorithm that we describe next (see Chapter 9 of [7] for more detail).

Given a link $L_f(\mathbf{w}, d)$, we define its *fractional weights* to be

$$\left(\frac{d}{w_0}, \cdots, \frac{d}{w_n} \right) \equiv \left(\frac{u_0}{v_0}, \cdots, \frac{u_n}{v_n} \right), \tag{22}$$

where

$$u_i = \frac{d}{\gcd(d, w_i)}, \qquad v_i = \frac{w_i}{\gcd(d, w_i)}. \tag{23}$$

We denote by $(\mathbf{u}, \mathbf{v})$ the tuple $(u_0, \ldots, u_n, v_0, \ldots, v_n)$. By (23), we may go between $(\mathbf{w}, d)$ and $(\mathbf{u}, \mathbf{v})$. We will sometimes write $L_f(\mathbf{u}, \mathbf{v})$ for $L_f(\mathbf{w}, d)$.

Definition 5.6 (Orlik's algorithm). *Let $\{i_1, \ldots, i_s\} \subset \{0, 1, \ldots, n\}$ be an ordered set of s indices, that is to say, $i_1 < i_2 < \cdots < i_s$. Let us denote by I its power set (consisting of all of the 2^s subsets of the set), and by J the set of all proper subsets. Given a $(2n+2)$-tuple $(\mathbf{u}, \mathbf{v}) = (u_0, \ldots, u_n, v_0, \ldots, v_n)$ of integers, we define inductively a set of 2^s positive integers, one for each ordered element of I, as follows:*

$$c_\emptyset = \gcd(u_0, \ldots, u_n),$$

and if $\{i_1, \ldots, i_s\} \in I$ is ordered, then

$$c_{i_1, \ldots, i_s} = \frac{\gcd(u_0, \ldots, \hat{u}_{i_1}, \ldots, \hat{u}_{i_s}, \ldots, u_n)}{\prod_J c_{j_1, \ldots, j_t}}. \tag{24}$$

Similarly, we also define a set of 2^s real numbers by

$$k_\emptyset = \varepsilon_{n+1},$$

and

$$k_{i_1, \ldots, i_s} = \varepsilon_{n-s+1} \sum_J (-1)^{s-t} \frac{u_{j_1} \cdots u_{j_t}}{v_{j_1} \cdots v_{j_t} \operatorname{lcm}(u_{j_1}, \ldots, u_{j_t})}, \tag{25}$$

where

$$\varepsilon_{n-s+1} = \begin{cases} 0 & \text{if } n-s+1 \text{ is even,} \\ 1 & \text{if } n-s+1 \text{ is odd,} \end{cases}$$

respectively. Finally, for any j such that $1 \leq j \leq r = [\max\{k_{i_1, \ldots, i_s}\}]$, where $[x]$ is the greatest integer less or equal than x, we set

$$d_j = \prod_{k_{i_1, \ldots, i_s} \geq j} c_{i_1, \ldots, i_s}. \tag{26}$$

$\square$

Orlik's torsion conjecture [32] is stated in terms of the integers computed by this algorithm:

Conjecture 5.7. *For a link $L_f(\mathbf{u}, \mathbf{v})$ with fractional weights $(\mathbf{u}, \mathbf{v})$, we have*

$$H_{n-1}(L_f(\mathbf{u}, \mathbf{v}), \mathbb{Z})_{\mathrm{tor}} = \mathbb{Z}/d_1 \oplus \cdots \oplus \mathbb{Z}/d_r.$$

This conjecture is known to hold in certain special cases [34, 36].

Proposition 5.8. Conjecture 5.7 *holds in the following cases:*

1. *In dimension 3, that is to say, when $n = 2$.*
2. *For Brieskorn–Pham polynomials (20).*
3. *For $f(\mathbf{z}) = z_0^{a_0} + z_0 z_1^{a_1} + z_1 z_2^{a_2} + \cdots + z_{n-1} z_n^{a_n}$.*

It is also known to hold for certain complete intersections given by generalized Brieskorn polynomials [36].

Before his tragic accident, by using Kollár's Theorem 6.1 below, Kris proved that Conjecture 5.7 holds in dimension 5 also. We present his argument in the appendix.

6 Extremal Sasakian Metrics in Dimension Five

We start by reviewing briefly some general facts about Sasakian geometry on simply connected 5-manifolds (cf. Chapter 10 of [7]).

6.1 Sasakian Geometry in Dimension Five

Sasakian geometry in dimension five is large enough to be interesting while remaining manageable, at least in the case of closed simply connected manifolds where we can use the Smale–Barden classification.

Up to diffeomorphism, Smale [43] classified all closed simply connected 5-manifolds that admit a spin structure and showed they must be of the form

$$M = kM_\infty \# M_{m_1} \# \cdots \# M_{m_n} \tag{27}$$

where $M_\infty = \mathbb{S}^2 \times \mathbb{S}^3$, kM_∞ is the k-fold connected sum of M_∞, m_i is a positive integer with m_i dividing m_{i+1} and $m_1 \geq 1$, and where M_m is $\mathbb{S}^5$ if $m = 1$, or a 5-manifold such that $H_2(M_m, \mathbb{Z}) = \mathbb{Z}/m \oplus \mathbb{Z}/m$, otherwise. The integer k in this expression can take on the values $0, 1, \ldots$, with $k = 0$ corresponding to the case where there is no M_∞ factor at all. It will be convenient to set the convention $0M_\infty = \mathbb{S}^5$ below, which is consistent with the fact that the sphere is the neutral element for the connected sum operation. The m_is can range in $1, 2, \ldots$.

Barden [2] extended Smale's classification to include the non-spin case also, where we must add the nontrivial $\mathbb{S}^3$-bundle over $\mathbb{S}^2$, denoted by X_∞, and certain rational homology spheres X_j parameterized by $j = -1, 1, 2, \ldots$ that do not admit contact structures, and can be safely ignored from consideration when studying the Sasakian case. Notice that $X_\infty \# X_\infty = 2M_\infty$, so in considering simply connected Sasakian 5-manifolds that are non-spin, it suffices to take a connected sum of the manifold (27) with one copy of X_∞, or equivalently, replace at most one copy of M_∞ with X_∞.

What Smale–Barden manifolds can admit Sasakian structures? A partial answer to this question has been given recently by Kollár [24]. In order to present this, it is convenient to describe the torsion subgroup of $H_2(M, \mathbb{Z})$ for the manifold M in terms of elementary divisors rather than the invariant factors in (27). The group $H_2(M, \mathbb{Z})$ can be written as a direct sum of cyclic groups of prime power order

$$H_2(M, \mathbb{Z}) = \mathbb{Z}^k \oplus \bigoplus_{p,i} (\mathbb{Z}/p^i)^{c(p^i)}, \tag{28}$$

where $k = b_2(M)$ and $c(p^i) = c(p^i, M)$. These nonnegative integers are determined by $H_2(M, \mathbb{Z})$, but the subgroups $\mathbb{Z}/p^i \subset H_2(M, \mathbb{Z})$ are not unique. We can choose the decomposition (28) such that the second Stiefel–Whitney class map

$$w_2 : H_2(M, \mathbb{Z}) \to \mathbb{Z}/2$$

is zero on all but one of the summands $\mathbb{Z}/2^j$. If we now assume that M carries a Sasakian structure, by Rukimbira's approximation theorem [37], M admits a Seifert fibered structure (in general, nonunique) with an associated Sasakian structure, or a *Sasaki–Seifert structure* as referred to in [7]. Kollár [24] proved that the existence of a Seifert fibered structure on M imposes constraints on the invariants k and $c(p^i)$ of (28) above. Namely, for the primes that appear in the expression, the cardinality of the set $\{i : c(p^i) > 0\}$ must be less than or equal to $k + 1$. For example, both M_2 and M_4 admit Sasakian structures but $M_2 \# M_4$ does not. Other obstructions involving the torsion subgroup of $H_2(M, \mathbb{Z})$ apply as well [7, 24].

The case of positive Sasakian structures is particularly attractive since they provide examples of Riemannian metrics with positive Ricci curvature. Kollár [25] (see also [26] in these Proceedings) has shown that positivity greatly restricts the allowable torsion groups. He proved that if a closed simply connected 5-manifold M admits a positive Sasakian structure, then the torsion subgroup of $H_2(M, \mathbb{Z})$ must be one of the following:

1. $(\mathbb{Z}/m)^2$, $m \in \mathbb{Z}^+$,
2. $(\mathbb{Z}/5)^4$,
3. $(\mathbb{Z}/4)^4$,
4. $(\mathbb{Z}/3)^4$, $(\mathbb{Z}/3)^6$, or $(\mathbb{Z}/3)^8$,
5. $(\mathbb{Z}/2)^{2n}$, $n \in \mathbb{Z}^+$.

Conversely, for each finite group G in the list above, there is a closed simply connected 5-manifold M with $H_2(M, \mathbb{Z})_{\text{tor}} = G$, and remarkably enough, all of these manifolds can be realized as the links of isolated hypersurface singularities.

Let us recall that a quasi-regular Sasakian structure can be viewed as a Seifert fibered structure. In dimension five, Kollár [25] has shown how the branch divisors D_i of the orbifold base determine the torsion in $H_2(M, \mathbb{Z})$. For simply connected 5-manifolds, this result is as follows:

Theorem 6.1. *Let M^5 be a compact simply connected 5-manifold with a quasi-regular Sasakian structure $\mathcal{S}$, let $(\mathcal{Z}, \sum_i (1 - \frac{1}{m_i}) D_i)$ denote the corresponding projective algebraic orbifold base, with branch divisors D_i and ramification index m_i, and let k be the second Betti number of M^5. Then the integral cohomology groups of M^5 are as follows:*

i	0	1	2	3	4	5
$H^i(M^5, \mathbb{Z})$	$\mathbb{Z}$	0	$\mathbb{Z}^k$	$\mathbb{Z}^k \oplus \sum_i (\mathbb{Z}/m_i)^{2g(D_i)}$	0	$\mathbb{Z}$

where $g(D)$ is the genus of the Riemann surface D (notice that $\mathbb{Z}/1$ is the trivial group).

Table 1 Simply connected spin 5-manifolds admitting Sasaki–Einstein metrics.

Manifold M	Sasaki–Einstein
$kM_\infty, k \geq 0$	Any k
$8M_\infty \# M_m, m > 2$	$m > 4$
$7M_\infty \# M_m, m > 2$	$m > 2$
$6M_\infty \# M_m, m > 2$	$m > 2$
$5M_\infty \# M_m, m > 2$	$m > 11$
$4M_\infty \# M_m, m > 2$	$m > 4$
$3M_\infty \# M_m, m > 2$	$m = 7, 9$ or $m > 10$
$2M_\infty \# M_m, m > 2$	$m > 11$
$M_\infty \# M_m, m > 2$	$m > 11$
$M_m, m > 2$	$m > 2$
$2M_5, 2M_4, 4M_3, M_\infty \# 2M_4$	Yes
$kM_\infty \# 2M_3$	$k = 0$
$kM_\infty \# 3M_3, k \geq 0$	$k = 0$
$kM_\infty \# nM_2, k \geq 0, n > 0$	$(k, n) = (0, 1)$ or $(1, n), n > 0$
$kM_\infty \# M_m, k > 8, 2 < m < 12$	

The right column indicates the restriction that ensures that the manifold on the left carries a Sasaki–Einstein metric.

6.2 Existence of Extremal Metrics

Here we reproduce a table [7] (see Table 1) that lists all simply connected spin 5-manifolds that can admit a Sasaki–Einstein metric (and which are, therefore, Sasaki extremal). In the first column, we list the type of manifold in terms of Smale's description (27), and in the second column we indicate restrictions on these under which a Sasaki–Einstein structure is known to exist. Any Smale manifold that is not listed here cannot admit a Sasaki–Einstein metric, but could, in principle, admit an extremal metric. We list the manifolds but not the number of deformation classes of positive Sasakian structures that may occur, a number that varies depending upon the manifold in question. For example, there are infinitely many such deformation classes on the 5-sphere $\mathbb{S}^5$ and on $k(\mathbb{S}^2 \times \mathbb{S}^3)$, as well as on some rational homology spheres. On the other hand, there is a unique deformation class of positive Sasakian structures on the rational homology spheres $2M_5$ and $4M_3$ [25, 26], and they both admit extremal metrics.

6.3 Brieskorn–Pham Links with No Extremal Sasakian Metrics

We now provide a table (Table 2) with examples of Brieskorn–Pham links $L_f(\mathbf{a})$ whose extremal sets are all empty. These are obtained by using Proposition 5.4, the required estimate on the index being easily checked. We list the link together

Table 2 Some examples of 5-dimensional Brieskorn–Pham links whose associated space of Sasakian structures, $\mathcal{S}(\xi)$, does not have extremal Sasakian metrics.

Manifold M	Link $L_f(\mathbf{a})$	Dim κ
M_1	$L(2,3,3,6l+1), l \geq 2$	1
M_1	$L(2,3,3,6l+5), l \geq 2$	1
M_1	$L(2,2,2,2l+1), l \geq 2$	2
M_∞	$L(2,2,2,2l), l \geq 3$	2
$2M_\infty$	$L(2,3,3,2(3l+1)), l \geq 2$	1
$2M_\infty$	$L(2,3,3,2(3l+2)), l \geq 2$	1
$4M_\infty$	$L(2,3,3,6l), l \geq 3$	1
$6M_\infty$	$L(2,3,4,12l), l \geq 3$	1
$8M_\infty$	$L(2,3,5,30l), l \geq 3$	1
$(k-1)M_\infty$	$L(2,2,k,2lk), l \geq 2, k \geq 3$	2
$4M_2$	$L(2,3,5,15(2l+1)), l \geq 2$	1
$2M_\infty\#M_2$	$L(2,3,4,6(2l+1)), l \geq 2$	1

The last column lists the dimension of their Sasaki cones.

Table 3 Five-dimensional links of isolated hypersurface singularities that do not carry Sasakian extremal structure.

Manifold M	Weight vector $\mathbf{w}$	d
$3M_\infty\#M_2$	$(2,2(4l+3),6l+5,2(6l+5)), l \geq 2$	$4(6l+5)$
$7M_\infty$	$(1,4l+1,3l+1,2(3l+1)), l \geq 2$	$4(3l+1)$
M_4	$(4,3(2l+1),4(2l+1),4(3l+1)), l \geq 5$	$12(2l+1)$

with the range for its parameters, the underlying manifold, and the dimension of the Sasaki cone κ of the corresponding CR structure. There are infinitely many such links (or Sasaki–Seifert structures) on each of the manifolds listed.

More examples can be had by considering general weighted homogeneous polynomials f instead. In those cases, we go back to the notation $L_f(\mathbf{w},d)$ for the link, $\mathbf{w}$ its weight vector, and d its degree. Their systematic study should be possible by using the classification of normal forms of Yau and Yu [46]. Here we content ourselves by giving a few examples. In all of these cases, the Sasaki cone is one-dimensional.

In the first row of Table 3, the existence of positive Sasakian structures on the Smale–Barden manifold $3M_\infty\#M_2$ is reported here for the first time. As indicated in the table, for $l \geq 2$ these links do not carry extremal Sasakian metrics. It is unknown if an extremal Sasakian metric exists when $l = 1$, and generally speaking, it is not known whether an extremal Sasakian metric exists at all on $3M_\infty\#M_2$. A similar situation occurs for the link $L_f((4,5,12,20),40)$, which gives a positive Sasakian structure on the manifold $3M_\infty\#M_4$, and which is not yet known to carry extremal Sasakian metrics. Thus, it is natural to pose the following:

Question 6.2. *Are there positive Sasakian manifolds that admit no extremal Sasakian metrics?*

6.4 Toric Sasakian 5-Manifolds

The study of contact toric $(2n + 1)$-manifolds goes back to [1], where it was shown that toric contact structures split into two types, those where the torus action is free, and those where it is not, with the free action case easily described, and the non-free case described via a Delzant type theorem when $n \geq 2$. A somewhat refined classification was obtained later [28], and the non-free case (again, for $n \geq 2$) was characterized in [3] by the condition that the Reeb vector field belongs to the Lie algebra of the $(n + 1)$-torus acting on the manifold. These toric actions were called actions of *Reeb type*, and it was proved that every toric contact structure of Reeb type admits a compatible Sasakian structure. Here we content ourselves with describing the Sasaki cone for a simple but interesting example. For more on toric Sasakian geometry, we refer the reader to [12, 15, 17].

Example 6.3. The Wang–Ziller manifold $M_{k_1,k_2}^{p_1,p_2}$ [45] is defined to be the total space of the $\mathbb{S}^1$-bundle over $\mathbb{CP}^{p_1} \times \mathbb{CP}^{p_2}$ whose first Chern class is $k_1\alpha_1 + k_2\alpha_2$, α_i being the positive generator of $H^2(\mathbb{CP}^{p_i}, \mathbb{Z})$ with $k_1, k_2 \in \mathbb{Z}^+$. These manifolds all admit homogeneous Einstein metrics [45], and also admit homogeneous Sasakian structures [6]. However, they are Sasaki–Einstein only if $k_1\alpha_1 + k_2\alpha_2$ is proportional to the first Chern class of $\mathbb{CP}^{p_1} \times \mathbb{CP}^{p_2}$, that is to say, only when $k_1\alpha_1 + k_2\alpha_2$ and $(p_1 + 1)\alpha_1 + (p_2 + 1)\alpha_2$ are proportional.

We are interested in $M_{k,l}^{p,q}$ as a toric Sasakian manifold and treat the five-dimensional case $M_{k_1,k_2}^{1,1}$ only. We shall also assume that $\gcd(k_1, k_2) = 1$, for then $M_{k_1,k_2}^{1,1}$ is simply connected, and diffeomorphic to $\mathbb{S}^2 \times \mathbb{S}^3$. If k_1 and k_2 are not relatively prime, the analysis of $M_{k_1,k_2}^{1,1}$ can be obtained easily from the simply connected case by taking an appropriate cyclic quotient. We shall also assume that if $(k_1, k_2) \neq (1, 1)$, then $k_1 < k_2$, for otherwise we can simply interchange the two $\mathbb{CP}^1$s.

We wish to find the Sasaki cone for $M_{k_1,k_2}^{1,1}$. For that, we identify this manifold with the quotient $\mathbb{S}^3 \times \mathbb{S}^3/\mathbb{S}^1(k_1, k_2)$, where the $\mathbb{S}^1$ action on $\mathbb{S}^3 \times \mathbb{S}^3$ is given by

$$(\mathbf{x}, \mathbf{y}) \mapsto (e^{ik_2\theta}\mathbf{x}, e^{-ik_1\theta}\mathbf{y}),$$

and points in $\mathbb{S}^3$ are viewed as points in $\mathbb{C}^2$. Alternatively, $M_{k_1,k_2}^{1,1}$ can be obtained by a Sasakian reduction [21] of the standard Sasakian structure $\mathcal{S}_0$ on $\mathbb{S}^7$ by the $\mathbb{S}^1(k_1, k_2)$ action, the moment map $\mu : \mathbb{S}^7 \to \mathbb{R}$ being given by

$$\mu = k_2(|z_0|^2 + |z_1|^2) - k_1(|z_2|^2 + |z_3|^2).$$

The zero level set is the product $\mathbb{S}^3(R_1) \times \mathbb{S}^3(R_2)$ of spheres whose radii are given by

$$R_i^2 = \frac{k_i}{k_1 + k_2}. \tag{29}$$

The Sasakian structure on $M_{k_1,k_2}^{1,1}$ induced by this reduction is denoted by $\mathcal{S}_{k_1,k_2} = (\xi,\eta,\Phi,g)$, where for ease of notation, we refrain from writing the tensor fields with the subscripts also.

Let us consider coordinates (z_0,z_1,z_2,z_3) in $\mathbb{C}^4$, where $z_j = x_j + iy_j$. Then the vector fields

$$H_i = y_i\frac{\partial}{\partial x_i} - x_i\frac{\partial}{\partial y_i}, \ 0 \le i \le 3,$$

restrict to vector fields on $\mathbb{S}^7$ that span the Lie algebra $\mathfrak{t}_4$ of the maximal torus $T^4 \subset \mathrm{SO}(8)$. The base space $\mathbb{CP}^1 \times \mathbb{CP}^1$ of $M_{k_1,k_2}^{1,1}$ is obtained from the zero level set $\mathbb{S}^3(R_1) \times \mathbb{S}^3(R_2)$ as the quotient by the free T^2-action given by $(\mathbf{x},\mathbf{y}) \mapsto (e^{i\theta_1}\mathbf{x}, e^{i\theta_2}\mathbf{y})$. Let $\tilde{X}$ denote the vector field on $\mathbb{CP}^1$ generating the $\mathbb{S}^1$-action defined in homogeneous coordinates by $[\zeta_0,\zeta_1] \mapsto [e^{i\phi}\zeta_0, e^{-i\phi}\zeta_1]$, and denote by $\tilde{X}_1$ and $\tilde{X}_2$ the vector field $\tilde{X}$ on the two copies of $\mathbb{CP}^1$. We let X_1 and X_2 denote the lifts of these vector fields to $\mathfrak{aut}(\mathcal{S}_{k_1,k_2})$.

The Reeb vector field ξ of $\mathcal{S}_{k_1,k_2}$ is that induced by the restriction of $H_0 + H_1 + H_2 + H_3$ to the zero level set $\mathbb{S}^3(R_1) \times \mathbb{S}^3(R_2)$ of the moment map μ (see Theorems 8.5.2 and 8.5.3 in [7]). The Lie algebra $\mathfrak{t}_3(k_1,k_2)$ of the 3-torus of $\mathcal{S}_{k_1,k_2}$ is spanned by the Reeb vector field ξ, and the two vector fields X_1 and X_2 described above. We have:

Lemma 6.4. *The Sasaki cone* $\kappa(\mathcal{D}_{k_1,k_2},J)$ *for the Wang–Ziller manifold* $M_{k_1,k_2}^{1,1}$ *is given by the set of all vector fields*

$$Z_{l_0,l_1,l_2} = l_0\xi + l_1X_1 + l_2X_2, \ (l_0,l_1,l_2) \in \mathbb{R}^3,$$

subject to the conditions

$$l_0 > 0, \qquad |k_1l_1 + k_2l_2| < (k_1+k_2)l_0.$$

Proof. The Sasaki cone $\kappa(\mathcal{D}_{k_1,k_2},J)$ is equal to the set of $\{Z \in \mathfrak{t}_3 \mid \eta(Z) > 0\}$ of positive vector fields (see Theorem 2.2). The vector fields X_1 and X_2 induced on the quotient $M_{k_1,k_2}^{1,1} \approx (\mathbb{S}^3(R_1) \times \mathbb{S}^3(R_2))/\mathbb{S}^1(k_1,k_2)$ by the restrictions of $H_0 - H_1$ and $H_2 - H_3$ to $\mathbb{S}^3(R_1) \times \mathbb{S}^3(R_2)$. So we have

$$\eta(Z_{l_0,l_1,l_2}) = \eta(l_0\xi + l_1X_1 + l_2X_2) = l_0 + l_1(|z_0|^2 - |z_1|^2) + l_2(|z_2|^2 - |z_3|^2),$$

with $l_0 > 0$. The desired result now follows if we use the expression (29) for the radii. $\square$

For $(l_0,l_1,l_2) = (1,0,0)$, we obtain the original Kähler structure on the base $\mathbb{CP}^1 \times \mathbb{CP}^1$, with Kähler metric $k_1\omega_1 + k_2\omega_2$ where ω_i is the standard Fubini–Study Kähler form on the corresponding $\mathbb{CP}^1$ factor. The scalar curvature of this metric is constant, and therefore, the Reeb vector field ξ on $M_{k_1,k_2}^{1,1}$ is strongly extremal. Moreover, by the openness theorem [13], there exist an open neighborhood of ξ in $\kappa(\mathcal{D}_{k_1,k_2},J)$ that is entirely contained in the extremal set $\mathfrak{e}(\mathcal{D}_{k_1,k_2},J)$. It would be of interest to determine how big a neighborhood this can be, and how it compares to the entire Sasaki cone. Although each $M_{k_1,k_2}^{1,1}$ carries a Sasakian structure $\mathcal{S}_{k_1,k_2}$ with an extremal Sasakian metric of constant scalar curvature, the only one with a Sasaki–Einstein metric is $M_{1,1}^{1,1}$. This follows from the fact that $c_1(\mathcal{D}_{k_1,k_2}) = 2(k_2 - k_1)x$, where x is the positive generator of $H^2(\mathbb{S}^2 \times \mathbb{S}^3, \mathbb{Z})$.

7 Appendix

We begin describing a graphical procedure that generalizes the now well-known Brieskorn graphs (see Theorem 10.3.5 and Remark 10.3.1 of [7] for a detailed discussion).

To a given link $L_f(\mathbf{w}, d) = L_f(\mathbf{u}, \mathbf{v})$ defined by a weighted homogeneous polynomial f, of weight vector $\mathbf{w}$ and fractional weight vector $(\mathbf{u}, \mathbf{v})$, we associate a graph as follows:

Definition 7.1. *Consider the following rational Brieskorn polytope*

$$G(\mathbf{u}, \mathbf{v}) = \quad$$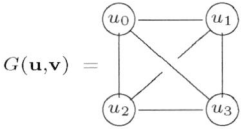

where we think of $G(\mathbf{u}, \mathbf{v})$ as a tetrahedron, and

1. *we label the vertices with pairs (u_i, α_i), $i = 0, 1, 2, 3$, where $\alpha_i = 1 - 1/v_j - 1/v_k - 1/v_l$ for a set of distinct indices $\{i, j, k, l\}$,*
2. *we label the edges with the numbers $\alpha_{ij} = \gcd(u_i, u_j)/v_j v_j$,*
3. *we label the faces with numbers $\alpha_{ijk} = \gcd(u_i, u_j, u_k)/v_i v_j v_k$,*
4. *we label the interior of the tetrahedron with the pair*

$$(t, \tau) \equiv \left(\frac{u_0 u_1 u_2 u_3}{v_0 v_1 v_2 v_3 \mathrm{lcm}(u_1, u_1, u_2, u_3)}, -1 + \frac{1}{v_0} + \frac{1}{v_1} + \frac{1}{v_2} + \frac{1}{v_3} \right).$$

Furthermore,

5. *we define the reduced indices m_i to be the factor of u_i that is not a factor of any of the u_js associated to the remaining 3 vertices,*
6. *we define numbers g_i so that $2g_i + \alpha_i$ is computed from the face opposite to the vertex (u_i, α_i) as the product of the edge numbers divided by the face number minus the sum of the edge numbers,*
7. *we define κ so that $\kappa + \tau - t$ is the sum of the six edge numbers minus the sum of four products of edge numbers divided by the face numbers, one such term for each face.*

Using this associated graph, we can reformulate Orlik's Conjecture 5.7 in dimension five as follows:

Conjecture 7.2. *The second homology group over $\mathbb{Z}$ of a five-dimensional link $L_f(\mathbf{u}, \mathbf{v})$ is given by*

$$H_2(L(\mathbf{a}), \mathbb{Z}) = \mathbb{Z}^\kappa \oplus (\mathbb{Z}/m_0)^{2g_0} \oplus (\mathbb{Z}/m_1)^{2g_1} \oplus (\mathbb{Z}/m_2)^{2g_2} \oplus (\mathbb{Z}/m_3)^{2g_3},$$

where κ, m_i, g_i are the integers given in Definition 7.1.

That κ is the second Betti number of the link was known long ago [31]. There remains to describe the torsion.

Before going any further in this direction, we need to reformulate Kollár's Theorem 6.1 for 5-dimensional links $L_f(\mathbf{w}, d)$. First, let us recall the following two simple conditions of quasi-smoothness for curves [22] and surfaces [23]:

Proposition 7.3 ([22]). *Let $C_d \subset \mathbb{CP}^2(w_0, w_1, w_2)$ be a curve of degree d defined by a homogeneous polynomial f. Assume that $d > a_i$, a_i integer, $i = 0, 1, 2$. Then C_d is quasi-smooth if for all $i \in \{0, 1, 2\}$ the following conditions are satisfied:*

1. *the polynomial f has a monomial $z_i^{a_i} z_j$ for some j (here we allow $i = j$) of degree d*
2. *the polynomial f has a monomial of degree d that does not involve z_i.*

Proposition 7.4 ([22, 23]). *A general hypersurface $X_d \subset \mathbb{CP}^3(w_0, w_1, w_2, w_3)$ defined by a homogeneous polynomial f is quasi-smooth if, and only if, all of the following three conditions hold*

1. *For each $i = 0, \ldots, 3$, there is an index j such that f has a monomial $z_i^{m_i} z_j \in H^0(\mathbb{CP}^3(\mathbf{w}), \mathcal{O}(d))$. Here $j = i$ is possible.*
2. *If $\gcd(w_i, w_j) > 1$, then f has a monomial $z_i^{b_i} z_j^{b_j} \in H^0(\mathbb{CP}^3(\mathbf{w}), \mathcal{O}(d))$.*
3. *For every i, j either f has a monomial $z_i^{b_i} z_j^{b_j} \in H^0(\mathbb{CP}^3(\mathbf{w}), \mathcal{O}(d))$, or f has monomials $z_i^{c_i} z_j^{c_j} z_k$ and $z_i^{d_i} z_j^{d_j} z_l \in H^0(\mathbb{CP}^3(\mathbf{w}), \mathcal{O}(d))$ with $\{k, l\} \neq \{i, j\}$.*

We are then ready for the following proposition.

Proposition 7.5. *Let $L_f(\mathbf{w}, d)$ be a link defined by a quasi-smooth weighted homogeneous polynomial $f = f(z_0, z_1, z_2, z_3)$, with weights $\mathbf{w} = (w_0, w_1, w_2, w_3)$, and of degree d. Let $L_f \to X_f \subset \mathbb{CP}^3(w_0, w_1, w_2, w_3)$ be the associated Seifert fibration. Consider the sets*

$$D_i = X_f \cap \{z_i = 0\} \subset \mathbb{CP}^2(w_0, \ldots, \hat{w}_i, \ldots, w_3).$$

Note that D_i need not be an orbifold, so pick $i \in A \subset \Omega = \{0, 1, 2, 3\}$ such that D_i is a quasi-smooth (orbifold) Riemann surface. Then D_i can be singular only if $\gcd(w_0, \ldots, \hat{w}_i, \ldots, w_3) = 1$, and we have that

$$H_2(L_f, \mathbb{Z}) = \mathbb{Z}^\kappa \oplus \sum_i (\mathbb{Z}/m_i)^{2g(D_i)},$$

where $\kappa = b_2(L_f) = b_2(X_f) - 1$, and the ramification indices are given by $m_i = \gcd\{w_0, \ldots, \hat{w}_i, \ldots, w_3\}$. In particular, let us set $d_i = d/m_i$, and let us normalize the weight vector by $\tilde{\mathbf{w}}^i = \frac{1}{m_i}(w_0, \ldots, \hat{w}_i, \ldots, w_3)$. Then $D_i \subset \mathbb{CP}^2(\tilde{\mathbf{w}}^i)$ is an orbifold Riemann surface of degree d_i embedded in the weighted projective 2-space with weights $\tilde{\mathbf{w}}^i$, and its genus $g(D_i)$ is given by the formula [33, 35]

$$g(\Sigma_{(\tilde{\mathbf{w}},d)}) = \frac{1}{2}\left(\frac{d^2}{\tilde{w}_0\tilde{w}_1\tilde{w}_2} - d\sum_{i<j}\frac{\gcd(\tilde{w}_i,\tilde{w}_j)}{\tilde{w}_i\tilde{w}_j} + \sum_i\frac{\gcd(d,\tilde{w}_i)}{\tilde{w}_i} - 1\right). \tag{30}$$

Proof. We only need to show that if D_i is singular, then we must have $m_i = 1$ so that $D_i \subset X_f^{reg}$ lies inside the smooth part of X_f. This will follow from the quasi-smoothness condition on f.

Suppose then that $m_0 = \gcd(w_1, w_2, w_3) > 1$, and that D_0 is singular. We shall derive a contradiction.

Since $m_0 > 1$, we have $\gcd(w_i, w_j) > 1$ for all $1 \leq i < j \leq 3$. Since $X_f \subset \mathbb{CP}^3(w_0, w_1, w_2, w_3)$ is quasi-smooth, $f(\mathbf{z})$ must contain the following monomial terms (other than the terms involving z_0):

$$z_1^{a_1}z_j, \quad z_2^{a_2}z_k, \quad z_3^{a_3}z_l, \quad j,k,l \in \{1,2,3\},$$

and

$$z_1^\alpha z_2^\beta, \quad z_2^\gamma z_3^\delta, \quad z_3^\tau z_1^\rho.$$

Note that, for example, $z_1^{a_1}z_0$ is not possible as this would mean that

$$d = w_0 + w_1 a_1,$$

and m_0 divides d and w_1 but not w_0. Hence, the polynomial $g(z_1, z_2, z_3) = f(0, z_1, z_2, z_3)$ will contain these six monomial terms. Note that the degree of g is $\tilde{d} = d/m_0$, and it has weights $\tilde{w}_i = w_i/m_0$. By Proposition 7.3, it is quasi-smooth. This contradicts the fact that D_0 is assumed to be singular. $\qquad\square$

7.1 Proof of Orlik's Conjecture in Dimension 5

In order to prove Conjecture 7.2, we only need to show that the (κ, m_i, g_i)s of Definition 7.1 are the same as the ones in Proposition 7.5. We need to pass between the $(\mathbf{u}, \mathbf{v})$-data and the $(\mathbf{w}, d)$-data.

It suffices to prove the needed statement for $i = 0$. We first observe that $(m_0, k_0) = (c_{123}, k_{123})$. In terms of the $(\mathbf{u}, \mathbf{v})$-data, we have

$$m_0 = \frac{u_0\gcd(u_0,u_1,u_2)\gcd(u_0,u_1,u_2)\gcd(u_0,u_2,u_3)\gcd(u_0,u_1,u_3)}{\gcd(u_0,u_1,u_2,u_3)\gcd(u_0,u_1)\gcd(u_0,u_2)\gcd(u_0,u_3)},$$

and

$$2g_0 = -1 + \left(\frac{1}{v_1} + \frac{1}{v_2} + \frac{1}{v_3} \right)$$

$$- \left(\frac{u_1 u_2}{v_1 v_2 \mathrm{lcm}(u_1, u_2)} + \frac{u_1 u_3}{v_1 v_3 \mathrm{lcm}(u_1, u_3)} + \frac{u_2 u_3}{v_2 v_3 \mathrm{lcm}(u_2, u_3)} \right)$$

$$+ \frac{u_1 u_2 u_3}{v_1 v_2 v_3 \mathrm{lcm}(u_1, u_2, u_3)} .$$

We recall now that

$$u_i = \frac{d}{\gcd(d, w_i)}, \qquad v_i = \frac{w_i}{\gcd(d, w_i)}, \qquad \frac{u_i}{v_i} = \frac{d}{w_i}.$$

Let us begin with $2g_0$. The first term is no problem as

$$\frac{1}{v_i} = \frac{\gcd(d, w_i)}{w_i},$$

as needed.

Since f is quasi-smooth, $\gcd(w_i, w_j)$ must divide the degree for all $i < j$. By this, we have the following

$$\frac{u_i u_j}{v_i v_j \mathrm{lcm}(u_i, u_j)} = \frac{\gcd(u_i, u_j)}{v_i v_j} = \frac{\gcd(u_i, u_j) \gcd(d, w_j) \gcd(d, w_i)}{w_i w_j}$$

$$= \frac{\gcd\left(u_i(\gcd(d, w_j) \gcd(d, w_i)), u_j(\gcd(d, w_j) \gcd(d, w_i))\right)}{w_i w_j}$$

$$= \frac{\gcd\left(d \gcd(d, w_j), d \gcd(d, w_i)\right)}{w_i w_j} = \frac{d \gcd\left(\gcd(d, w_j), \gcd(d, w_i)\right)}{w_i w_j}$$

$$= \frac{d \gcd(w_i, w_j)}{w_i w_j},$$

where the last equality holds because $\gcd(w_i, w_j)$ divides d. Alternatively (an argument that is simpler), we could get the same result by showing that if $\gcd(w_i, w_j) | d$, then

$$\mathrm{lcm} \left(\frac{d}{\gcd(d, w_i)}, \frac{d}{\gcd(d, w_i)} \right) = \frac{d}{\gcd(w_i, w_j)}.$$

The last term is handled similarly,

$$\frac{u_1 u_2 u_3}{v_1 v_2 v_3 \mathrm{lcm}(u_1, u_2, u_3)} = \frac{d^2 \gcd(w_1, w_2, w_3)}{w_1 w_2 w_3},$$

because under the assumption that $\gcd(w_i, w_j) | d$ for all $i \neq j$, we have that

$$\mathrm{lcm}\left(\frac{d}{\gcd{(d,w_i)}}, \frac{d}{\gcd{(d,w_i)}}, \frac{d}{\gcd{(d,w_k)}}\right) = \frac{d}{\gcd{(w_i,w_j,w_k)}}.$$

Combining all of these terms, we get that

$$2g_0 = -1 + \sum_{i=1}^{3} \frac{\gcd{(d,w_i)}}{w_i} - d \sum_{1 \leq i < j \leq 3} \frac{\gcd{(w_i,w_j)}}{w_i w_j} + \frac{d^2 \gcd{(w_1,w_2,w_3)}}{w_1 w_2 w_3}.$$

If we now re-scale

$$\tilde{\mathbf{w}} = \frac{(w_1,w_2,w_3)}{\gcd{(w_1,w_2,w_3)}}, \quad \tilde{d} = \frac{d}{\gcd{(w_1,w_2,w_3)}},$$

we see that for $0 \leq i < j < k \leq 3$, we have that $k_{ijk} = 2g(\Sigma_{(\tilde{\mathbf{w}},\tilde{d})})$, as desired.

It remains to show that $m_0 = \gcd{(w_1,w_2,w_3)}$, which we shall leave as an exercise.

The conjecture follows now by Proposition 7.5. This proposition shows that when any of the $m_i > 1$, then the corresponding g_i is the genus of a quasi-smooth curve D_i, and therefore, $2g_i$ must not only be integral but also equal to twice that genus as it was shown. When $m_i = 1$, it does not matter what $2g_i$ is as it does not enter into the torsion formula in either Kollár's theorem or in Orlik's algorithm.

Acknowledgments During the preparation of this work, the first two authors were partially supported by NSF grant DMS-0504367.

References

1. A. Bangaya & P. Molino, *Géométric des formes de contact complètement intégrables de type toriques*, Séminaire G. Darboux de Géométrie et Topologie Différentielle, 1991–1992, Montpellier.

2. D. Barden, *Simply connected five-manifolds*, Ann. Math. (2) 82 (1965), pp. 365–385. MR 32 #1714.

3. C.P. Boyer & K. Galicki, *A note on toric contact geometry*, J. Geom. Phys. 35 (2000), 4, pp. 288–298. MR 2001h:53124.

4. ———, *Einstein metrics on rational homology spheres*, J. Diff. Geom. (3) 74 (2006), pp. 353–362. MR MR2269781.

5. ———, *New Einstein metrics in dimension five*, J. Diff. Geom. (3) 57 (2001), pp. 443–463. MR 2003b:53047.

6. ———, *On Sasakian-Einstein geometry*, Internat. J. Math. (7) 11 (2000), pp. 873–909. MR 2001k:53081.

7. ———, *Sasakian Geometry*, Oxford Mathematical Monographs, Oxford University Press, 2008, MR2382957.

8. ———, *Sasakian geometry, hypersurface singularities, and Einstein metrics*, Rend. Circ. Mat. Palermo (2) Suppl. (2005), 75, suppl., pp. 57–87. MR 2152356.

9. C.P. Boyer, K. Galicki & J. Kollár, *Einstein metrics on spheres*, Ann. Math. (2) 162 (2005), no. 1, pp. 557–580. MR 2178969(2006j:53058).

10. C.P. Boyer, K. Galicki, J. Kollár & E. Thomas, *Einstein metrics on exotic spheres in dimensions 7, 11, and 15*, Experiment. Math. 14 (2005), no. 1, pp. 59–64. MR 2146519 (2006a:53042).

11. C.P. Boyer, K. Galicki & P. Matzeu, *On eta-Einstein Sasakian geometry*, Commun. Math. Phys. (1) 262 (2006), pp. 177–208. MR 2200887.

12. C.P. Boyer, K. Galicki & L. Ornea, *Constructions in Sasakian Geometry*, Math. Zeit. (4) 257 (2007), pp. 907–924. MR2342558.

13. C.P. Boyer, K. Galicki & S.R. Simanca, *Canonical Sasakian metrics*, Commun. Math. Phys. (3) 279 (2008), pp. 705–733. MR2386725.

14. E. Calabi, *Extremal Kähler metrics II*, in Differential geometry and complex analysis (I. Chavel & H.M. Farkas eds.), Springer-Verlag, Berlin, 1985, pp. 95–114.

15. K. Cho, A. Futaki & H. Ono, *Uniqueness and examples of compact toric Sasaki–Einstein metrics*, Comm. Math. Phys. (2) 277 (2008), pp. 439–458. MR2358291 (2008j:53076).

16. A. Futaki, *An obstruction to the existence of Einstein Kähler metrics*, Invent. Math., 73 (1983), pp. 437–443. MR 84j:53072.

17. A. Futaki, H. Ono & G. Wang, *Transverse Kähler geometry of Sasaki manifolds and toric Sasaki–Einstein manifolds*, preprint arXiv:math. DG/0607586, (unpublished).

18. J.P. Gauntlett, D. Martelli, J. Sparks & W. Waldram, *Sasaki–Einstein metrics on $S^2 \times S^3$*, Adv. Theor. Math. Phys., 8 (2004), pp. 711–734.

19. J.P. Gauntlett, D. Martelli, J. Sparks & S.-T. Yau, *Obstructions to the existence of Sasaki–Einstein metrics*, Comm. Math. Phys. (3) 273 (2007), pp. 803–827.

20. A. Ghigi & J. Kollár, *Kähler–Einstein metrics on orbifolds and Einstein metrics on Spheres*, Comment. Math. Helvetici (4) 82 (2007), pp. 877–902. MR2341843 (2008j:32027).

21. G. Grantcharov & L. Ornea, *Reduction of Sasakian manifolds*, J. Math. Phys. (8) 42 (2001), pp. 3809–3816. MR 1845220 (2002e:53060).

22. A.R. Iano-Fletcher, *Working with weighted complete intersections*, Explicit birational geometry of 3-folds, London Math. Soc. Lecture Note Ser. 281, pp. 101–173, Cambridge Univ. Press, Cambridge, (2000).

23. J.M. Johnson & J. Kollár, *Kähler–Einstein metrics on log del Pezzo surfaces in weighted projective 3-spaces*, Ann. Inst. Fourier (Grenoble) (1) 51 (2001), pp. 69–79. MR 2002b:32041.

24. J. Kollár, *Circle actions on simply connected 5-manifolds*, Topology (3) 45 (2006), pp. 643–671. MR 2218760.

25. ———, *Einstein metrics on five-dimensional Seifert bundles*, J. Geom. Anal. (3) 15 (2005), pp. 445–476. MR 2190241.

26. ———, *Positive Sasakian structures on 5-manifolds*, these Proceedings, Eds. Galicki & Simanca, Birkhauser, Boston.

27. C. LeBrun & S.R. Simanca, *On the Kähler Classes of Extremal Metrics*, Geometry and Global Analysis (Sendai, Japan 1993), First Math. Soc. Japan Intern. Res. Inst. Eds. Kotake, Nishikawa & Schoen.

28. E. Lerman, *Contact toric manifolds*, J. Symp. Geom. (4) 1 (2002), pp. 785–828, MR 2 039 164.

29. D. Martelli, J. Sparks & S.-T. Yau, *Sasaki–Einstein manifolds and volume minimisation*, Comm. Math. Phys. (3) 280 (2008), pp. 611–673. MR2399609.

30. J.W. Milnor, *Singular points of complex hypersurfaces*, Annals of Mathematics Studies 61, Princeton University Press, Princeton, N.J., 1968. MR 39 #969.

31. J. Milnor & P. Orlik, *Isolated singularities defined by weighted homogeneous polynomials*, Topology 9 (1970), pp. 385–393. MR 45 #2757.

32. P. Orlik, *On the homology of weighted homogeneous manifolds*, Proceedings of the 2nd Conference on Compact Transformation Groups (Univ. of Mass., Amherst, Mass., 1971), Part I, Lect. Notes Math. 298, Springer-Verlag, Berlin, 1972, pp. 260–269. MR 55 #3312.

33. ———, *Seifert manifolds*, Lect. Notes Math. 291, Springer-Verlag, Berlin, 1972, MR 54 #13950.

34. P. Orlik & R.C. Randell, *The monodromy of weighted homogeneous singularities*, Invent. Math. (3) 39 (1977), pp. 199–211. MR 57 #314.

35. P. Orlik & P. Wagreich, *Isolated singularities of algebraic surfaces with C^* action*, Ann. Math. (2) 93 (1971), pp. 205–228. MR 44 #1662.

36. R.C. Randell, *The homology of generalized Brieskorn manifolds*, Topology (4) 14 (1975), pp. 347–355. MR 54 #1270.

37. P. Rukimbira, *Chern-Hamilton's conjecture and K-contactness*, Houston J. Math. (4) 21 (1995), pp. 709–718. MR 96m:53032.

38. S.R. Simanca, *Canonical metrics on compact almost complex manifolds*, Publicacões Matemáticas do IMPA, IMPA, Rio de Janeiro (2004), 97 pp.

39. ———, *Heat Flows for Extremal Kähler Metrics*, Ann. Scuola Norm. Sup. Pisa CL. Sci., 4 (2005), pp. 187–217.

40. ———, *Precompactness of the Calabi Energy*, Internat. J. Math., 7 (1996) pp. 245–254.

41. ———, *Strongly Extremal Kähler Metrics*, Ann. Global Anal. Geom. 18 (2000), no. 1, pp. 29–46.

42. S.R. Simanca & L.D. Stelling, *Canonical Kähler classes*. Asian J. Math. 5 (2001), no. 4, pp. 585–598.

43. S. Smale, *On the structure of 5-manifolds*, Ann. Math. (2) 75 (1962), pp. 38–46. MR 25 #4544.

44. T. Takahashi, *Deformations of Sasakian structures and its application to the Brieskorn manifolds*, Tôhoku Math. J. (2) 30 (1978), no. 1, pp. 37–43. MR 81e:53024.

45. M.Y. Wang & W. Ziller, *Einstein metrics on principal torus bundles*, J. Diff. Geom. (1) 31 (1990), pp. 215–248. MR 91f:53041.

46. S.S.-T. Yau and Y. Yu, *Classification of 3-dimensional isolated rational hypersurface singularities with $\mathbb{C}^*$-action*, Rocky Mountain J. Math. (5) 35 (2005), pp. 1795–1809. MR 2206037(2006j:32034).

Printed in the United States of America